The study of structural analysis and design is a central subject in civil, aeronautical, and mechanical engineering. This book presents a modern and unified introduction to structural analysis, with a strong focus on how structures actually behave.

The unifying theme is the application of energy methods, developed without the formal mathematics of the calculus of variations. The energy approach makes it possible to articulate the logical relationship between equilibrium and compatibility; emphasize the unity of structural analysis, particularly for indeterminate structures; and identify the roles of idealization and discretization in structural modeling. Thus, energy methods also serve as a prelude to the main ideas behind modern computational approaches to structural analysis and design.

Overall, the author intends to convey a style of thinking about and modeling structures and their behavior, and to introduce the intellectual roots from which most computer tools derive. As an aid to upper-level undergraduate students in mastering this material, the text includes numerous worked examples, as well as homework problems.

Structural Modeling and Analysis

The Natchez Trace Parkway Bridge, located near Franklin Tennessee, was completed in 1994. It earned the 1995 Outstanding Civil Engineering Achievement Award of Merit from the American Society of Civil Engineers. It was the first arch bridge built in the United States from precast concrete segments, as shown in the figure immediately preceding Chapter 1. As such, this bridge combines original technical design with breathtaking aesthetic vision. (Photo courtesy of Figg Engineering Group.)

Structural Modeling and Analysis

Clive L. Dym
Harvey Mudd College

CAMBRIDGE UNIVERSITY PRESS
Cambridge, New York, Melbourne, Madrid, Cape Town, Singapore, São Paulo

Cambridge University Press
The Edinburgh Building, Cambridge CB2 2RU, UK

Published in the United States of America by Cambridge University Press, New York

www.cambridge.org
Information on this title: www.cambridge.org/9780521495363

First published 1997
This digitally printed first paperback version 2005

A catalogue record for this publication is available from the British Library

Library of Congress Cataloguing in Publication data

Dym, Clive L.
Structural modeling and analysis / Clive L. Dym.
p.   cm.
Includes bibliographical references and Index.
ISBN 0 521 49536 9
1. Structural analysis (Engineering)   2. Structural design.
I. Title.
TA645.D95   1997
624.1′7 – dc21   96-45567
CIP

ISBN-13 978-0-521-49536-3 hardback
ISBN-10 0-521-49536-9 hardback

ISBN-13 978-0-521-02007-7 paperback
ISBN-10 0-521-02007-7 paperback

Contents

List of Photographs*

*Photographs indicated by an unadorned chapter number appear on the page immediately preceding the chapter.

Preface

1 Still Another Introduction to Structural Analysis?

I studied civil engineering as an undergraduate at the Cooper Union, earning my B. C. E. in 1962. I took three courses in structures and two on the strength of materials, as well as two design courses that heavily featured structural design. The curriculum totaled 143.5 credits. Recently, engineering schools have begun to reduce their total credit hours in an emerging trend toward a total of 120 credits. Further, civil engineering curricula have been reoriented away from a traditional emphasis on structural and geotechnical engineering toward more general programs that include environmental engineering, transportation engineering, systems analysis, and computer-aided engineering. Consequently, the number of required courses in structural analysis taken by civil engineering undergraduates has declined steadily, reaching a point where, typically, civil engineering majors take only one *required* course in structures, usually after (only) one course in the strength of materials.

In the face of such changes, you might expect that the way in which we teach structures has also shifted markedly. However, beyond extending the classical approaches to this subject to include elementary structures programs on attached floppy disks, the basic textbooks on structural analysis remain largely unchanged. That is, those books that present the first course in structures are not terribly different than those that were available when I was an undergraduate. To be sure, there are many books on computational techniques, especially on finite element methods, but they are intended for advanced undergraduate and graduate courses. There are one or two elementary books that purport to support computer-based modeling, but in my view they won't provide the reader with a sound basis – about both theory and behavior – for learning much about structures. I think we can do better, and this book is my contribution to the continuing dialogue on how we should transmit the beginnings of an extraordinary body of knowledge.

We clearly need a shift away from traditional approaches to the subject. Structural analysis is often presented as a collection of tools – often seemingly unrelated – for handling a set of fairly specific problem types. A major dividing line on the problem-type axis is the distinction between statically determinate and statically indeterminate structures. Further, the tools are often presented in an order reflecting their respective degree of difficulty of application, rather than in an order reflecting a coherent view of the discipline. In fact, structural analysis is taught in two or three different disciplines, including civil engineering, aeronautical engineering, and mechanical engineering. (It is also taught in architecture programs, but that really is a horse of another color.) Perhaps reflecting long-standing differences in the different engineering disciplines, traditional approaches to structural analysis are often presented as distinct from their logical underpinnings in mechanics, especially engineering or applied mechanics. There are, I feel, better ways to do this.

I haven't said much about computers, yet there is no doubt that the ways in which structures are analyzed and designed are dramatically different than what was done in my professional youth. Computers are ubiquitous both in the classroom and in the design office. However, in terms of introducing structural analysis in a first course, the major response to this change in the working environment has been the inclusion of elementary computer programs within a shrinking structural curriculum. One common result of this is more time spent on generating numbers and less time spent on understanding what meaning – if any – to attach to the numbers that are generated with these programs. In short, I believe that as computers become ever more powerful, it is even more important to teach *basic structural modeling*, with a heavy emphasis on *understanding behavior*, as well as on interpreting results in terms of the limitations of the models being applied. In fact, I have heard it argued that the generation of numerical analyses for particular cases is, in the real world, a task increasingly performed by technicians – and not by professional engineers. As numerical analysis becomes both more common and significantly easier, the premiums will be earned by those who know *which* calculations to perform and *what meaning* to attach to the subsequent results.

As I've already indicated, there have been many, many books written about structures. I have included a bibliography of recent and classical textbooks and tradebooks on structural mechanics, structural analysis, and structural engineering, organized into five broad categories: classical civil engineering structures, energy methods, finite element and matrix methods, special topics, and general structural behavior.

Any teacher of structures could, of course, choose a selection of books from several categories to generate material for a single course. However, it seems pretty clear that there is no single book that, in a one semester course, could: present a unified approach to analysis tools; successfully integrate structural modeling and analysis with elements of structural behavior; provide a solid basis for further learning for those going on to further work in structural engineering; and, finally, make structural analysis a useful and more interesting subject for those students whose primary focus is not structural engineering. This book represents my attempt, classroom tested here at Harvey Mudd, to write a book that achieves these ends.

2 What I am Trying to Do in This Book

Meeting the four goals just outlined is a formidable challenge, especially within the constraint of a one-semester introduction to a complex subject. The approach I take includes (1) choosing a unifying theme and (2) limiting the scope of the implementation to a manageable feast.

The unifying theme is the application of energy methods. I believe these methods can be developed and applied in a relatively straightforward way, without the heavy machinery of the formal mathematics of the calculus of variations. An energy approach allows us to develop a logical relationship between equilibrium concepts and compatibility concepts. Further, energy approaches also provide a sound base for developing approximate solutions for structural analysis (e.g., Rayleigh–Ritz methods) as a prelude to introducing the central ideas behind numerical approaches (e.g., finite element methods (FEM)). In fact, it is possible to go further in this direction and identify *indirect* energy methods with the *idealization* part of the modeling process and *direct* energy methods with *idealization and discretization* in structural modeling.

I also use the energy-based approach to emphasize the unity of structural analysis, particularly in the consideration of indeterminate structures. While there are important differences between determinate and indeterminate structures that emerge as a consequence of design considerations (e.g., stress and deflection limitations, redundancy, stability, etc.), their analysis should reflect behavioral and computational differences, not that they are different branches of structural analysis (as is often the case with traditional approaches).

Some will feel that energy methods themselves require enough background and sophistication that they are not readily accessible to undergraduates. However, I argue from my own classroom experience that energy methods can be readily introduced by focusing first on elementary structural models such as simple springs and discrete structures (e.g., the bent beam as a pair of rigid links joined by a rotational spring). Further, as I've said before, energy approaches can be introduced without the heavy machinery of the calculus of variations by introducing small systematic variations in an informal and intuitive manner, and then focusing on the pragmatics of applying the *delta* or "δ operator." The subject can thus be introduced correctly and without too much jargon.

Trying to do all of this in one-semester is tough, very tough. This severe constraint is reflected by (generally) limiting considerations to linearly elastic trusses, linearly elastic beams, and frames with symmetrical crosssections, responding in bending about a principal axis. This class of structures is sufficiently complex and practically interesting to illustrate the major points. It should also provide an adequate base for studying more complex structures in further courses. In addition, not all of the material presented need be covered in a one-semester, first course on structures. For example, the general statements of energy methods (Chapter 6) can be summarized or even deleted, as can some of the material in Chapters 4 and 7 where classical approximate solutions for the extensions of bars and the bending deflections of beams are demonstrated.

I will also emphasize two other points in this book. First, I will normally do examples and solve problems in terms of variables, with numerical values inserted irregularly and then only in the very last steps. I do this to emphasize the validation and interpretation of results in terms of (1) dimensional consistency and (2) the range of values of proper dimensionless ratios (e.g., the thickness-to-length ratio of a long, slender beam). This approach also lays the foundation for doing "back of the envelope calculations" in support of both analysis and design, particularly as an aid in the evaluation of computer results.

Second, I will try to emphasize structural behavior in terms of reasonableness of deflected shapes and of force and stress distributions (e.g., shear and moment diagrams). Again, I want to reinforce the notion that an intuitive feel for how a structure will respond can be developed, and that this intuition can often be expressed in sketches or simple formulas that help one interpret more complex analyses (or experiments).

There is an important point here that is intended particularly for student readers; it will use language that you will see again in what follows. The models of various structural elements that I will present are done with the intention of conveying a *style of modeling and thinking about structures and their behaviors*. Thus, specific numerical results are less important here than *knowing what sorts of things to look for*, including the right dimensions, the right dimensional ratios, and the presence and absence of terms embodying specific kinds of behavior. Certainly getting the magnitudes right is important, which is where the numerical results come in. All of the results I have given can be done for specific cases on a computer with one program or another. And there are certainly countless problems you can find where you can do it all numerically. However, the kinds of approaches emphasized here are intended to convey a flavor of what we always need to look for whenever we are doing any numerical work, but most especially when we are using computer-based tools, sometimes called "black boxes." Remember, a computer can't tell you whether the axial force *should* be greater than or smaller than the moment, it can only give you numbers. *You* have to apply some engineering judgement to see if you want to accept those numbers.

As with many other engineering courses, mastery of the material is developed and reflected in the ability to solve problems. There are many examples worked out in the main text, forming an integral part of the narrative, and they are also highlighted in a special format. The homework problems given include examples of the physical systems being modeled, often showing results that there was no time to get to in the main text. Therefore, it is very important for you, as reader, to *do* as many of the given problems as you can, and still more that you can find elsewhere. Remember, you can never do too many problems!

Finally, a word on notation, always a difficult issue. Whereas many aspects of structural engineering notation are fairly standard (e.g., E for the modulus of elasticity), many others vary – and oft times they simply conflict. This is particularly true for deflection and displacement quantities, and the concern is made even more difficult because I use the traditional δ for the variational operator, although it is often used to denote a deflection. So, the policy will be as follows. In Chapter 3, where I discuss discrete elements (i.e., springs), I denote extensions by ξ, whereas in Chapter 4, where I discuss one-dimensional structural elements, I use Δ for displacements and deflections. Later on, in an attempt to be consistent also with other works on structural mechanics, I use lower case "deltas" with subscripts, (e.g., δ_{Ch} or δ_T) to denote particular structural deflections – with one exception: δ_D is introduced in Eq. (4.9) as the Dirac delta function.

3 How This Book Is Organized

This book is organized as follows. In Chapter 1 I outline some important aspects of structural behavior, focusing on what purposes structures and structural elements are designed to achieve. I also discuss in general terms considerations of design for both strength and stiffness, as well as notions of load paths, redundancy, and safety. In Chapter 2 I review some fundamental structural models, focusing here mostly on *idealization*, although I complete the chapter by bringing *discretization* under the modeling umbrella as well. I also review the meaning of determinacy. I devote Chapter 3 to introducing the minimum potential energy principle for the elastic spring and for discrete models of structures. This will constitute the first introduction to the δ operator and to energy principles. These ideas are reinforced and extended in Chapter 4, which is devoted to a discussion of axially-loaded members or bars. These structural types provide a simple but robust platform for introducing potential and complementary energy principles, including the Castigliano theorems, the second of which is used to derive the standard unit-load calculation for the displacements of truss joints. Bars also offer the opportunity to introduce both indirect and direct energy approaches, the latter of which allows me to reinforce the idea of discretization and to introduce both matrix notation and the notion of interpolation. In Chapter 5 I describe how Castigliano's theorems are used to analyze those assemblages of bars called trusses, and I give a brief review of the classical methods of sections and of joints.

In Chapter 6, building on the examples presented in Chapters 3, 4, and 5, I provide formal statements of the energy principles in three dimensions (in Cartesian coordinates) and in matrix form. (And, as noted earlier, this material can be summarized or even deleted in a first course.) In Chapter 7 I use energy-based tools to derive the basic models for elastic beams and I introduce and apply some ideas about discretization and the direct displacement approach. In Chapter 8 I use energy-based tools, including Castigliano's second theorem, to calculate (and use) beam deflections, including examples of beams supported by elastic restraints. In Chapter 9 I introduce frames as assemblages of beams, again making frequent use of Castigliano's second theorem. Last, in Chapter 10 I present an overview of the force (flexibility) and displacement (stiffness) methods of analyzing structures, describe how they are represented in matrix format, and outline how these methods and formats are used in computational-based approaches to structural analysis and design.

A comment on the topic of modern computational-based approaches to structural engineering. While I summarize that in Chapter 10, you will find that it is a topic I highlight and discuss at several points in the book (e.g., Sections 4.6, 6.5, and 7.3). However, I am *not* going to tell you how to write programs or format input, nor do I provide programs, disks, or user's manuals. This is because, as I've already said, a goodly chunk of what this book is about is providing a basis for understanding the intellectual roots from which most of these computer tools derive, what these tools can do, why they do it in the ways they do, and most important, *how their results can be assessed for good use*. That is, the emphasis is on knowing the *why* of whatever calculations or computations are being done.

4 A Note on Style

Writing style is an important issue, even for a writer of technical textbooks or monographs. For example, my older brother, Harry, is a mathematician who writes with an economy and elegance of style that I can only admire. I try to choose words as carefully as topics and equations, and I wanted this book to be informal, even conversational, hoping that this would be more effective than a more formal pedagogic style. Thus, I intend the book to "sound" as I think I do in a relaxed, informal classroom setting, and I hope it works for you!

תושלב״ע

Acknowledgments

I have long been interested in structures and structural engineering, starting even before I was introduced to their analysis and design by Professor Anthony E. Armenàkas while studying civil engineering at Cooper Union. My interest in the variational approaches was inspired and encouraged by Professors Joseph Kempner of Polytechnic University (nee "Brooklyn Poly") and Nicholas J. Hoff and Jean Mayers of Stanford University. (Nicholas was my doctoral advisor; he earned his own doctorate under Stephen P. Timoshenko at Stanford in 1942, so I am part of a very distinguished tradition in applied mechanics. This lineage also shows a Russian immigrant to America mentoring an Hungarian immigrant, who in turn mentored an immigrant born in England of Eastern European parents.) Later I wrote two graduate textbooks on variational approaches to solid mechanics with Irving H. Shames, a long-time University Professor at the State University of New York at Buffalo and now at George Washington University. Professor Steven J. Fenves encouraged me to think about new approaches to teaching structures to undergraduates while we were colleagues at Carnegie Mellon University in the early 1970s. I am grateful to all, but especially to Joe, Nicholas, Irv, and Steve.

I have profited enormously from a careful, critical reading of the manuscript as it unfolded by a friend and colleague at Harvey Mudd College, Professor Harry E. Williams. Harry has an uncanny knack for reminding one of what the issues and principles really are, and for providing support and intellectual rigor and honesty at the same time.

Two other Harvey Mudd colleagues have also been helpful. Philip D. Cha has read parts of the manuscript and provided useful feedback. Phil also answered "urgent" questions, as has Ziyad H. Duron, with whom I have had useful talks about what students really ought to know about structures.

Professors Peter A. Chang (University of Maryland), Steve Fenves, and Victor Kaliakin (University of Delaware) have each read the penultimate version of this book and provided detailed and constructive reviews. As a result of what I learned from these very civil engineers, I made major changes that have significantly improved the book's organization and format. However, I alone remain responsible for errors, oversights, and infelicities.

There are twenty-three photographs of structures in this book. They are partly about structural history and structural function, yet they are also a small personal reflection on some structures that I find beautiful. The photographers of each are identified, but I do want to thank several people who provided both inspiration and very practical help. Professor David P. Billington of Princeton University wrote a wonderful book, *The Tower and The Bridge*, gave me access to his photographs, and intoduced me to his close colleague, J. Wayman Williams of Basking Ridge, New Jersey. Wayman is both a structural engineer and a photographer of structures, and he provided me with photographs, ideas, and very enthusiastic encouragement. Emeritus Professor Colin O'Connor of the University of Queensland, Australia, author of the evocative book *Roman Bridges*, kindly provided me with prints. Professor E. C. Ruddock of the University of Edinburgh directed me to the Royal Commission on the Ancient and Historical Monuments of Wales, who provided the photograph of a beautiful Welsh bridge.

Florence Padgett of the Press is once again my wonderfully supportive editor. This is our fourth collaboration, including two projects at another publisher (to remain nameless), and our professional relationship has blossomed into a valued friendship.

Florence also encouraged me to involve my younger daughter, Miriam, an artist, in my work. As one result, Miriam did the cover art for an earlier book on engineering design. For this book, she designed the entire cover and drew the figures that appear throughout. I am very grateful to Miriam for her work.

Finally, Joan Elizabeth Wilson Anderson provided encouragement, friendship, love, and oft-needed distraction while this book was being completed and produced. It is a great pleasure to acknowledge what this has meant to me.

Structural Modeling and Analysis

The Natchez Trace Parkway Bridge, here under construction. As noted (cf. Frontispiece), it was the first arch bridge built in the United States from precast concrete segments. Thus, it philosophically is akin to the masonry structures that we show later in the book because it is built up of discrete, blocklike segments. Of course, the designers had the dual advantage of better materials (e.g., the precast, steel-reinforced segments) and of better analysis techniques, the subject which with this book is concerned. (Photo by J. Wayman Williams.)

1

Structural Mechanics: The Big Picture

This book is about structural mechanics, that is, about the modeling and analysis of how structures behave in the real world. As a discipline, the material we cover is in part an extension of coursework titled "Statics" and "Strength of Materials," and it can also be viewed as a subset of the "Theory of Elasticity," a subject typically taken as a graduate course. This very brief epistemological lesson is meant to remind the reader that what we are about to cover derives from larger principles, some versions of which will be familiar.

1.1 Structures and One-Dimensional Structural Elements

We begin by looking at some structures, the real, physical objects that are our focus. We show pictures of several interesting structures in Figs. 1.1–1.4 (and elsewhere in the book, including some abstract structural elements on the book's cover). We also start by connecting with an ancient building material, stone. There is some appeal in the simple-minded idea that masonry structures – such as Egypt's pyramids, China's Great Wall, and the Mayan temples of Central America – represent some version of a pile of bricks, one on top another. Each of these structures seems rooted to its foundations by its own weight. In fact, as we see in the two Roman structures shown in Fig. 1.1, masonry arches have been used to construct relatively light and airy structures. Other masonry structures – such as the Salisbury Cathedral of Figs. 1.3(a) and 1.4(b) – show great artistry in their flying buttresses and their majestically open clerestory spaces. However, as can be seen in Fig. 1.4(b), there are no interior floors in these graceful cathedrals because it is impossible to build a flat span over a space with a material that cannot support tensile forces. As with the Coliseum and other stone structures, the design of such graceful structures is an accomplishment based on recognizing the arch as a structural form that most effectively uses stone, a material that functions well only in compression.

In Fig. 1.2 we show pictures of three beautiful bridges (and several others are shown in the frontispiece and the title pictures of Chapters 1–3, 5, 6, and 8–10). Our imagination is fueled by bridges that span both broad reaches and very deep gorges. Think of the archetypal pedestrian bridge, hung from rope or cables, that has been a focus of stories ranging from Thornton Wilder's classic novel, The *Bridge of San Luis Rey*, to the epic adventures of Indiana Jones.

But bridges have evolved, and interestingly enough, they moved away from cable suspension through arches of various materials, including stone (Fig. 1.2(a)), iron, steel (Fig. 1.2(b)), and concrete (Fig. 1.2(c)). Only since the end of the last century have we returned to the notion of suspending a structure from cables, resulting in well-known suspension bridges such as New York's George Washington bridge (pictured immediately

Figure 1.1a.
Figure 1.1. These two Roman structures – a water-carrying aqueduct and a famous stadium – were built of stone almost two thousand years ago. They use the properties of stone as expressed in the form of an arch to lasting and beautiful advantage. They are (a) the Pont du Gard in Remoulins, France (photo by Colin O'Connor from *Roman Bridges*, Cambridge University Press, 1993; courtesy of the author); and (b) the Coliseum in Rome (photo by Clive L. Dym).

preceding Chapter 8) – which now carries a second deck for which the bridge was designed, although it wasn't added until almost thirty years later – and cable-stayed bridges such as the Chesapeake and Delaware Canal Bridge (pictured immediately preceding Chapter 2).

Another facet of structural development is shown in Fig. 1.3, wherein we see some snapshots of the evolution of the "skyscraper" – a term whose very meaning has also evolved with our ability to build taller and higher. While the cathedrals of the middle ages were tall structures, they were essentially one-story structures. Designers had learned to support tall and slender walls with flying buttresses, but they were unable to build floors to span the spaces between their tall walls. The development of the skyscraper had to wait in part for the development of a material (rolled steel) and of configurations (beams and trusses) that would enable floors for buildings. One of the intermediate steps was Gustave Eiffel's tower of iron in Paris (Fig. 1.3(b)), the shape of which conforms to an optimal expectation of design for wind. The Hancock Center (Fig. 1.3(c)) is emblematic of the modern skyscraper not only because of its great height and volume, but also because of its presentation of its structural form as a clear expression of that structure's function.

The structures we have shown so far can be summarized almost as one dimensional, or at least planar structures with loads anticipated to act in the plane of the drawings (which

Figure 1.1b.

are all elevations). There are clearly three-dimensional aspects to these structures and their loads, but their behavior can be viewed or modeled in one- or two-dimensional ways. On the other hand, there are structural types wherein large surface areas (other than vertical walls) are evident and for which special models and design and analysis techniques are required. In the civil engineering domain, these *surface structures* are typically roofs, and we show a few in Fig. 1.4.

Figure 1.2a.
Figure 1.2. Three beautiful bridges, each of which expresses the beauty of the arch with a different material. They are (a) the Puente del Diablo, built by the Romans in Martorell, Spain (photo by Colin O'Connor from *Roman Bridges*, Cambridge University Press, 1993; courtesy of the author); (b) Robert Maillart's Salginatobel Bridge, built near Schiers, Switzerland, in 1930 (photo courtesy of M.-C. Blumer-Maillart); and (c) Othmar Amman's Bayonne Bridge between Bayonne, New Jersey, and Staten Island, New York, completed in 1931 (photo by J. Wayman Williams). The bridges represent a similar aesthetic, although their technological details are quite different. For example, the Salginatobel is a three-hinged concrete arch, whereas the Bayonne is a two-hinged steel arch.

These structures are special because they are meant to be relatively thin, curved surfaces so that their weight can be supported. Note, however, that we mean thin in comparison with the dimensions they span, not necessarily in an absolute sense. Clearly the masonry and concrete roofs shown in Fig. 1.4 are not going to be all that thin. By way of contrast, the walls of aircraft, spacecraft, or submarines are really pretty thin because, in these cases, weight is much more at a premium than it is for a typical civil engineering structure.

One of the things we can learn from this very brief history is that there are models of structural behavior that can be expressed in terms of some simple structural elements. As a starting point, we can note that structures can be seen as being of one of two basic types. One type has the principal direction or shape of the structure aligned or virtually coincident with the direction of the load it is supporting, the most familiar example being a cable or rope supporting a weight or structure at its end. Such structures are often called *funicular structures*, and they include structural devices that work in compression as well as in pure tension, for example, the compression bars in a truss. Arches are also considered to be funicular structures because the compressive thrusts that they exhibit in supporting

Figure 1.2b.

loads lines up with the (curved) shape of the arch. The point is that for these structural elements – and they can be one- or two-dimensional – the sense of the principal structural action is coincident with both the shape of the structure and the direction of the applied loads.

In the second type of structure, the load is applied perpendicular or normal to the line or plane of the structure. The structural action that supports the load occurs in the line or plane of the structure, although we will see that deflections due to the loads occur normal to the plane of the structure (i.e., in the same direction as the loading). Beams and bridges clearly work this way, as do the surface structures that we have shown in Fig. 1.4.

Figure 1.2c.

The net effect of this dichotomy is that we need different kinds of models for analyzing different kinds of structural behavior. We have shown some one-dimensional versions of the models in Fig. 1.5, and we will discuss them in greater detail in Section 2.1.

Something else we can learn from our brief history – and remember that we did not intend to provide a comprehensive history of structures – is that structural engineering comprises both art and science. We should recognize that structural engineering has evolved over time, as we have mastered new technologies, including those of materials, fabrication and assembly, tools for analysis and design, and better insight into the behavior of structures and the ways we can model that behavior. Thus, a first course in structures such as this one is just the first step into a field with a long and interesting history, as well as a future filled with serious and equally interesting challenges.

1.2 The Conceptual "Elements" of Structural Mechanics

Let us imagine that we want to design a structure, say, a suspension bridge, a high-rise building, or a football stadium. What does it mean to "design" such structures? What ideas, models, techniques, or calculations do they have in common? Is there a theory that binds them together?

There is such a theory, and we can outline it in general terms as a set of six physical quantities, unified by three fundamental sets of basic principles (expressed in well-known equations) and to which are applied some common design criteria or behavioral goals, which we want the structure to meet. This book is about the elements of this theory, and the six *basic physical quantities* that we use to describe various aspects of structural behavior are as follows.

Figure 1.3a.

Figure 1.3. Three different expressions of the human urges to reach (and "scrape") the sky, each representing a different technological time. These tall buildings are (a) Salisbury Cathedral, Salisbury, England, built around the twelfth century (photo by Clive L. Dym); (b) Gustave Eiffel's Tower, Paris, France, finished in 1889 (photo by David P. Billington); and (c) the Hancock Center, Chicago, Illinois, completed in 1970 (photo by J. Wayman Williams). Fazlur Kahn of Skidmore, Owings, and Merrill was the structural designer of the Hancock Center, and he designed a very tall, framed tube that clearly shows in its X-bracing the structural form that carries all of the wind and most of the gravity loads acting on the building.

Figure 1.3b.

1.2.1 Loads

These are the forces that the structure is expected to carry or support. They may include the dynamic forces that result from a train traversing a bridge, the forces produced by the winds whistling by a tall building, and the forces produced as stadium fans respond in near unison to a play on the field. These *external loads* are regarded as *givens*, that is, as conditions of use that the structure is expected to accomodate. We can often specify these external loads pretty well, in which case we can say we are dealing with *deterministic*

Figure 1.3c.

problems. However, there are also loads that can only be estimated in a *probabilistic* sense, such as those produced by earthquakes or by wave forces on structures in the ocean. (We shall have more to say about loads in Section 1.3.4.) Finally, as a matter of notation, we will typically label externally applied forces as P.

1.2.2 Reactions

These are the forces that are required to hold the structure in place as it is subjected to all of the applied, external loads. These reactions are, typically, transmitted

Figure 1.4a.

Figure 1.4. Roofs have often proved an interesting challenge to structural engineers and architects, and certainly their lives have been made easier by advancing technologies. Here we illustrate the timeless beauty of some roof forms, including (a) a segment of the roof of the Emperor Herod's Palace in Caesarea, Israel (photo by Clive L. Dym); (b) an interior view of the clerestory (or clearstory) walls and roof structure of Salisbury Cathedral, Salisbury, England (photo by Clive L. Dym); and (c) the Little Sports Palace, Rome, Italy, designed by Pier Luigi Nervi and built in 1957 (photo by David P. Billington). Nervi used two sets of ribs in his design, an innovation that provided greater stiffness – thus extending the area that could be spanned – for an otherwise relatively thin shell roof.

Figure 1.4b.

through the foundations that situate the structure on and in the ground. We calculate these reactions by applying Newton's second law to the entire structure, acting as a rigid body. If such a calculation does allow the complete determination of all of the required reactions, the structure is said to be statically determinate, deriving from the fact that for static loading, we are simply solving Newton's second law for static equilibrium, as we do in "Statics":

$$\Sigma \boldsymbol{F} = 0$$

Typically, however, there are more reactions than can be determined from the limited number of equations of static equilibrium, so that more extensive considerations enter into determining the reactions. In these cases we must account for the deformations or deflections of the structure and the internal forces (next section) that cause these deflections as part of the calculations needed to determine the reactions. Not surprisingly, such configurations of structures and their reactions are referred to as *statically indeterminate*.

In either event, both for determinate and indeterminate structures, we will normally label reactions as R.

1.2.3 Internal Forces

These are the forces internal to the structure, carried by the various elements that make up the entire structure. Thus, for a high-rise building, we would include the forces in the columns, in the floor systems, in the window panes and in the rest of the building's cladding, and so on. This set would also include the forces and moments transmitted at

Figure 1.4c.

connections, whether from columns to floor joists or window frames to wall elements, whether transmitted through bolted joints or through welded connections.

We use a variety of symbols for forces and forcelike quantities, and they are more reflective of history than an organized taxonomy. Thus, for springs we talk of internal forces F_s, for bars and trusses the axial force can be P or F_i, and for beams we talk of force resultants, including moments M and shear forces V.

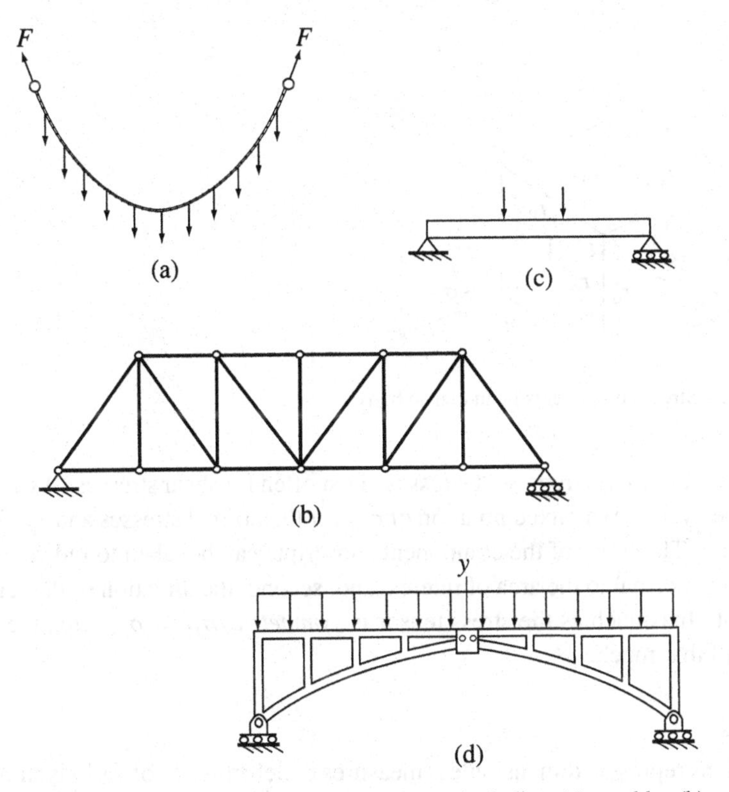

Figure 1.5. Basic one-dimensional elements, including (a) a cable, (b) axially loaded members, (c) a beam, and (d) an arch.

1.2.4 Stresses

The stresses are the detailed, point-by-point distributions of the internal forces over the relevant geometry of the structural members. Thus, for uniaxially loaded bars or rods, the stress is normal to the bar's cross section and uniformly distributed over it. For a bent beam, the most important stress is also normal to the beam's cross section, but it is distributed linearly over that cross section so as to produce the resultant bending moment needed to maintain equilibrium. (These two elements should be familiar from "Strength of Materials," but we will review them in this book as well.)

The notation we use for stresses derives from the fact that they are idealizations of what happens when we divide one vector, a force, by another vector, the area on which the force acts. The vector nature of the area stems from the *unit normal vector* we construct perpendicular to the area to define its "direction," as we show in Fig. 1.6. Thus, we need to identify three components of the force and three for the area, and then in some sense we need to divide the two sets of components. There is no mathematical law for dividing vectors, but we can think of the stress as being a *second-order tensor*, with six symmetric components, that is,

$$\begin{bmatrix} \sigma_{xx} & \sigma_{yx} & \sigma_{zx} \\ \sigma_{xy} & \sigma_{yy} & \sigma_{zy} \\ \sigma_{xz} & \sigma_{yz} & \sigma_{zz} \end{bmatrix} \tag{1.1}$$

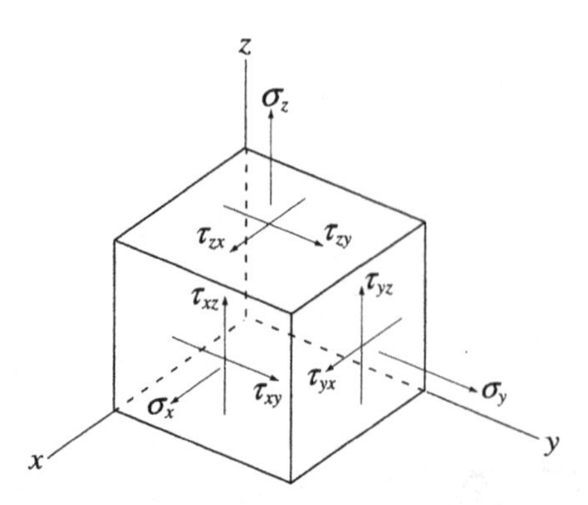

Figure 1.6. Stress components in an elastic body.

We also use the notation τ_{xy} to represent stresses, most often for shear stresses. As a result, we often see (and even use) a mixed notation of σ_{xx} for the normal stresses and τ_{yz} for the shear components. The order of the component subscripts can be taken to indicate, first, the direction of the normal to the area of interest and, second, the direction of the relevant force component. Inasmuch as the stress tensor is *symmetric*, $\sigma_{xy} = \sigma_{yx}$, etc., the order doesn't matter all that much.

1.2.5 Strains

The strains represent dimensionless measures of deformation of various structural elements. As is true of stress, strain is also a second-order tensor with components that can be arrayed just like Eq. (1.1). The strain tensor includes the three *normal strains* that reflect the (dimensionless) change in length of a strained element divided by the element's original length (see Fig. 1.7). In Cartesian coordinates the normal strains are

$$\begin{aligned} \varepsilon_{xx} &= \frac{\partial u}{\partial x} \\ \varepsilon_{yy} &= \frac{\partial v}{\partial y} \\ \varepsilon_{zz} &= \frac{\partial w}{\partial z} \end{aligned} \tag{1.2}$$

The strain tensor also includes the three *shear strains* that reflect the changes in the angles of a strained parallelapiped (see Fig. 1.7)

$$\begin{aligned} \gamma_{xy} &= \frac{\partial u}{\partial y} + \frac{\partial v}{\partial x} = 2\varepsilon_{xy} = 2\varepsilon_{yx} \\ \gamma_{xz} &= \frac{\partial u}{\partial z} + \frac{\partial w}{\partial x} = 2\varepsilon_{xz} = 2\varepsilon_{zx} \\ \gamma_{yz} &= \frac{\partial v}{\partial z} + \frac{\partial w}{\partial y} = 2\varepsilon_{yz} = 2\varepsilon_{zy} \end{aligned} \tag{1.3}$$

Note that in Eqs. (1.3) we have written strain in terms of two variables, one of which is used to describe the shear strains in their true tensorial character ε_{xy}, the other being the

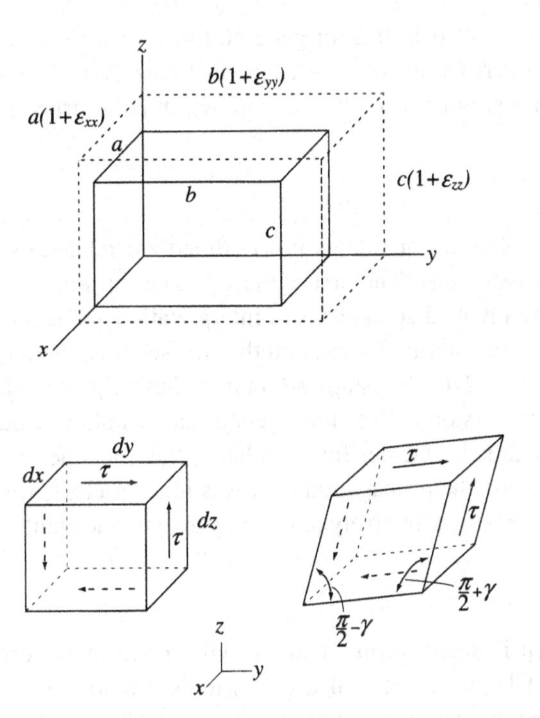

Figure 1.7. Normal and shear strains in an elastic body.

engineering shear strains, γ_{xy}, which are more commonly used in structural engineering calculations.

It is clear that strains are dimensionless, in part because they have dimensions of length/length, in part because of the way they are defined. This is important in providing both another point to check on the correctness of any calculations and a scale for measuring the magnitude of the phenomena we are modeling.

1.2.6 Displacements and Deflections

These are the more evident indicators of the change in dimensions and perhaps shape of a structure as it responds to the applied loads. For structural systems, the deflected shape of the structural system taken as a whole might be modeled rather differently than we might model the structural elements that make up the system. Thus, we might see bending in a simple beam, bending for a bridge whose major members are truss frameworks, or bending of a tall and slender high-rise building. Clearly, the ensemble model may turn out to be an elegant average of complex arrangements of elements.

The basic "unit" of displacement is implicit in the definitions of normal and shear strain just given, Eqs. (1.2) and (1.3). We think of a continuum model and identify the displacement at any point in the continuous solid as having components in the three coordinate directions that can, in general, depend on all three coordinates. Thus, the displacement is a vector of the general form

$$\{u\} = \left\{ \begin{array}{c} u(x, y, z) \\ v(x, y, z) \\ w(x, y, z) \end{array} \right\} \tag{1.4}$$

Remember, though, that while Eq. (1.4) is useful for general, theoretical formulations, we typically describe the deflection response of a structure in terms of more specific idealizations that conform to some geometric model of how we think a structure will behave.

1.2.7 Equilibrium

We have said that the six physical quantities just outlined are related by three basic physical principles or sets of equations. The first of these is equilibrium.

The *equations of equilibrium* are invoked at all levels of interpretation and calculation of the external loads, the reactions, the internal forces, and the stresses (i.e., the physical quantities outlined in Sections 1.2.1–1.2.4). For static structures, these equations are all variant representations of the static version of Newton's second law as already quoted. When dynamic effects need be included, whether for calculating the dynamic response to a time-varying load or for analyzing the propagation of waves in a structure, we must include the proper inertia terms, so that we effectively must analyze dynamic equilibrium:

$$\Sigma F = m\frac{dv}{dt}$$

Now we will apply equilibrium in many forms, and we will derive these forms as we need them. For example, in Chapter 3 we will use Newton's law to describe the equilibrium of the forces acting on discrete models of structures, in Chapter 4 we will write the equilibrium equation for the axial stress in axially loaded bars, while in Chapter 5 we will derive the equilibrium equations for a three-dimensional element as part of a more general discussion.

1.2.8 Constitutive Laws

The *constitutive laws* define the material of which a structure is made as they relate the stresses to the ensuing strains in terms of material properties. The most familiar of these is Hooke's law, which defines linear elastic materials for which stresses are linearly related to strains. This model of material behavior is quite useful in structural engineering and we will use it exclusively in this book.

Further, we will use Hooke's law to describe elastic materials that also have two other properties. We will consider our structures to be made of materials that are *homogeneous*, by which we mean that the material properties are not functions of the coordinates, and *isotropic*, by which we mean that they are independent of our orientation (or of a rotation of axes) at any point in the body or structure. Introducing these restrictions for our models of material behavior does preclude a complete discussion of all structures under all circumstances (cf. Chapter 2), however, it also means that we need introduce only two elastic constants (Young's modulus E and Poisson's ratio ν) for materials that do cover a substantial part of the structural waterfront.

1.2.9 Compatibility

Compatibility conditions are also referred to as continuity or consistency conditions on the strains and the deflections (i.e., physical quantities of Sections 1.2.5 and 1.2.6) because they assure that the deformed geometry of a loaded structure is a single-valued, continuous function. Thus, in ways that also signal a desire not to violate common sense,

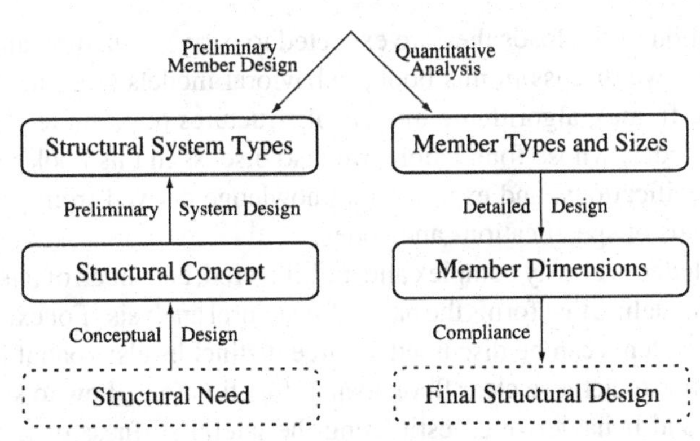

Figure 1.8. A pictorial view of the structural engineering problem. (Courtesy of Steven J. Fenves.)

as a structure deforms under a load we want to ensure that: two originally separate points do not move to the same end point, holes do not emerge as the structure deforms, and elements that are connected together initially remain appropriately connected as the structure deforms under the applied loading.

We note that in the preceding discussion we have imposed certain limitations in our formulation of basic structural mechanics, namely, that we are dealing only with the static response of structures made of linear elastic materials. The unified approach that forms the backbone of our approach can be extended to more general cases, both theoretically and in many numerical implementations, but we believe that the limitations imposed here provide a complex enough subject for a first course in modeling and analyzing structural behavior.

1.3 Structural Analysis and Structural Design

Now the basic principles just described are applied in both structural *analysis* and structural *design*. To illustrate this, let us consider a fairly typical structural engineering problem (cf. Fig. 1.8). We start by identifying a *structural need*, whether it is for a mill building or a concert hall. Then we choose a *structural concept*, perhaps a simple steel frame and steel roof truss for the mill building, and likely something considerably more complicated for the concert hall. We then move to *preliminary design* during which we estimate approximate sizes for the principal structural members, the object being to see whether the type of structural system that we have chosen is practically feasible. We then try to realize the rest of the structure by estimating the types and sizes of the remaining members (e.g., in the mill building, purlins for the roof truss and floor joists as needed). Then we assemble the *detailed design* in which we calculate actual dimensions and placements for all members and their connections. In the last step we check to ensure that our design meets all statutory requirements, including both applicable building codes and the design specifications that are promulgated by various professional associations (to be discussed in greater detail in Section 1.4).

We employ many kinds of knowledge to complete a structural design, including: classical mechanics (e.g., Newton's laws); structural mechanics (e.g., models of columns and beams); geometry of structures (e.g., relating the geometry of members and assemblages of members to the orientation of the loads they are expected to carry); structural analysis techniques, including those we discuss in this book; behavioral models (e.g., modeling the stiffness of a complete frame); algorithmic models of structures (e.g., finite element method (FEM) computer codes, whose foundations we also discuss in this book); structural design codes and specifications; and experiential knowledge derived from practice and encoded in various kinds of specifications and codes.

Structural design knowledge is clearly complex and multilayered, and much of it is quite beyond the mathematical modeling that forms the basis of structural analysis. For example, the behavior of structural systems can be discussed at three distinct levels: spatial layout (e.g., where to place columns to achieve clear floor spans), function (e.g., how to support different kinds of loads), and behavior (e.g., estimating the lateral stiffness of frames). And while we cannot use analysis models to perform these structural design tasks, we also cannot complete a structural design without having serious analysis capability because we can't verify the applicability of a structural concept, size the various members, detail the location and design of connections, and so on, without such analysis capability. Perhaps another way of saying it as that without analyzing structures prior to their construction, we would be forced to depend on trial and error as our guiding principle of structural design!

In this book we will focus on a unified approach to developing basic mathematical models for the mechanics of structures. Further, we are focusing almost exclusively on one language or representation for structural engineering knowledge, namely, mathematics. The approaches we will develop can be used to: derive detailed models for specific structural elements (e.g., the behavior of truss or beam elements), develop case-specific analyses (e.g., buckling or loss of stability of slender columns), deduce "back of the envelope" formulas to model various complex phenomena (e.g., the beams like response of tall buildings), and lay the foundations for the numerical programs that are so widely used in structural engineering (e.g., FEM codes).

We should also observe that the view of design presented in Fig. 1.8 is less than complete because it does not picture the many iterative loops that inevitably occur in the design of complex structures. That is, the figure does not show the extensive calculations that go on when we refine the loads and member dimensions over the course of a design as we home in on better detail. This is especially true in the design of indeterminate structures because member properties are assumed ab initio, and then a process of iterative refinement is undertaken to converge to a final structural configuration by repeated analysis.

1.4 Design Criteria, Codes, Specifications, and Loads

We have noted that the analysis models we will begin to develop are integrally tied into the design function. Thus, it is worthwhile to define the two broad paths to specific structural designs and to the kinds of design criteria that span the kinds of structural behavior that must be modeled in structural design. We will also discuss some aspects of the design codes that represent formulations of design criteria to meet various legal and professional requirements.

1.4.1 Design for Stiffness and Design for Strength

We can identify and highlight two broad classes of design goals by looking at two commonplace situations. Consider the chair you use when you sit at a dining room table, the table itself while you eat, or the desk you will use while working out your structures homework. One feature that all these simple structures – and yes, they are structures – have in common is that they support weights or loads significantly greater than those to which they are normally exposed. That is, your desk will typically support a collection of people standing on its top, and likely many more people than you would want on your desk. Why is this so? Are desks designed to be super strong, to the point of being uneconomic designs, achieving a level of structural performance not normally needed? In fact, these structures are designed for *stiffness*, not for load-carrying capacity, so that when we write on a desk it doesn't flex under the pressure of our work, and when we move to sit on a chair we can identify a destination for our backsides!

Conversely, we often construct a makeshift platform of a board – say, a 2 × 10 – propped up between two widely spaced stepladders to use when painting the upper part of a wall. We also know that this board is fairly flexible if it's long enough, so we can feel ourselves moving up and down as the board flexes. While this movement is unsettling, it is usually true that the board itself doesn't break or otherwise fail, so that it does safely carry our weight while flexing. Although this particular structure has not been formally designed, we intuitively recognize that it has been designed for *strength*, to sustain our weight, and not to minimize how much flexing we will experience.

As an example of these two broad types of design criteria, consider the design of a simple linear spring. Inasmuch as this book deals only with linear elastic structures, we will find that structural elements can often be modeled as linear springs whose spring constants represent aggregated models of the geometry and material properties of the structural elements. However, for a simple linear spring, the constitutive law is given as (see also Chapter 3)

$$F_s = k\xi \tag{1.5}$$

Here we use the symbol ξ to define the relative extension of the spring, that is, the difference in the displacements of the spring's two ends, and F_s and k represent the force in the spring and the spring stiffness, respectively. In designing for stiffness we would choose a value of k so that the spring's relative extension would not exceed a prescribed value while under the influence of a given force P, that is,

$$\xi = \frac{P}{k} < \xi_{prescr} \tag{1.6}$$

On the other hand, if we were designing for strength, we would simply require that the force in the spring be such that

$$F_s = k\xi < P \tag{1.7}$$

Thus, in designing for strength, we need only ensure that the maximum spring force F_s be less than the applied external force P.

We note emphatically that these simple cases are by no means indicative of the full range of challenging design problems that await the structural engineer. Even for the simple spring, one can wonder about the possibility of designing for stiffness and strength at the same time – although just a moment's reflection will suggest that the stiffness design is likely to dictate a more restrictive constraint on the single design parameter k, so that designing for stiffness will almost certainly guarantee meeting a requisite strength requirement. However, for more complex structures subjected to many more loads, the very meaning of strength and stiffness design becomes more complicated. This is due in part to an increasingly sophisticated understanding of structural and material behavior and in part to a changing emphasis on how far – and in what ways – we can push structures before we regard them as failed. For example, we can now utilize the ductile behavior of high-strength steels more effectively by allowing parts of structures to deform *plastically*, that is, beyond the linear elastic range typically defined by Hooke's law and Young's modulus. In fact, our increased understanding has generally made structural design both more interesting and much harder. It is just because there are so many structural elements, arrayed in so many configurations, that there are also many interesting – and tough – challenges in trying to develop better – even optimal – structural designs with regard to a variety of important parameters, including weight and cost.

1.4.2 Some Behavioral Design Criteria

Now, in our simple model of the last section we have not even begun to exhaust the kinds of physical behavior for which structures are designed. For example, if the spring had a mass m attached, we might be interested in its dynamic response, in which case we might need to meet a restriction on the natural frequency of the spring-mass system modeled as an oscillator, for example,

$$\omega = \sqrt{\frac{k}{m}} < \omega_{prescr} \tag{1.8}$$

Again, with just a casual suggestion about the properties of our spring, we see that there are other kinds of physical behaviors and phenomena that might well dictate important aspects of a structural design. Thus, we now identify some commonly used *behavioral design criteria* that suggest different measures of behavior that are seen in structural design. Expressed in terms of the six basic physical quantities listed in Section 1.2, such *behavioral criteria* include the following.

Strength criteria. These criteria require that *stresses* in structural elements be below certain specified values that reflect *limit states* of structural behavior wherein certain behaviors are limited. For example, *yielding* is the limit state that restricts structural behavior to remain entirely elastic, so that all stresses are below the yield points of the stress–strain curves of the structure's material (viz., Fig. 2.15). *Buckling* is the limit state that restricts structural behavior to avoid a loss of stability that occurs at critical loads that define the existence of alternate or multiple equilibrium states.

Stiffness criteria. These criteria are typically related to functions perceived in relation to structural *deformation* or *deflections*. For example, there may be an upper limit to floor deflections in an otherwise flexible, high-rise office building, or we might restrict the magnitude of a floor's response to some disturbance in a sensitive laboratory environment.

Fatigue and crack criteria. These criteria are applied to both *stresses* and *deflections*, and they are concerned with avoiding failure due to repetitive motion, such as in intense vibration environments. Familiar examples are the failure of ductile metals (like paper clips) when bent repeatedly and the propagation of cracks in both brittle and ductile materials, especially near sharp edges, bolt holes, and other discontinuities. For example, we often see cracks in both steel and reinforced concrete bridges.

Stability criteria. These criteria are typically expressed in terms of the *externally applied loads*, requiring them to be below certain values to avoid structural failure through a loss of stable equilibrium. One very common example is the buckling failure of columns under compressive axial loads, but losses of stability are conceptually quite similar to the difference between considering a ball *stable* when it is placed within a bowl and *unstable* when it is placed on top of an inverted bowl.

Vibration criteria. These criteria are various in nature and are applied to *stresses* and *deflections* (from our list of basic physical quantities of interest), but also to unlisted quantities such as the *natural frequency* of a structure, the *spectral content* of an external load, and other dynamic variables that are beyond the scope of our present work.

Indeed, in what follows we will limit ourselves to thinking about strength and stiffness as important characterizations of structural behavior. The balance of the list is meant only as a reminder that while we can (and will!) do interesting and important problems with the techniques developed in this book, there is much, much more to the art of structural engineering.

1.4.3 Design Specifications and Codes, and Fabrication Specifications

We could once more look at our picture of the structural engineering problem (Fig. 1.8) and ask: Where and how do design criteria enter? Further, how are they expressed to meet concerns – such as professional practices and legal requirements – beyond the technical criteria? The answers to these questions are involved because of the involvement of many issues and of many stakeholders, that is, people who have a stake in the outcome of a design. Further, each of these stakeholders will see the design of a structure from a different perspective. Consider the design of a hospital, for example. The owner of the hospital – perhaps a medical school of a large university, perhaps the health bureau of a large urban county – poses a multifaceted structural need (cf. Fig. 1.8), that is, that the new hospital will be esthetically pleasing to the eye, that it will meet the needs of both patients and medical staff, and that it can be built within a certain budget. The architect's concerns range from conceptual (i.e., the concern with esthetics), down through preliminary system design (i.e., the number of beds, the location of operating rooms and of special-purpose diagnostic equipment, the proximity of these facilities), and to some aspects of very detailed design (e.g., the number of fire exits and exit signs). The structural engineer does conceptual design to develop alternate framing plans on which to hang the architect's esthetic vision; preliminary systems design to figure out how to support the various rooms, corridors, service cores, and the like; and detailed design when working out such details as piping intersections and member connections. The army of contractors that assemble such a complex building must understand clearly the designers' intentions and feel confident that the structure can actually be built as designed. And, of course, relevant civil governing bodies will typically impose restrictions in terms of the safety of the hospital for its occupants and users, and in terms of its impact on its local environment. In each case, the stakeholders have targets at which to aim their own thinking about the project.

Now, it is fairly obvious that there are different ways in which designs are represented to and interpreted by those involved. The hospital owner who might be interested in the comparative economics of precast concrete slab versus steel frame construction, is not likely to care whether a steel frame is welded or bolted, but certainly does want to know that the hospital structure will withstand earthquakes and other potential disasters. In fact, what is going on in this hospital design process is the *translation* or conversion of the owners's (or client's) wishes from a set of fairly abstract qualities (e.g., esthetics, safety, etc.) into a set of targets against which to measure the performance of the structure being designed, including size, footprint, capacity, parking, and other gross attributes of the (in this case) hospital. However, the design targets also include *design codes* and *design specifications*, which serve as the codifications of both "best practice" and relevant legal requirements that pertain to a hospital (and to other structures). These codes and specifications are included in the "compliance" phase of Fig. 1.8. The structural engineering part of this translation process is often identified with the statement that the structure is designed to have the *resistance* necessary to withstand the specified *loads* that must be carried. The structural designer must therefore identify the loads that occur due to the hospital's intended operation in the given environment and choose the configuration and materials needed to conform with good design practice about how to resist (or support) those loads.

Thus, the design targets at which structural engineers aim their thinking include three different articulations of expected structural behavior: design specifications, design codes, and loads. However, there is considerable overlap in the ways that these terms are used, so we will try to provide some clarity. Design codes are typically promulgated by governmental or quasi-governmental bodies to reflect what has emerged as successful building practice. Thus, in a sense they represent the emergence of such good practice as seen through the eyes of public officials as historians. Some of the codes that structural designers must take into account include local and regional building codes for residential construction and the *Uniform Building Code* (UBC) promulgated by the International Conference of Building Officials. There are also codes that stem from insurance and liability requirements and are incorporated into regional codes (e.g., the *Life Safety Code Handbook* (LSC) of the National Fire Protection Association) as a matter of public policy.

Design specifications are typically produced by professional organizations to reflect what is considered to be the best state of the art, that is, the best understanding of how things work in a specific professional context. Some of the codes are produced by those skilled with specific materials (e.g., the American Institute of Steel Construction's (AISC) manuals for steel design and the American Concrete Institute's (ACI) codes for concrete) and some by experts in certain kinds of environments or behavior (e.g., the seismic code of the Structural Engineers Association of California (SEAOC) of California, called the SEAOC's *Recommended Lateral Force Requirements and Commentary*).

Loads refer to the various environments to which a structure may be subjected and to the specific instances or realizations of these environments. Some codes that describe environments and loads include the *American National Standard Minimum Design Loads for Buildings and Other Structures*, put forward by the American National Standards

Institute (ANSI), and the *Standard Specifications for Highway Bridges*, published by the American Association of State Highway and Transportation Officials (AASHTO). We will detail some of the kinds of loads (or loading environments) in Section 1.4.4, and we also present a sampler bibliography of design codes and design specifications that are in use in structural engineering in America.

As written documents, design specifications and codes actually reflect three distinct approaches, as we will describe immediately below. We note that they are often written with loads as the "prescribed variable," but they can also be cast in terms of required physical attributes or desired physical behavior.

Prescriptive specifications and codes. Specific dimensions and other attributes of the designed structural element are specified so as to completely define all of the acceptable configurations. In other words, acceptable physical objects are prescribed. For example, "A wall stud in a wood frame house is acceptable as a column if made from Grade A fir of nominal 2 × 4 cross section and of length that does not exceed 96 inches."

Procedural specifications and codes. Specific procedures are given for calculating desired attributes or behavior. In other words, acceptable calculations are described. For example, "A column in a wood frame house is safe if the axial compressive load it supports is such that $P \leq P_{allow}$, where $P_{allow} = \pi^2 EI/L^2$."

Performance specifications and codes. The desired behavior is just expressed as such. In other words, acceptable behavior (or a range of behavior) is described. For example, "A column in a wood frame house is considered safe if it will support an 800-pound gorilla."

Our point here is simply that the structural analysis that is done as part of the structural design process can be interpreted in several different lights as the design unfolds. This provides further motivation for thinking ahead about structural calculations, so that they can be used to produce answers to the proper questions.

In a strongly related way, we should also remember that engineers *do not* typically build what they design. Rather, they produce *fabrication specifications* from which their design can be manufactured or assembled. These fabrication specifications are expressed in drawings and blueprints and in accompanying volumes of detailed "specs" that including parts lists, materials specifications, assembly instructions, and so on. Clearly, a successful structural design is a building whose performance meets (or exceeds) the given specifications and satisfies (or exceeds) the client's expectations. However, it is important to remember that fabrication specifications may evolve or be expressed in greater detail as the design unfolds. Even more important, because the construction or assembly is going to be done by others, often without the benefit of any interaction with the designer(s), the fabrication specifications created by the structural designer(s) must be: *complete*, so that the fabrication specification explicitly applies in any possible situation; *unique*, so that the fabrication specification yields one and only one result in any possible situation; and *correct*, so that the built structure is exactly the one intended by the author(s) of the specification.

Lest the reader think that this is just an academic issue, consider one of the most infamous and costly – especially in human terms – structural failures in American history. The 1981 collapse of Kansas City's Hyatt Regency Hotel occurred because a contractor, unable to procure threaded rods that were long enough to suspend a second-floor walkway from a roof truss, used shorter rods and hung it instead from the intermediate fourth-floor walkway. However, the fourth-floor walkway supports were not designed to carry the second-floor walkway in addition to its design loads, so a major disaster occurred. Had the designer been able to signal to the contractor the *design intention* of suspending the second-floor walkway directly from the roof truss, this accident might never have happened. And had the contractor thought to ask why the longer rods were called for, the designer's intent to hang the walkway from the roof truss could have been made explicit, rather than implicit, and so a major tragedy might have been averted.

In fact, the subject of preserving and making explicit the *designer's intent* is a serious topic of design research today, and in many design domains. It is usually wiser (and safer) *not to assume* intentions from a drawing or blueprint, any more than an instructor can always make fair assumptions from a partially completed examination paper. Thus, it is a point to keep in mind in doing both design and analysis, that anything out of the ordinary ought to be so marked to ensure that it receives proper attention.

1.4.4 Design Specifications and Design Loads

Finally, we address the issue of the kinds of loads that a building or other kind of structure is expected to support or resist during its expected lifetime. As we have already noted, these loads represent part of the translation of the desires and expectations of the building's owner. The loads also represent a characterization of the environment in which a building will operate, as will become clear in the short list that follows.

Dead loads. Dead loads constitute the weight of the building or structure, together with any and all objects or devices that are permanently attached. Thus, dead loads include the weight of the structural members, all walls – including both exterior cladding and interior partitions, plumbing equipment and pipe runs, heating and air conditioning equipment and ducts, elevators, and so on.

Often, and with experience, dead loads are cast in terms of weight per unit area. For some kinds of loads, for example, the weight of a concrete slab of a given thickness, such statements of dead load are evident and straightforward. In other cases, some averaging is done so as to "smear" the weight of a set of discrete members, for example, over the entire floor area they might be supporting.

Live loads. Live loads are transient or temporary loads that result from the building's use. These are also "smeared" or aggregated areal distributions, and their magnitudes can vary enormously with the building's use. For example, apartment buildings and classrooms are designed for live loads of about 2 kN/m^2 (40 lb/ft^2), while dance halls and auditoria are designed for a bit less than 3 kN/m^2, and heavy manufacturing facilities for about 12 kN/m^2. These load estimates are *static* loads, and so dynamic or time-varying loads that occur as a result of machinery operation of vehicular traffic must be dealt with separately.

Wind loads. The first of the "environmental loads," wind loads result from a structure's exposed surface acting as an obstruction to the flow of a fluid, air, that we identify as the wind. These loads can be substantial, depending in significant part on geographical location and on building characteristics, especially if we're designing a tall, slender building. In fact, the aerodynamic effects are very hard to model for skyscrapers, and serious consequences that range from perceptible discomfort to structural damage can occur if wind loads are not properly taken into account. As a result, many such buildings are aerodynamically tested in very large, specially designed wind tunnels. Loads due to events such as tornadoes are not included in typical wind load codes, so they must be investigated separately if tornado resistance is thought to be important for the building at hand.

Seismic loads. Seismic loads are another – and perhaps the most widely known – of the environmental loads. Designing for seismic events is very hard, because we do not have anything like a complete understanding of the kinds of loads that occur during earthquakes, so we cannot clearly identify appropriate design considerations and criteria. In part this is because earthquakes are dynamic and uncertain in their occurrence, and the kind of motion that is imparted to buildings depends on both the nature of the ground at the earthquake's epicenter and the nature of the ground under the building being designed. Further, there is still ongoing debate about the correct metrics for assessing seismic response, with there being wide agreement only about fundamentals such as trying to incorporate a maximum of flexibility in the structure and ductility in the joints, and in designing a structure to accommodate significant intrafloor shear forces in taller buildings.

Rain and snow loads. These are more environmental loads, and clearly highly dependent on geography. Rain is particularly important for structures with flat roofs because sudden rainstorms and/or inadequate drainage can produce significant loads due to the water "ponding" on a roof. Similarly, snow can comprise a significant load, particularly in Northern climates, even when a building might have a steep roof design to minimize the buildup of snow loads. Ponding has also been known to occur under snow loads, especially as melting begins at the bottom of a roof's snow pack.

This list clearly does not cover all of the kinds of loads that could exist for all kinds of structures. For example, we have not described the loads that occur on bridges that carry automobile and truck traffic or railway traffic. There are special codes for bridge design for such structures. Nor have we mentioned the hydrostatic pressures that a dam must resist. The point is that there are many loads that need to be taken into account, depending on the structure's purpose and its geographic location.

Further, for any given structure, loads will have to be taken in combinations that reflect an experienced perception of reality. For example, it is unlikely that both a seismic event and a hurricane will occur simultaneously, so a weighted or modified sum of loads can be used. The design approach can also influence the way in which loads are combined for design purposes. For example, depending on geography and purpose, a combination of dead, live, and snow loads might be thought more probable than a combination of dead, live, and earthquake or wind, so that the former combination would be taken as a more critical design load. Again, the point is that design is both complex and complicated, and we will not do any serious structural design in this first course. However, it is also impossible to do any serious structural design without having a fundamental understanding of how structures are modeled and how they behave – and achieving such understanding is our goal for this first course.

[illegible] and Delaware [illegible] and beautiful example of a [illegible] [illegible] (Photo courtesy of the [illegible])

The Chesapeake and Delaware Canal Bridge at St. Georges, Delaware, completed in 1995, is a recent and beautiful example of a cable-stayed bridge. Such bridges differ from "classical" suspension bridges in that the roadway is hung from the tower(s) with only a single line of cables, a development that is made possible by using a box girder as the basic deck element. Thus, torsional resistance to lateral bridge loads is provided by deck that is a deeper and stiffer than what is seen in the conventional suspension bridge. (Photo courtesy of Figg Engineering Group.)

2

Structural Models and Modeling

Is is clear from our discussion in Chapter 1 that we want to be in the business of modeling structures and their behavior. Thus, we will devote this chapter to describing in greater detail what we mean when we talk about structural models, and we will do this within the context of the modeling process.

2.1 Modeling Structural Elements: Bars, Beams, Arches, Frames, Surfaces

We begin by extending our brief structural history of Section 1.2 to describe specific models of structural elements in greater detail. We have noted that there are two broad categories of structures: those whose principal direction or line of action is coincident with the loads to which they are subjected and those whose principal direction or action is perpendicular to the direction of the applied loads. Within these broad confines we can further subdivide the domain of structural elements according to the number of physical dimensions needed to properly account for the behavior of interest – a number that is at least one, often two, but certainly no more than three! Thus, structures are often viewed as one- or two-dimensional structures, depending on the relative magnitudes of a structure's physical dimensions and the number of (independent) spatial variables in the equation(s) of equilibrium in the structure's mathematical model.

Now, the directionality and dimensionality issues do interact in our attempt to identify structures with neat categories. This is a consequence of the fact that it is easier to create simplified, *idealized* models of structural elements or types than it is to ensure that all structures act in just such simple ways. However, we will also suggest that our idealized models are widely representative of basic structural forms, so that these models provide a vocabulary for identifying, describing, and analyzing more complicated *structural forms* and thus more complex structures.

There are cases where one- and two-dimensional idealizations won't do. For example, stress analysis of holes and cracks require more detailed analysis, even when they occur in the gusset plates of trusses made up of one-dimensional bars. Such analyses are, however, considered more as aspects of three-dimensional elasticity theory than of structural analysis, and so they are not discussed here.

2.1.1 One-Dimensional Basic Structural Elements

In the first category we discussed, in which the loads and their attendant principal structural responses have the same directions, we are usually talking about structural elements that can be modeled as *one-dimensional structures*. These structures act in pure tension, as ropes or some of the bars in a truss, or they act in pure compression, again in truss bars and in columns – although columns deserve a category of their own because they

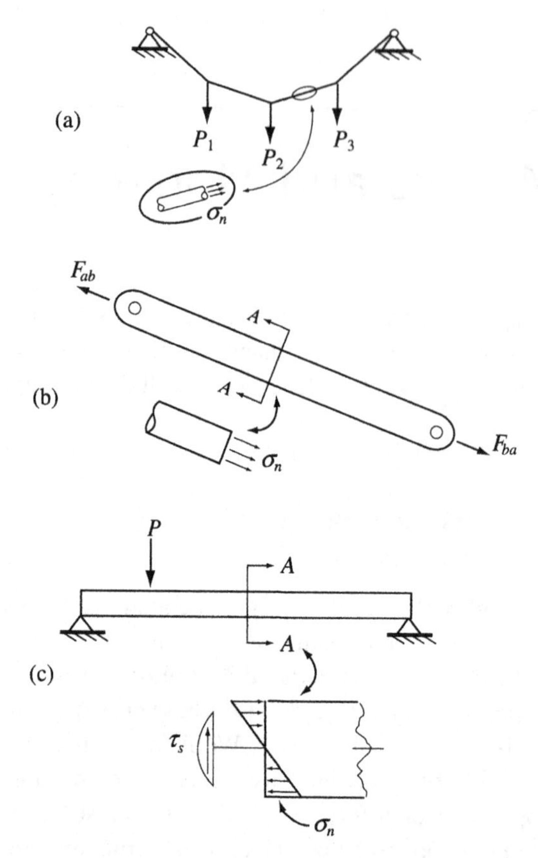

Figure 2.1. Some one-dimensional structural elements and their dominant stresses: (a) cables, (b) bars, and (c) beams.

exhibit unusual behavior because of limits on the loads they can carry that are dictated by nonlinear geometrical concerns.

In Figs. 2.1 and 2.2 we show the basic one-dimensional structural elements and their dominant or major stresses. For this discussion, we will denote such dominant stresses by σ_n for normal stresses and by τ_s for shear stresses. We also note that each of these elements is far longer than it is deep or wide. Stated otherwise, the depth h and the width b of each of these elements are small compared to its length L, and "small" typically means less than one-tenth.

We begin with the cables of Fig. 2.1(a) wherein the loads are often directed either perpendicular to or at other nonzero angles with the cable, and the cable itself generally has "kinks" at points where the concentrated loads are placed. (Where the load is aligned with the cable, we have the simple case of a rope in tension.) The resulting stresses in a cable or rope are normal stresses that act along the cable's lines or segments, always in tension. Remember that you can't push on a rope. Such cables are clearly mainstays of classic pedestrian bridges, and experience and intuition suggest that the cables of such bridges will need both horizontal and vertical support at the ends.

In Fig. 2.1(b) we show bars, for which we assume that loads can be applied only at the endpoints, not along their lengths. Thus, each bar can be viewed as a single segment of

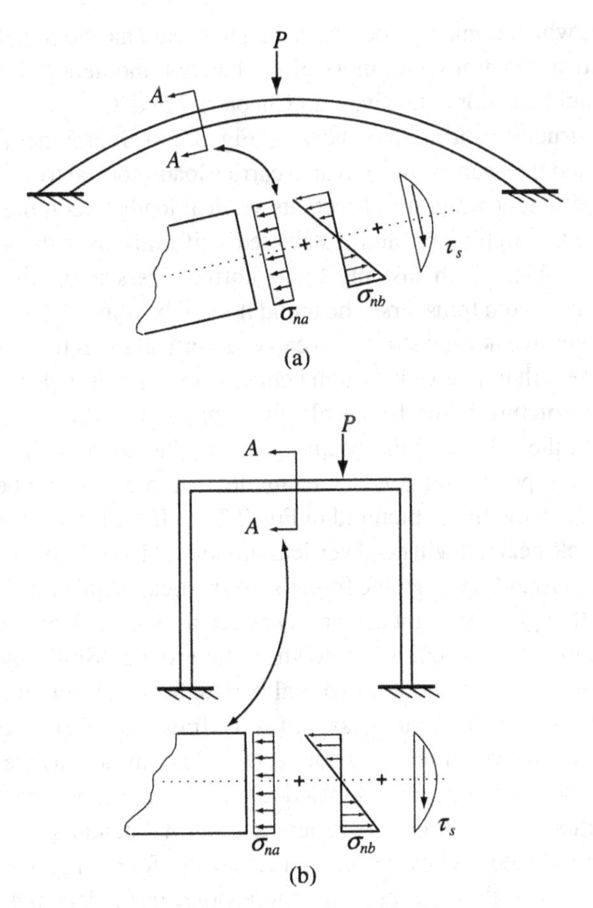

Figure 2.2. More one-dimensional structural elements and their dominant stresses: (a) arches, and (b) frames.

the cables of Fig. 2.1(a), although bars are more useful if we allow them to support both tensile and compressive normal stresses. This is because bars taken individually are not of much use, while assembled into truss structures they become very useful, as we will see in Chapter 5. Thus, bars and cables form a class of one-dimensional structures wherein the dominant stresses are normal stresses acting along the line(s) of the structure and the loads are applied, respectively, in directions that may depart significantly from those lines (for cables) and in arbitrary directions at the endpoints (for bars) at which they are joined to other bars to make up trusses.

The next one-dimensional structure we examine is the class of beams (cf. Fig. 2.1(c)). Here the loads are almost invariably normal to the axis of the beam, and the beam supports these loads by a combination of normal bending stresses in the axial direction and transverse shear stresses perpendicular to the the beam's axis. Recall that a beam cannot be supported solely by the transverse shear stresses because it is impossible to satisfy moment equilibrium without the bending stresses (cf. the beam section A–A of Fig. 2.1(c)). Beams are useful because they transfer loads from locations along their span to the reactions that are typically located at the beam's ends. While doing this, they *bend* or deflect transversely. Thus, the beam is clearly the kind of structure intended to support loads perpendicular to its dominant geometrical line, and its visible behavior is its vertical deflection. However, unlike the cable of Fig. 2.1(a), which arguably does the same, the beam has no horizontal reactions at its ends, only vertical reactions and, more often than not, moments. We will discuss beam analysis and beam behavior extensively in Chapters 7 and 8.

The third one-dimensional structure, the arch, is shown in Fig. 2.2(a). It provides an interesting contrast to the cable and the beam. The arch also carries loads normal to its basic direction; however, it is most efficient when it redirects its vertical loads to compressive normal stresses directed along the arch's axis and distributed uniformly over the arch's thickness. However, unlike the cable, which has only a pure normal stress across its cross section, a less-than-ideal arch uses both transverse shear and normal bending stresses – in addition to the dominant compressive normal stress – to carry its vertical loads to the reactions at the ends of the arch. We will not delve into arch behavior any further in this book.

The final one-dimensional structure is the frame. In the same way that an arch has behavior in common with both the cable and the beam, the frame has its roots in beam behavior combined with barlike aspects. Imagine, for example, that three slender beams are used as building blocks to erect the frame pictured in Fig. 2.2(b). If the topmost frame element is something like a simple beam, it will need vertical supports at its ends to transmit its vertically applied load to the ground (through the frame's two vertical members). Thus, the vertical legs of the frames, though they might act largely as beams, must exhibit barlike (or columnlike) behavior to carry the vertical loads down to the ground. Similarly, the horizontal bar atop the frame will combine both horizontal barlike behavior with its own beamlike bending. Thus, the transverse or shear stresses of some frame members become the axial stresses of the members to which they are connected. And, in fact, as we will show in Chapter 9, frame members are generally modeled as beams that support uniform axial, barlike stresses, in addition to their transverse shear and normal bending stresses. Further, although we have introduced arches before frames in the foregoing, we will devote Chapter 9 to frames because they are used far more often in modern building design. Arches are more typically used for some bridges, and occasionally for some special effects in buildings, such as the proscenium arches of theaters and concert halls. Frame elements, on the other hand, are literally the building blocks of modern office and apartment building design.

2.1.2 Two-Dimensional Basic Structural Elements

Clearly not all structures are one dimensional or made up solely of one-dimensional components. There are several important two-dimensional or *surface* structures, both planar and curved. Although they are beyond our current concerns, it is worth describing such surface structures at least briefly.

The first surface structure is a two-dimensional counterpart of the cable, called the *membrane*. A membrane supports vertical loads through purely tensile normal stresses (cf. Fig. 2.3(a)), except that the two-dimensional nature of the membrane renders it unable to kink like a cable. Perhaps the most common form of a membrane is the trampoline, but membranes are often used as roofs or domes over stadia and are partially supported by the air pressure contained by the roof. In fact, membranes are not often used in buildings because they require enormous tensile stresses and forces in order to maintain a virtually flat surface under loads applied perpendicular to that surface, which in turn means that there needs to be some massive structure erected to support those in-plane stresses and forces. In fact, there are parallels between the membrane support requirements and the

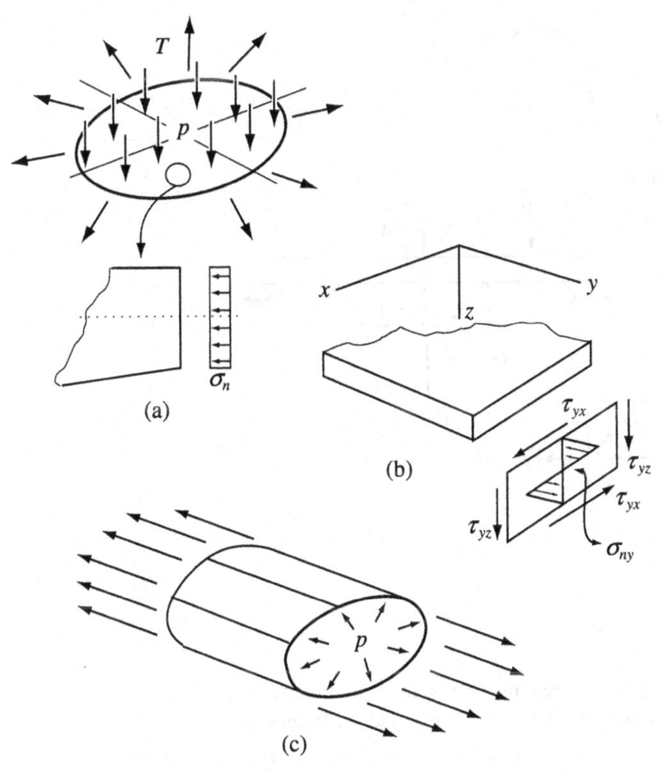

Figure 2.3. Two-dimensional structural elements and their dominant stresses: (a) membranes, (b) plates, and (c) shells.

comparable situations for cables and arches. It is no surprise, for example, that arches are often contained between immobile river banks or rock faces, or that they are built upon massive haunches – all to support the large compressive stresses and forces that result from the redirection of the loads carried by the arch.

The next surface structure of interest is the *plate*, shown in Fig. 2.3(b), which is a two-dimensional analog of the beam. The basic behavior is just like that of beams in that transversely applied vertical loads are converted into normal bending and transverse shear stresses, although the extensions from beam theory are somewhat complicated by the nature of the two-dimensional modeling needed to describe the bending of a plate structure. Plates are used in a variety of structures, mostly to model aspects of the *cladding* or covering of a structure, including everything from roofs to windows, and also to model coverings over holes in the ground, whether manholes or construction trenches.

In Fig. 2.3(c) we illustrate *shell* structures, or simply *shells*, which reflect a combination of behaviors that arise because shells are *curved* surface structures. In fact, shells are often described as curved plates, but their behavior is more complicated because the curvature in a shell produces an interesting coupling of the in-plane normal stresses and the transverse shear stresses. Of particular note is the fact that the in-plane normal stresses have two components, one uniform, like the axial stresses in bars and arches, and a second that varies linearly through the thickness, like the bending stresses in beams (see Chapter 7) and plates. Again, shell models are more complicated because of the curvature effects,

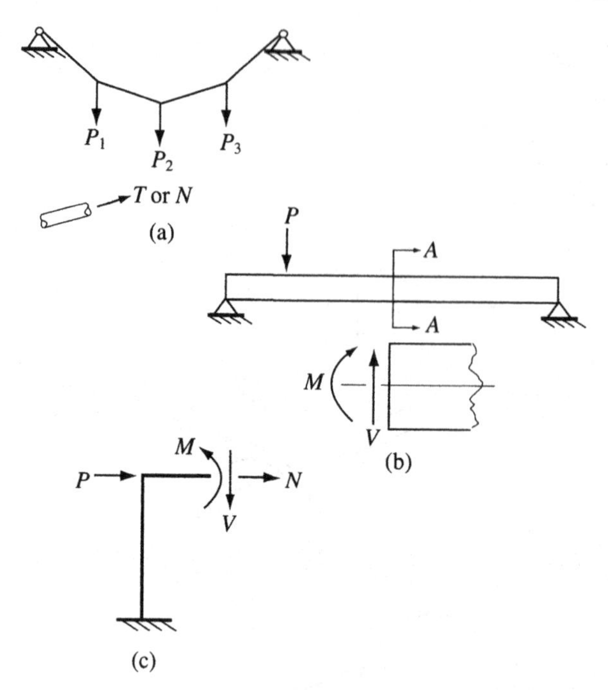

Figure 2.4. Integrated force resultants for one-dimensional structures: (a) cables and bars, (b) beams, and (c) frames and arches.

and their attendant behavior is also more complicated – and often more interesting. Shells are used in storage structures, such as grain silos and oil tanks, and in dramatic roof structures, such as the hyberbolic paraboloids and other curved surfaces that often provide graceful roofs for modern houses of worship. Shell structures are also evident in aircraft, in spacecraft, and in objects as mundane as trash cans.

2.1.3 Stress Resultants for One-Dimensional Structural Elements

Before leaving this overview of basic structural elements, we want to provide an advance look at a descriptive device that is essential to modeling these elements. It will also significantly enhance our ability to model *supports* (in Section 2.2).

We have chosen to display the responses of the structural elements just described in terms of the dominant stresses that occur in the cross sections of each particular structure. However, the analytical models of these structures are cast in terms of force stress resultants (also called force resultants) that represent integrated values of the dominant stresses, with the integrations being over the cross-sectional areas that are perpendicular to the axes of these elements. We will show these in more rigorous detail as we discuss each of these structural types in Chapters 4, 7, and 9, but for now it is useful to summarize them.

For cables and bars, we simply integrate the uniform normal stress over the area and define an axial stress resultant $N(x)$ as (cf. Fig. 2.4(a))

$$N(x) \equiv \iint_{A=hb} \sigma_n dz dy \tag{2.1}$$

Thus, $N(x)$ represents the *axial force resultant*, tensile or compressive, produced by the dominant normal stress across the bar or cable. As a matter of notation, for cables and ropes the notation $T(x)$ is often used, reflecting the fact that these structures are always in tension. Similarly, when bars are assembled in trusses, the force in each bar is typically denoted by F_{ij}, where the subscripts stand for the numbers of the joints that the bar connects. These bar forces can be either tensile or compressive.

For the beam, the dominant stresses are the normal bending stress σ_n and the transverse shear stress τ_s. These are integrated differently because the bending and shear stresses are distributed differently over the cross section (cf. Section 7.2.2). The normal bending stress is itself distributed linearly through the thickness of the beam (it is constant across the width, as are all beam variables); however, its net average across the thickness is zero because of this linear distribution. On the other hand, the integral of the first moment of the bending stress about the centroid of the beam's cross section produces the *moment* or *bending stress resultant* $M(x)$, defined as (see Fig. 2.4(b))

$$M(x) \equiv \int_h \sigma_n z b dz \tag{2.2}$$

The transverse shear stress generally varies quadratically over a beam's cross section (again, see Section 7.2.2). However, we are not interested in the details of that shear stress distribution, only in the resultant (or simple average). Thus, we define the *shear force* or *shear stress resultant* $V(x)$ as (see Fig. 2.4(b))

$$V(x) \equiv \iint_{A=hb} \tau_s dz dy \tag{2.3}$$

Now, for frames and arches, we expect that all three of the aforementioned force and moment resultants would be present, in the most general case, for reasons we have already noted. However, as we will show in Chapter 9, while axial, shear, and bending resultants are invariably present in frames to satisfy static equilibrium, their individual effects on the deformation of a frame will be substantially different. Similarly for arches, where we have noted that the ideal – or most efficient – arch has only a compressive normal stress distributed across its thickness, there will be both shear and bending effects. For the arch, however, these three resultants are often coupled in interesting ways by the very curvature of the arch. However, to be most general or inclusive, it is best to assume that both frame and arch cross sections are acted upon by the complete set of force and moment resultants, as per Fig. 2.4(c).

2.2 Modeling Structural Supports

Now, although we have just offered a few comments about the reactions needed to support some of the one- and two-dimensional structures we described, we haven't said very much. As this is obviously a very important topic, we will turn to it next. How do we support a structure? And, equally relevant for our purposes, how do we model such supports?

It is clear that to support a structure means that we both want to keep it from moving or going anywhere, and that we want to ensure that the loads applied to the structure are

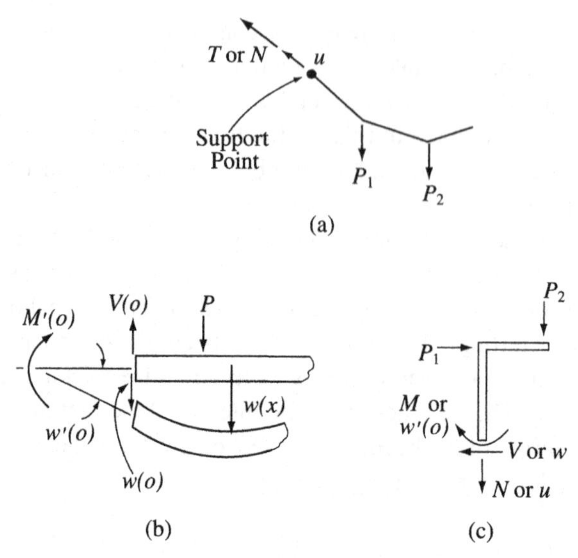

Figure 2.5. Force-displacement duals at structural supports: (a) a rope or cable, (b) a beam, and (c) a frame.

constructively carried down into a foundation attached to the ground that, for the most part, is considered the immovable object on which our structure can depend for support. The reason that the modeling aspect becomes important is that the foregoing statement – although fairly abstract – is (generally) quite correct. We need to express the need for support reactions – or for permissible deflections or displacements – in the language used to describe the behavior of the structures of interest. This means that the kinds of reactions or supports are best described here in rather general terms, with appropriate details left for the development of the individual structural models.

There are two basic themes for describing support reactions. The first is that the number of generalized force reactions needed at each support is exactly equal to the number of independent resultants (force and moment) required to write a complete set of equilibrium equations for the structural model of interest. We say generalized force reactions to accommodate the fact that some of our force resultants are moments and, thus, strictly speaking, they are not forces.

The second theme is that each generalized reaction force has a generalized displacement *dual*, that is, a displacement quantity that can be prescribed *instead of* that force – although not in addition to the force. The displacement dual is that generalized displacement through which the corresponding reaction force would move to do work.

Thus, as a very simple example, consider the rope or cable shown in Fig. 2.5(a). Inasmuch as this cable requires only a statement of equilibrium regarding its uniform axial stress (or its net axial force), we need stipulate only one reaction force, or its dual, at each end. At the top end, since it represents the suspension point from which we are hanging the cable and its load, there is only one reaction force, and it must balance the tension in the cable. At the other end, we could require the force we want the cable to carry, say, a weight W, or we could require that the cable be stretched by a known amount, say, Δ. However, we cannot stipulate both W and Δ because these two quantities are duals of one another. As we will see in general terms in Chapter 5, their product is directly proportional to the work done at the cable end, and we cannot independently prescribe both the force and the displacement needed to do work there. Stated in the context of a very simple yet familiar example, we cannot independently require specified values for a force applied to extend a linear spring and to the displacement that would result.

For a bar, whether alone or as a member in a truss, the situation is very much the same as for the cable or rope. Although the components of forces applied to a bar at its ends may take on any direction, their net effect or resultant must be colinear with the bar, and we will see in Chapter 4 that the net deformation of a bar is simply the extension or compression of its length.

The situation of a beam is more complicated because there are two independent generalized forces needed to describe what is happening inside the beam as a result of the external loads, a moment and a shear. Thus, at each end of a beam, or at each support, there are two choices that can be made (cf. Fig. 2.5(b)). We prescribe either a shear force $V(x)$ or its corresponding dual, the transverse or bending deflection of the beam, *and* we prescribe either a moment $M(x)$ or its corresponding dual, the slope of the bending deflection of the beam. How this emerges will become much clearer in Section 7.2.4, but for now we can say that the beam support alternatives can be stated as

$$\text{At} \quad x = x^* \quad \text{prescribe either} \quad V(x) \quad \text{or} \quad w(x) \tag{2.4a}$$

and

$$\text{At} \quad x = x^* \quad \text{prescribe either} \quad M(x) \quad \text{or} \quad w'(x) \tag{2.4b}$$

where $x = x^*$ is a boundary or support point of a beam and $w(x)$ denotes the transverse or bending deflection of the beam.

For a simple beam with supports at both ends, say, $x = 0$ and $x = L$, there are two possibilities at each end and, thus, four possible permutations of ways we can support a two-ended beam. We have sketched some of these in Fig. 2.6, alongside some drawings of how such kinds of supports are actually implemented in the real world of structural engineering. As you no doubt recall from your course in "Strength of Materials," there are two basic support cases. In the first, the *simple support* or *pinned* case, we require that both the bending displacement $w(x)$ and the moment $M(x)$ vanish, whereas for the *clamped* or *fixed* case, we require that both the bending displacement $w(x)$ and its slope $w'(x)$ vanish. There are several ways these supports are implemented, as we show in Fig. 2.6, but perhaps the most interesting aspects are only implicit in these pictures.

For example, for a simple girder bridge, such as those used for highway and railroad bridges, or something that evidently functions just like a beam within a framed structure, we often see just the sort of pins and girder connections shown in Fig. 2.6. However, it is appropriate to wonder whether or not these boundary conditions are simply limited to the shear and moment choices given in Eqs. (2.4), or whether or not we need to account for any horizontal motion and its axial force dual.

For bridges, it may be reasonable to ignore the horizontal force–displacement duality because expansion joints are purposely built in to ensure a freedom of horizontal motion due to temperature expansion. For framed structures, if a beam is situated atop two vertical legs, as with the frame of Fig. 2.2(b), then we clearly have to ensure that any horizontal motion of the beam is compatible with any transverse motion of the legs, and we have to

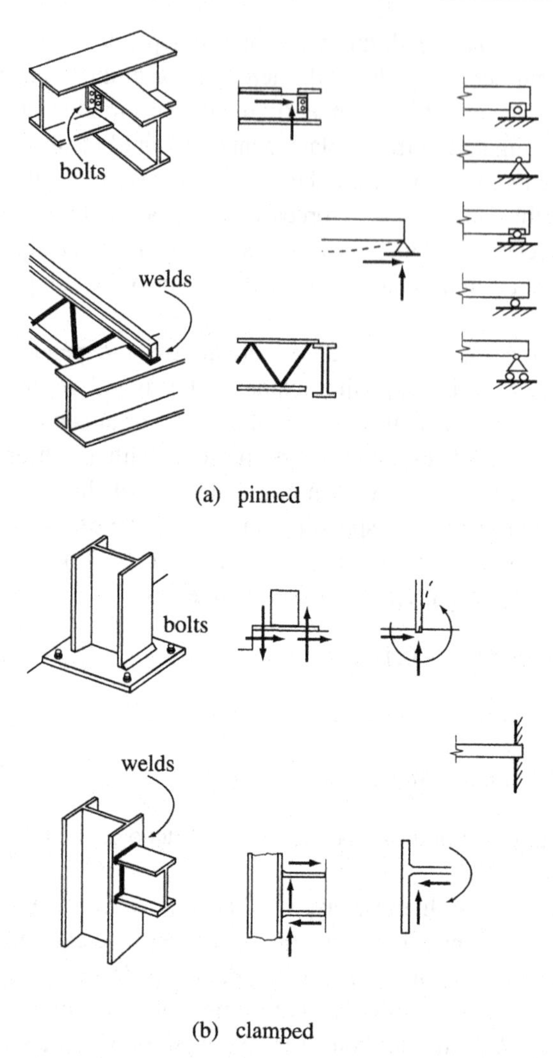

Figure 2.6. Classical support conditions at the ends of beams when they are standing alone and when they are attached to frames: (a) pinned and (b) clamped. (After Figure 3-8 of D. L. Schodek, *Structures*, Prentice-Hall, 1980.)

ensure the equilibrium considerations of the beam's axial load being transmitted between the two column legs.

But the most interesting aspect is that the beam–column frame connections shown in Fig. 2.6 cannot be as black and white as the choices offered by Eqs. (2.4) because, in general, we cannot literally expect deflections or slopes to be zero when beams are attached to columns that are not themselves perfectly rigid. That is, there is a fundamental modeling assumption contained both within our discussion of the cable of Fig. 2.5(a) and for the elementary beam supports, namely, that there is a device that will provide an absolutely immovable support against the bending displacement or the bending slope (or both) of a beam. This is a good model for a variety of practical circumstances, but it is not universally applicable. In particular, in the case of frame structures (cf. Fig. 2.5(c)), the relative stiffnesses of beams and of the columns to which they are attached become very important in analyzing frame behavior. We will offer more discussion of these relative stiffness effects in frames in Chapter 9, but we will also discuss (in Section 8.2.3) the behavior of beams supported on both translational and rotational springs just because we may have practical need of reflecting in our boundary conditions a support deflection due to ground settlement or an end-fixity condition, which might reflect an elastic connection rather than an entirely rigid wall encastering a beam.

And, of course, as we have said before and will often say again, we are doing our best to model reality, here the reality being supports as they would appear in actual construction. Thus, support conditions, both those discussed here and those applied in more complex structural calculations, are intended to reflect realistic support behavior whose effects we wish to take into account. So, for our physical and mathematical models, we should choose boundary and support conditions with an eye toward what's really out there in the world.

2.3 Indeterminate Structures, Redundancy, Load Paths, Stability

One of the issues that derives from the nature of the support conditions is their *determinacy* or, more accurately, their *statical determinacy*. That is to say, given a structure and a set of loads that act on that structure, we have to face the question of whether we can calculate the structure's support reactions simply by writing the equations of static equilibrium. It turns out that sometimes we can, and sometimes we can't. The basis of the answer is simple enough. If there are enough independent equations, we can calculate the unknown reaction forces. If there aren't enough equations, if there are more unknown reaction forces than available equilibrium equations, then we can't. How do we know? And does it matter much whether a structure is classified as statically determinate or as *statically indeterminate*?

2.3.1 On Counting the Degree of Indeterminacy

First, on how we know whether a structure is determinate, it is a relatively simple matter to count up the unknowns and compare that number with the number of available equations. We could do this in the abstract, for three-dimensional elastic solids, and we could do it for individual cases or models. For one-dimensional bars taken alone, for example, we have only one equilibrium equation, so it is easy enough to inspect a particular bar's configuration and assess its determinacy by counting the number of supports at which a reaction force is required (as opposed to bar ends where displacement is specified or permitted).

When bars are assembled into trusses the situation is more complicated. For *planar trusses*, that is, trusses that act in a plane, we can write three equilibrium equations (i.e., $\Sigma F_x = 0, \Sigma F_y = 0, \Sigma M_z = 0$) for the truss in its entirety. However, we can also write two equations (i.e., $\Sigma F_x = 0, \Sigma F_y = 0$) for each truss joint, and if we do that for every joint in the truss, we automatically include the three equations taken for the truss in its entirety. Thus, the three truss equations are subsumed in the complete set of joint equations because at the end of each bar the resultant forces are colinear with the bar, so that $\Sigma M_z = 0$ for the bar (see also Section 5.1). Thus, we have $2j$ independent equations for a truss with j joints. On the other hand, in terms of unknowns, we have r reaction forces and b bar forces, that is, the resultant tensile or compressive forces that occur in each bar. The test for the determinacy of a planar truss is, then,

$$\text{planar truss is determinate if} \quad r + b = 2j \tag{2.5a}$$

$$\text{planar truss is indeterminate if} \quad r + b > 2j \tag{2.5b}$$

We will discuss the meaning of the case $r + b < 2j$ in Section 2.3.3.

The kind of calculation we have done can clearly be extended to other models, for example, space (three-dimensional) trusses. For our purposes, the most important "other case" is the set of problems governed by models of beam behavior. Here, dealing with planar beams of the kind we will extensively detail in Chapters 7 and 8, we can recognize that once again we have three equilibrium equations (i.e., $\Sigma F_x = 0$, $\Sigma F_y = 0$, $\Sigma M_z = 0$), and that the number of potential force unknowns can be counted by simply looking at the support reactions. We also need to be a bit careful, however, to recognize that there is no net deformation (and, hence, no net axial force resultant) in elementary beam theory, so that the horizontal equilibrium equation (i.e., $\Sigma F_x = 0$) is typically useful only for calculating the single horizontal reaction required to keep a beam from rolling away on a horizontal plane. On the other hand, we must count the horizontal equilibrium equation as having greater import when we are analyzing frames because (1) the horizontal reactions of a frame have more physical meaning and impact and (2) the axial forces on the horizontal bars of a frame are the lateral loads that bend the frame's vertical legs (cf. Chapter 9).

Coming back to beams, now, at a pinned or simply supported support we need to account only for the vertical reaction (in shear), while at a clamped support we must account for both the shear and the moment. Thus, a beam clamped at both ends has four unknowns, and discounting the horizontal, we have only two equations (i.e., $\Sigma F_y = 0$, $\Sigma M_z = 0$), so the beam is indeterminate to the second degree.

It now becomes interesting to wonder why there should be more reactions than we can calculate with the equations of static equilibrium. Is it intentional? Is there something to be gained by incorporating "extra" reaction forces? After all, if we have just enough unknowns that we can calculate them all with the equations of equilibrium, do we not have a structure that is in equilibrium and will not go anywhere? That is, if all of the equations of static equilibrium are properly satisfied, are we not told by Newton's laws that we have a structure that is (statically) stable? Why do we need these extra, redundant forces? Are we simply interested in making it harder to analyze indeterminate structures (a view commonly held by students in structures courses)?

We will have occasion to visit the issue of degree of indeterminacy often as we do particular bar, truss, beam, and frame problems, and we will then provide some techniques for handling such problems. For now it is important to recognize that there are many problems for which the external static equations of equilibrium are insufficient for analyzing a structure, and that many structures are intentionally designed to be statically indeterminate. So, while we defer discussions of techniques for analyzing indeterminate structures, it is a good time to discuss the reasons why structures are designed to be that way.

2.3.2 On Indeterminacy and Why Redundancy Matters

The issue of indeterminacy does matter, but before explaining why, we want to introduce one more technical term. We have already noted that any reaction or supporting forces beyond the minimum number needed to satisfy the equations of static equilibrium

Table 2.1. Maximum deflections and moments for uniformly loaded beams with simple supports and with clamped supports.

	Simple supports	*Clamped supports*
$\dfrac{EIw_{\max}(L/2)}{q_0 L^4}$	5/384	1/384
$\dfrac{M_{\max}(L/2)}{q_0 L^2}$	3/24	1/24

are extra or, in an equivalent word, redundant. Thus, we will use the term *redundant* to refer to each of and all of the forces beyond the minimum number required by static equilbrium.

Why would we design redundants into structures? What purpose(s) do redundants serve? The answer has two parts that are, briefly, as follows. First, indeterminate structures are generally stiffer than their determinate counterparts, so that the generalized forces (and thus the stresses) and the deflections they experience will be less than the corresponding results for their determinate counterparts. Second, the existence of redundant reactions means that the failure of some of the supports could still leave a viable, determinate structure that remains in equilibrium and, thus, is still stable.

On the first point, we show in Table 2.1 the maximum bending deflections and stresses for two beams of length L that are supporting a uniform load-per-unit length q_0. (Note that Table 2.1 is a precursor to Table 7.3 and its corresponding discussion in Section 7.3.4.) We can see from these results that the maximum displacement is reduced by 80% and the maximum moment by 67%, so that the increase in stiffness caused by the transition from the determinate simply supported beam to the indeterminate fixed-ended beam is quite substantial. It is virtually impossible to quantify such stiffness gains in general in a meaningful way. However, it certainly is intuitively evident that by increasing the number of redundants we are simultaneously restricting a structure's ability to deflect and deform, and so we are stiffening that structure. For the uniformly loaded beams just considered, for example, it is clear that clamping the supports restricts the beam's flexibility in comparison with that of the simply supported beam. Thus, we have effectively increased the beam's stiffness by clamping its ends. It should, therefore, not be in the least surprising that our increasingly vast experience with structures confirms this idea.

On the second point, consider the two beams shown in Figs. 2.7(a,b), which we might imagine as bridges over a simple, yet deep, gap. Imagine that the right support of each washes away, resulting in the situations shown in Figs. 2.7(c,d). The first beam, originally simply supported (and thus statically determinate) has become a *mechanism* (see Section 2.3.3) with the loss of its right-hand support. That is, what was once an elastic beam has now become the equivalent of a *rigid bar* whose only support is the pin at the left support, and this pin is unable to supply sufficient constraint to keep the structure in a stable state. The rigid bar will simply rotate freely about its remaining support.

Now, as shown in Fig. 2.7(d), the second bridge, originally fixed at both ends, now resembles a diving board. That is, it is now a cantilever beam that carries a uniform load, so it remains a stable structure. It won't be as stiff as the original fixed-fixed beam, but it does support the original load. However, what we also see is that we we have provided an

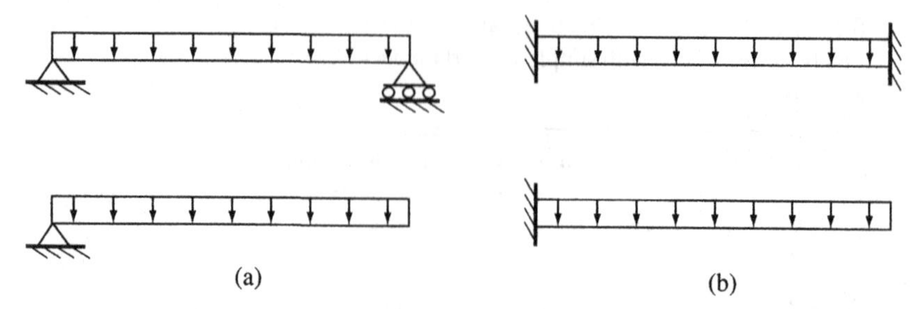

Figure 2.7. Loss of reactions in two simple beamlike bridges: (a) simply supported and (b) fixed at both ends.

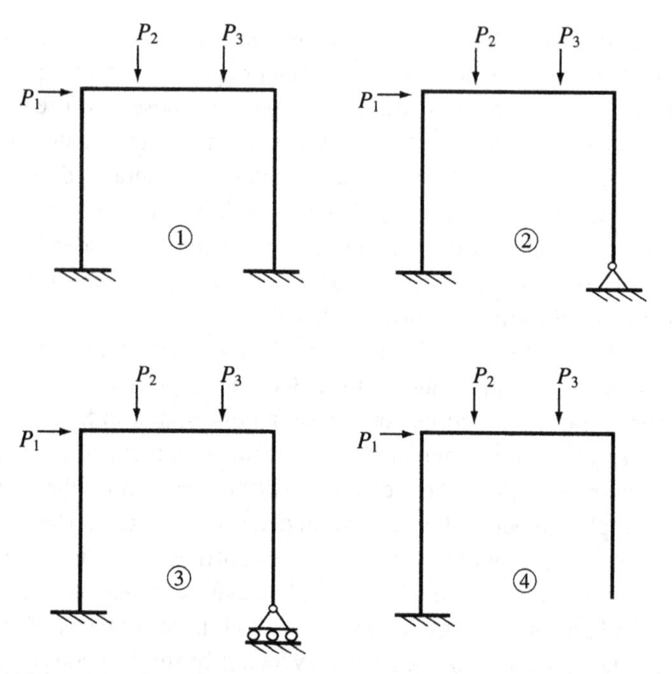

Figure 2.8. Alternate load paths created by the removal of redundants in a structural frame.

alternate *load path* by which the applied loads are carried through to the bridge's remaining foundation. The redundant reactions can be said to provide additional safety for the beam or bridge by providing a redundant load path.

Consider now the sequence of frames shown in Fig. 2.8. Each is stable, and each progresses from a highly redundant structure (indeterminate to the 3rd degree) to one that is clearly determinate. It is as if the right support had slowly weakened and eventually given way entirely. What remains is still a stable structure; it is just considerably more flexible, or less stiff, than the original structure, although it does support the same loads. The set of reactions on each frame will be different, with the changes occurring as another redundant reaction is removed. This progression of structures shows that we are providing alternate load paths by which the applied loads are carried down to the frame's foundation. Thus, as with the simple beam bridge, we are providing additional safety for the frame by providing redundant load paths.

Such redundancy shows up in other ways. For example, it might seem impossible to guard against leaks due to cracks or ruptures in the skins of storage tanks designed to hold liquids or gasses, just as it is impossible to contain within a burst balloon the pressurized gas that inflated it. However, it is now routine for oil to be transported by sea in tankers that are *double hulled*, so that should a tanker run aground or hit a reef, there is a redundant layer of protection that guards against oil spills. The environmental and other costs attached to major oil spills are so large that the extra expense of designing and building oil tankers to be redundant is considered a sound business decision.

Perhaps it is a commentary on our times, but a similar set of considerations is now being widely discussed in the context of building design. A recent and tragic spur to such discussions was the bombing of the Alfred P. Murrah Federal Building in Oklahoma City on 19 April 1995 (Fig. 2.9). It has been estimated that 80% of the 168 deaths that day resulted from structure failure, rather than as a direct result of the blast. Thus, the issue of extending building criteria to include structural measures against progressive collapse is a very pertinent topic, especially in terms of regulations for government office buildings, as well as for commercial buildings such as the World Trade Center. In our terms, the point is that structural protection against progressive collapse can be provided by incorporating designs of alternate load paths to minimize the effects of losing critical structural elements, no matter what causes these critical elements to fail.

As we have noted in Chapter 1, there are many considerations that enter into building design. For example, design considerations for a public building such as the Murrah Building include cost, aesthetics, public access, and the very appearance of public accessibility for a government office building in a democracy. And there are certainly other safety measures that can be, and are being, taken into account. Our point is simply that there are good reasons to design and build indeterminate structures, and these are some of the challenges of structural engineering that we can at least appreciate at this (early) point.

2.3.3 Two Important Aspects of Structural Stability

It is interesting to enter into a discussion of stability immediately after our comments on the role of redundancy in structural design. No doubt we all sense that we have some gut-level feeling or intuition of whether a structure is stable, or whether it might suddenly collapse at our feet. And certainly anyone could look at a picture of the remains of the Murrah Building after the explosion (cf. Fig. 2.9) and wonder whether that partially destroyed structure was itself stable, that is, whether it would stand for long on its own. But, in fact, there are two rather technical meanings attached to the notion of stability that we want to discuss here, and although at least one is related to the notion of redundancy, we have some other points to make.

The meaning of stability that is related to redundancy is also partially related to the test for the determinacy of a planar truss that was given in Eqs. (2.5). In the notation of those equations, we now ask: What does happen when $r + b < 2j$? This question is the converse of the test for determinacy, and its answer is that in this case the total number of reaction forces and bar forces is too small to satisfy equilibrium at all of the joints. Remember that the satisfaction of equilibrium in Newtonian mechanics implies that there is a sufficient number of *constraints* to keep the structure from moving. Further, these constraints are both *internal*, as manifested in bar forces, and *external*, as manifested

Figure 2.9. The remaining structure of the Alfred P. Murrah Federal Building, Oklahoma City, Oklahoma, after the explosion of 19 April 1995. (Photo courtesy of AP/World Wide.)

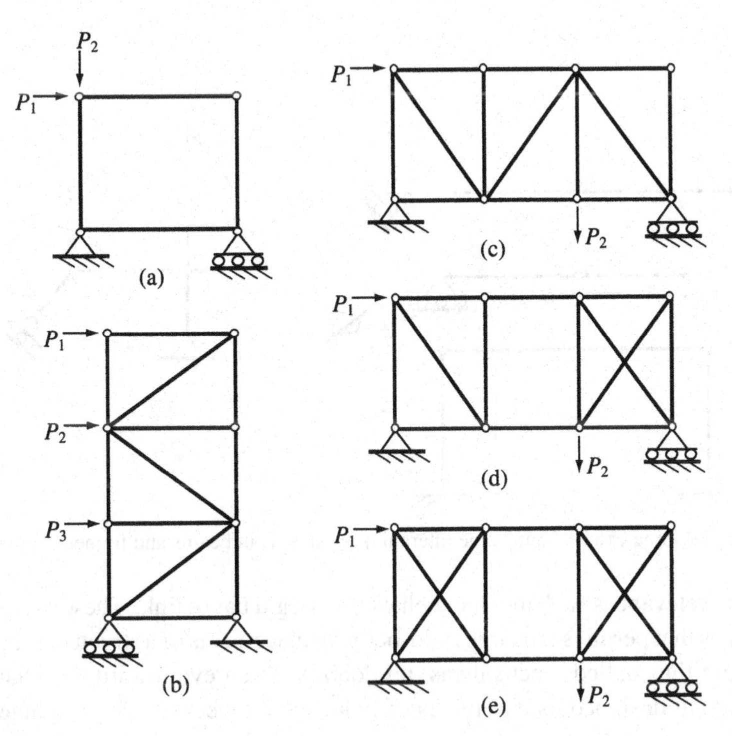

Figure 2.10. Determinacy and stability in trusses as created by adding or moving bars.

in reaction forces. We show a classic instance of this in Fig. 2.10(a), noting that this structure is underconstrained internally (and its attributes satisfy the given inequality because $r+b = 3+4 = 7 < 2j = 2(4) = 8$). On the other hand, the truss in Fig. 2.10(b) is both stable – remember that trusses are assemblies of triangles for good reason – and determinate, and for which $r + b = 3 + 13 = 16 = 2(8)$, which is twice the number of joints, so Eq. (2.5a) is satisfied. It is also easily seen that if another bar is added to the truss in Fig. 2.10(b), say, another diagonal across one of the two center panels, the new truss would be internally indeterminate because while we can still calculate the three reactions, we don't generate any more equations by adding just that bar, so now we can't calculate all of the bar forces (now numbering 14).

The relationship between determinacy and stability is sufficiently complex – especially for trusses – that simply applying formulas is not enough to determine determinacy and stability. Consider the sequence of trusses in Figs. 2.10(c–e). In the first two (Figs. 2.10(c,d)) we have the same numbers of reactions, bars, and joints ($r + b = 3 + 13 = 16 = 2j = 2(8)$), yet the first of these trusses is determinate and stable, the second is determinate and unstable. Further, if we add another bar, we can make this truss both unstable and indeterminate (Fig. 2.10(e)). In both of the last two trusses we have created a *mechanism* by virtue of making the center panel unable to remain rigid or, in another view, of making that panel (and thus the truss) unable to transmit a vertical load across that center panel. Thus, a structure can be unstable while being either determinate or indeterminate.

The idea of a mechanism derives from the fact that large rigid-body motions of some structural elements result from these hinges, and the elasticity of the members joined at the

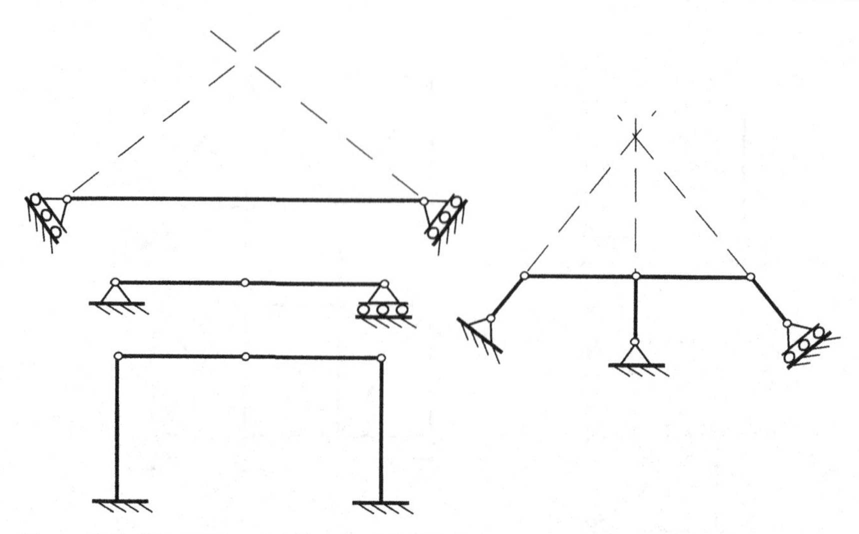

Figure 2.11. Some external and some internal mechanisms in beams and frames.

hinges becomes irrelevant as each member behaves as a rigid bar or link. These assemblies of links and hinges that permit such large rigid-body displacements of a structure, or pieces of a structure, are thus called mechanisms, no doubt with an eye toward "mechanical" mechanisms that are designed as ensembles of links and connectors such as hinges and springs.

Note that the mechanisms we have shown so far have all resulted from a lack of sufficient internal constraint against instability. It is also true, as you no doubt recall from elementary statics, that *external instability* can result if all of the supports of a truss are either *parallel* or *concurrent*. In fact, this is true for any kind of structure, whether it be a truss, a beam, a frame, an arch, or whatever. In these cases, the possibility of unstable structural behavior occurs because there is no external constraint against either a force directed at an angle to the parallel reactions or a net moment about the point of concurrency.

Internal mechanisms also appear in beams and in framed structures (see Fig. 2.11). Here the mechanisms typically occur because of the presence of *hinges*, that is, points of zero moment. This is not to say that all hinges produce mechanisms, only that hinges are either a signpost of an unintended mechanism or a device used to allow or insert unstable behavior in a structure.

The other kind of stability we want to discuss here is of a markedly different kind. Consider the vertically loaded bar in Fig. 2.12(a). We can calculate the axial normal stress and the net axial shortening of this one-dimensional bar (cf. Chapter 4). However, this structure is capable of another kind of behavior altogether, one in which its predominant motion is transverse or in bending, as shown in Fig. 2.12(b). In this case, this vertical bar (which can also be drawn horizontally, as in Fig. 2.12(c), with no loss of meaning) is considered to be a *column* for which *the onset of bending behavior* occurs at axial loads that are determined by solving an *eigenvalue problem* centered around the following differential equation:

$$EI\frac{d^4w(x)}{dx^4} + P\frac{d^2w(x)}{dx^2} = 0 \tag{2.6}$$

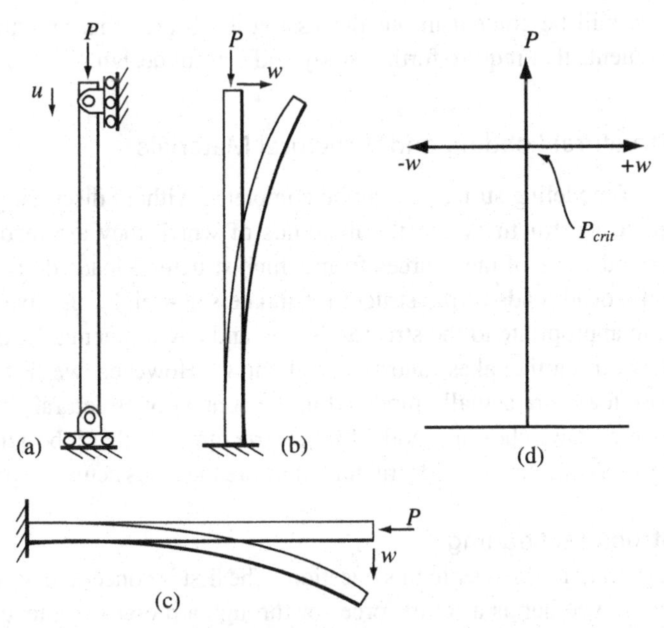

Figure 2.12. On column behavior: (a) an axially loaded bar, (b) a cantilevered column, (c) the cantilever column drawn horizontally, and (d) a graphical portrayal of (linearized) column behavior.

Now, while the variational techniques we espouse in this book can be applied to such problems with great elegance and power, we will not treat those mathematical details either now or in the balance of this book. It is important, however, that we recognize two facets of column behavior.

The first is that the mathematics of column behavior are inherently nonlinear. That is, even though Eq. (2.6) looks like (and is) a linear differential equation, the column's loss of stability is determined by a critical value of one of the coefficients in this equation, the value of the load $P = P_{crit}$. Thus, we cannot apply superposition as we typically would wherein we expect that the response of a structure (or any linear system) to two loads (applied as forcing functions on the right-hand sides of the equilibrium equation(s)) can be written as the sum of the responses to each load applied independently.

We illustrate the second facet of column behavior in Fig. 2.12(d), which represents a characteristic result of solving the eigenvalue problem implicit in Eq. (2.6). This figure shows that there is no bending response $w(x) = 0$ for loads smaller than the critical load ($P < P_{crit}$), and that we cannot determine the magnitude of the column's bending deflection when the critical load is reached ($P = P_{crit}$). This is one reflection of the truly nonlinear behavior of the column, which is only partially reflected in the *linearized version* of the problem given in Eq. (2.6) and in Fig. 2.12(d).

The practical import of this is, in brief, as follows. The critical load of a column can be regarded as an upper limit of the load that can be supported by the column before unwanted transverse or bending deflections (perpendicular to the column's axis) can occur. In actual structural practice, frame members often serve as *beam–columns*, carrying both transverse and axial loads, each of which produces bending deflections. In this case the kinds of elementary models we will use to analyze framed structures (in Chapter 9) have to be extended. However, we will be content in our discussion to observe that columns are important structural elements that require further study and careful design.

2.4 Modeling Structural Loading and Structural Materials

No discussion of modeling structures can be completed without discussing both the loads that are applied to structures and the materials of which they are made. In Section 1.4.4 we described some of the sources from which structural loads derive, and these sources include the dead loads of the structure's innate self-weight, the live loads resulting from the traffic appropriate to the structure's use, and environmental loads that include the effects of wind, earthquakes, and rain and snow. However, we have said nothing about how those loads are actually modeled in the context of structural analysis and design. That is, once we have chosen a model for the structure, whether it be a truss, a beam, an arch, a bridge, or a complex framed structure, how are the loads actually applied?

2.4.1 Modeling Structural Loading

There are two parts to the answer to this question. The first is concerned with how a set of loads is applied – whether as a set of forces or through a prescribed movement, such as are used in seismic analysis – and about how these loads are distributed in space and time. Wind loads, for example, can be felt as short-term gusts or as long-term steady winds, while seismic movements tend to last only briefly. Further, the aerodynamic forces resulting from wind blowing across a slender suspension bridge are going to produce a markedly different effect than the spatial distribution of wind pressure over a very tall and slender building, and the latter will be strongly influenced as well by whether the building is surrounded by other tall structures or is more of a stand-alone kind of skyscraper. The effects of some of these types of loads are so complex that part of structural modeling includes expensive physical modeling and testing. For example, actual scale models of skyscraper designs are now routinely tested in large wind tunnels, with appropriate inclusion of models of significant neighboring structures. On the other hand, some of the distributions can be successfully modeled in rough terms, for example, by postulating that wind pressure is distributed linearly over a building, from a zero value at its base to a maximum value at the roof.

The second part of the answer to the question about how loads are applied has to do with the connection between the model or idealization of a structure, on the one hand, and the kinds of loads, on the other. The clearest example of this interaction is encapsulated in a simple planar truss, whether it is used as a frame for supporting a roof or as one side of a railway or highway bridge. As we will confirm in Chapter 5, trusses can only accept loads at their joints or pins (because each of their bars is an axially loaded member). However, common sense tells us that roof trusses have to support loads spread out over the roof's surface and bridges have to support the traffic on the continuous railroad tracks or roadways that they are supporting. The answer here is a simple one, namely, that there are intervening structural elements (e.g., beams, purlins, joists; see Fig. 2.13) that pick up the spatially distributed loads and carry or "convert" them to the point loads of the trusses.

Now, complete load analyses are well beyond the kind of introduction to structural modeling, so what are we going to do here? We will typically use the most fundamental

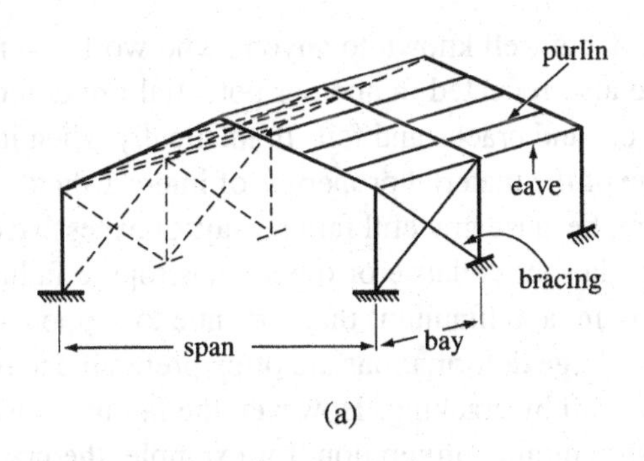

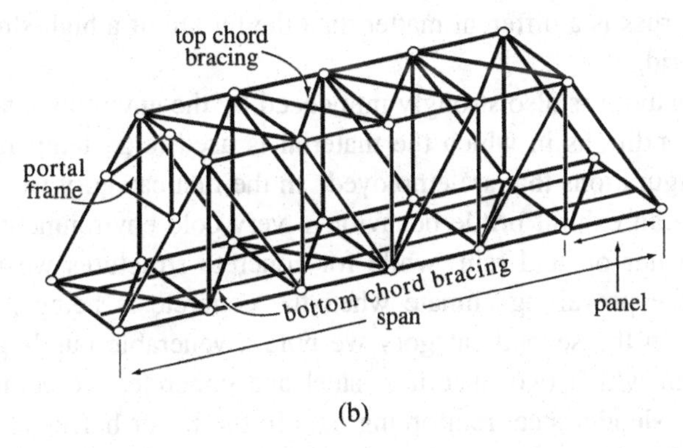

Figure 2.13. Load transfer in trusses in trusses of (a) roofs and (b) of bridges.

static models for our problems, all of which will ring familiar from "Strength of Materials": concentrated forces and moments, uniformly distributed line loads (per unit length of a one-dimensional structure), line loads with a triangular distribution, occasionally some other spatial polynomials, and, of course, loads that can be made up by superposing the fundamental loads.

2.4.2 Modeling Structural Materials

Like many of our other discussions, a complete discussion of structural materials is multidimensional. It clearly must include considerations of intrinsic material properties, such as the modulus of elasticity E, with dimensions of force per unit area, and the density ρ, having dimensions of mass per unit volume. However, knowledge of such basic properties is not a sufficient basis for choosing structural materials. Some of the general materials issues we need to understand include: different aspects of *material behavior*, such as elastic limits, ultimate strengths, and crack initiation characteristics; how *combinations of material properties* compare; how materials are used in varying *configurations*, because their geometry is often very important; the influence of the *environment*; and, finally, the *cost*, including both unit material costs and costs that arise from different fabrication and construction requirements.

Wood, for example, is a structural material whose strength-to-weight ratio is high, but whose strength limits are below what would be needed for, say, use in high-rise building design. Further, wood's properties are nonuniform, varying as they do with grain orientation and with cell structure, facts well known to anyone who works with wood as a hobby. These two properties are also reflected in another potential limitation of wood construction, namely, that it splinters and cracks and fails dramatically when its strength is exceeded. This suggests that one of the material properties of interest should be its behavior during degradation or failure, because graceful failure, say, of excessive deflection without rupture, is preferable to a flat-out collapse or other catastrophic failure. Thus, *ductile* materials (like steel, aluminum, and titanium) that continue to support significant loads while undergoing undesirably large deformations are often preferable to *brittle* materials (like concrete and stone) that fail by cracking. However, the failure considerations are also strongly affected by the structural configuration, for example, the cracking of a single bar in a wooden roof truss is a different matter than the failure of a high-strength cable on a long suspension bridge.

The choice of material behavior is also strongly influenced by the environmental circumstances and operating conditions in which the material is used (e.g., temperature) and by the geometrical configurations that are employed. In the first category we might need to recognize that steel can exhibit brittle behavior in very cold environments and it can corrode due to salt water or acid rain, while for concrete structures we might well have to worry about the prevailing climate when the concrete is being poured and during its curing time. In the second category we note a venerable building material, reinforced concrete, in which two materials, steel and concrete, are combined in a particular configuration, slender steel reinforcing bars in the lower halves of thick concrete beams or floor slabs, to take good advantage of their individual behavioral properties, the ductility of steel in tension and the effectiveness of concrete aggregate in compression.

One of the most interesting, and most challenging, materials issues is the relation between structural function, structural configuration, and structural material. For example, whether to design a bridge as a girder or an arch or a suspension bridge, and if an arch, whether it should be made of reinforced concrete or erected as a steel framework. Among the obvious variables that will influence these choices are the width of the gap that is being spanned, the aesthetic judgment of the owner or sponsor, and the availability of funding. Some of the complex interactions between function, configuration, and material are reflected in the two charts given in Figs. 2.14 and 2.15. Some of trade-offs that are more reflective of technical issues are more evident in these charts than are the issues of taste, aesthetics, and cost.

Clearly, there are a lot of materials issues, many more than we have discussed. Equally clearly, we will have to limit our materials modeling at this stage. The model we will use throughout is derived from a classic stress–strain curve, such as the one shown in Fig. 2.16. In that picture we note various stages of behavior for a mild, ductile steel. Of particular interest is that, as exemplified by a standard uniaxial test, most of the stress that a piece of steel can carry is reached during that portion of the curve where the stress and strain are linearly related. Thus, as our one-dimensional model of material behavior, we take

$$\sigma = E\varepsilon \tag{2.7}$$

This model will serve us well for most of our applications, save for the general development of variational approaches given in Chapter 5 and for a small number of instances where we want to allow for behavior that includes shear in an otherwise one-dimensional

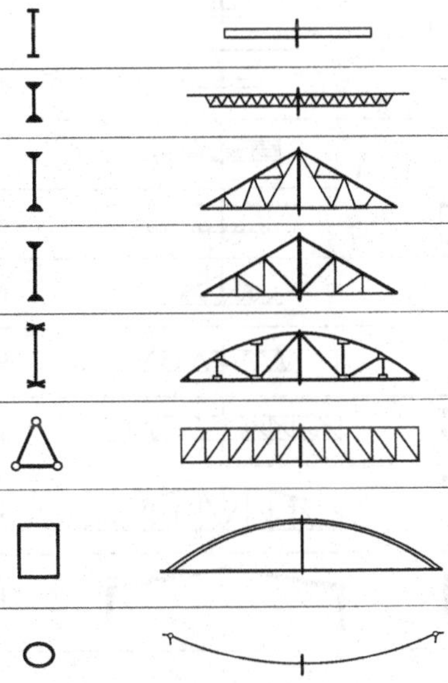

System	Cross-Section	Profile (Elevation)	Span Range (m)	Approximate Depth
Plate girders			$6<L<25$	$L/15$–$L/20$
Open-web joists			$3<L<36$	$L/18$–$L/22$
Fink truss			$10<L<?$	$L/4$–$L/5$
Howe truss			$10<L<33$	$L/4$–$L/5$
Bow-string truss			$18<L<36$	$L/6$–$L/10$
Special trusses			$23<L<?$	$L/4$–$L/15$
Arches			$18<L<?$	$L/3$–$L/5$
Cables			$22<L<?$	$L/5$–$L/11$

Figure 2.14. Configuration versus span tradeoffs for steel systems. (After Figure 15–7 of D. L. Schodek, *Structures*, Prentice-Hall, 1980.)

model. In such cases we insist that we are dealing with *isotropic elastic materials*, whose material properties do not change with orientation, and for which we need only one more elastic constant, that is, Poisson's ratio, denoted by υ. We will say a bit more about this as the occasion warrants.

2.5 Modeling: On Approximating Magnitudes; The Role of Dimensions

This book, like many engineering textbooks, focuses on solving problems within a restricted domain, here the limited area of introductory structural mechanics. The basic principles involved in this book are few – one might even argue, to extend an old Chassidic joke, that Newtonian mechanics is contained entirely within $\vec{F} = m\vec{a}$ and everything else is just a matter of detail – but the approach we take in this book to setting up and solving structures problems is the "real stuff" of this course. There are some key ideas we want to emphasize now, that have to do with modeling, with magnitudes and sizes, and with some advice on "when to plug in the numbers."

First, it is important to remember that when we describe or formulate a problem in words, draw a sketch (e.g., a free-body diagram), write down or derive a formula, and crank through to get some numbers, we are engaged in *modeling*. In each of these activities we are formulating and representing a model of the problem in a *modeling language*. And as we go from words to pictures to formulas to numbers, we must be sure that we are translating our problem correctly and consistently. We have to maintain our assumptions, and at the right level of detail. Our primary modeling language is mathematics, so we must be able to translate fluently into and from mathematics. Further, as practicing engineers, we must always remember that we are dealing with models of a problem – *models of*

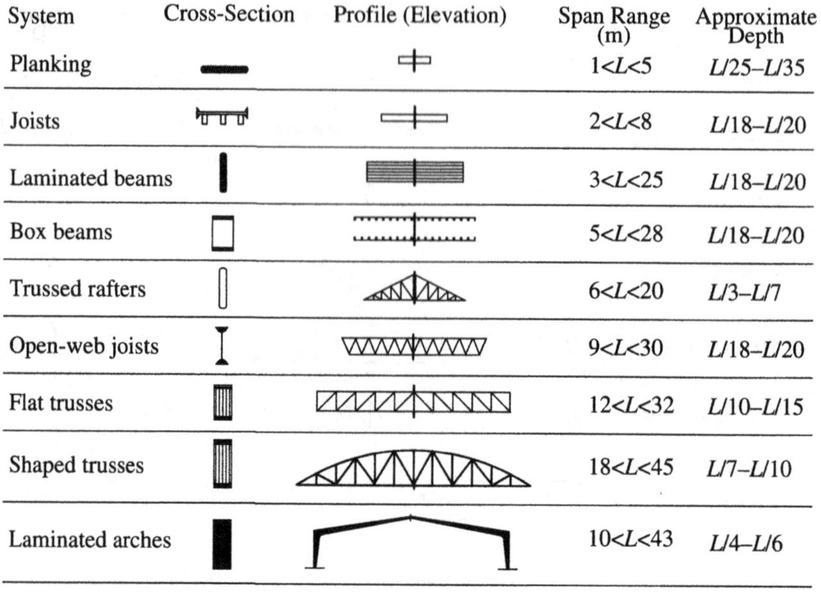

System	Cross-Section	Profile (Elevation)	Span Range (m)	Approximate Depth
Planking			$1<L<5$	$L/25$–$L/35$
Joists			$2<L<8$	$L/18$–$L/20$
Laminated beams			$3<L<25$	$L/18$–$L/20$
Box beams			$5<L<28$	$L/18$–$L/20$
Trussed rafters			$6<L<20$	$L/3$–$L/7$
Open-web joists			$9<L<30$	$L/18$–$L/20$
Flat trusses			$12<L<32$	$L/10$–$L/15$
Shaped trusses			$18<L<45$	$L/7$–$L/10$
Laminated arches			$10<L<43$	$L/4$–$L/6$

Figure 2.15. Configuration versus span tradeoffs for timber systems. (After Figure 15–3 of D. L. Schodek, *Structures*, Prentice-Hall, 1980.)

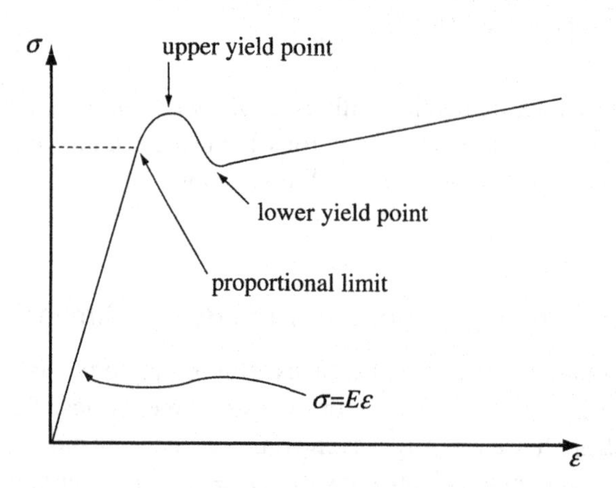

Figure 2.16. Stress–strain curve for mild, ductile steel.

reality. Thus, if our results do not match experimental data or intuitive expectations, we may well have a model that is simply wrong.

Second, we often *idealize* or approximate situations or objects when building models so that we can analyze the behavior of interest to us. We perform two kinds of idealizations, the first being physical, the second mathematical. One very common idealization is to say that something is small, such as the range of motion of the classical linear pendulum. To translate the verbal statement of a physical assumption into a mathematical idealization or model is to say that angles will be small, that is,

$$\sin\theta \cong \theta, \qquad \cos\theta \cong 1 \tag{2.8}$$

for some range of values $\theta \leq \theta_{pres}$ that is prescribed to be acceptable.

Now here is a classic case of the need to be careful in translating assumptions. The first approximation in Eq. (2.8) is common for small angles of θ, but the second one requires further thought about what small means – especially in relation to what. That's because in many physical models the cosinusoid is linked to or compared with unity. For example, the height that a pendulum swings above its datum can be written as

$$\begin{aligned} y &= l(1 - \cos\theta) \neq l(1 - 1) \\ &\cong l(1 - (1 - \theta^2/2!)) = l\theta^2/2 \end{aligned} \tag{2.9}$$

from which we see that our approximation for the cosinusoid would have to be different than what we first indicated (in Eq. (2.8)) because its value is being compared to unity, not to zero. If we were sloppy in translating this assumption when trying, for example, to evaluate the potential energy of the pendulum, we would have wrongly found that potential energy to be zero.

All of the mathematical models of structural behavior that we deal with in this book are *linear*. This is true because we start from an assumption that all of our structures can be modeled as having both a linear stress–strain law (cf. Section 2.4) and that the strains that we are analyzing are all sufficiently small that we can justifiably ignore the geometeric nonlinearities that are part of more precise models of the mechanical behavior of solids. Thus, as a practical matter, all of the material we will present has already passed the "small strain" test of

$$\varepsilon\left(1 + \tfrac{1}{2}\varepsilon\right) \cong \varepsilon \tag{2.10}$$

We show Eq. (2.10) in part as a reminder that when we say the strain ε is small, we must say in relation to what, and here $\varepsilon \ll 1$. (Remember that the strain ε is dimensionless.) We also show this result to confirm the fundamental assumption we have just stated for all our work, namely, that *all of the strains we will compute are very, very small when compared to unity*. This will also have implications for the relative magnitudes of particular structural deflections.

The fact that we can make the kinds of judgments we have described so far in this section also reflects a habit of thought, a mindset with which solutions to problems of all sorts can be sought – and it is a mindset that we will adopt with all of the problems we solve in this book, and that we urge you to adopt for solving structures problems and all engineering problems. The fundamental idea is that when we are making a calculation, we will proceed as far as possible with our *formulas left in symbolic terms*. That is, we will substitute numbers only as the very, very last step in performing calculations. We adopt this attitude because a symbolic result: encourages checking the dimensions of a result, which is an essential part of model verification; makes it easier to relate a mathematical model back to its physical idealization, and to reason about both; mkes it easier to use a consistent set of units for a problem, and to check the order-of-magnitude of the answer; and leaves open the door to using a model again because it is far more general in symbolic terms than it is when written in terms of problem-specific parameter values.

2.6 Modeling: Idealization and Discretization

Our final note on modeling is brief, intended only to foreshadow an issue that we introduced in the Preface and which we will discuss several times in the rest of this book (e.g., Sections 4.6, 6.5, and 8.2, and Chapter 10). First we note that we are not going to make extensive use of matrix algebra in this book, although we will often cast our results in matrix form, especially when it seems convenient and appropriate, and we have attached a brief and gentle introduction to matrix manipulation in the Appendix. And we also note that while we will often refer to structural calculations that are implemented in various styles of computer programs, we will not provide any guidance to writing such programs or using them in any software package or computational environment. So, what is our final note on modeling?

In the previous section we talked explicitly of the idealization part of the modeling process, wherein we develop an abstraction or an approximation of something that we wish to analyze. We do this, of course, because we can deal with only so many variables at any given moment, and even more importantly, the phenomenon or device we are modeling does not always require an infinite amount of detail to provide a useful understanding. Thus, we start modeling by seeking an *idealization*. For much of our work, the idealization will be expressed in differential equations and in formulas, that is, in the language of mathematical models.

However, as a practical matter, we do ultimately need numbers to analyze and design structures, for loads, for structural dimensions, for material properties, for particular behaviors, and for member sizes. As long as our problem is relatively small, we can use mathematical formulas to develop models of our structural problem and then to calculate the numbers we need. For large-scale structural engineering problems, however, we need more efficient ways of generating meaningful numbers. These more efficient ways use computer programs that further refine our mathematical models by *discretizing* them, that is, by representing them in terms of physically meaningful quantities that can be associated with discrete points in a body or a structure. There are several different ways of doing this sort of discretization of models that are originally developed in terms of the continuous mathematics of elastic bodies, although they generally have their roots in the notion of casting a continuous function in the form of a *Fourier series* whose amplitudes represent values of the relevant dependent variable at specific, discrete locations. We will illustrate this discretization process at several points in what follows, especially as we seek approximate solutions to problems involving both bars and beams.

The most widely used, modern form of discretization, which goes well beyond what is possible with Fourier series, is the finite element method. FEM approaches can be elegantly written in matrix form. Further, because matrix manipulation is easily automated, it too is a part of the FEM process. As a result, we not only cast many of our results in matrix form, we also close the book with a discussion of the underlying principles of matrix representations of structural calculations. Inasmuch as there are two basic paradigms that can be used, the force (flexibility) method and the displacement (stiffness) method, Chapter 10 provides parallel discussions of matrix formulations of both methods, although we also indicate which method is most often used, and why.

Now, given the introductory nature of this book, we should expect that more complex structures, loadings, and geometry will require solution processes far more complicated than what we will display. And there are extremely complicated issues that derive from attempts to achieve great accuracy for structural models that require thousands and or tens of thousands of unknowns, all issues involving both matrix mathematics and high-end computing. In addition, we do not pretend to touch many effects that are quite important in structural engineering, including nonlinear matrials and geometry, vibration and dynamics, structural stability, and so on. However, the processes we intend to demonstrate throughout can be taken as a model of current approaches to idealization and discretization in structural engineering.

The Forth Bridge, over the Firth of Forth near Edinburgh, Scotland, was completed in 1890. The cantilevered trusses are made of steel, and the bridge itself is huge. At 1710 feet each span exceeds the length of the Brooklyn Bridge by better than 100 feet, and it towers 342 feet above its masonry piers. Its massiveness is overwhelming when seen up close, but it "morphs" into a far more graceful picture when viewed from afar. (Photo by Clive L. Dym.)

3

Elementary Discrete Structural Models: Energy Approaches

In this chapter we will use a well-known example of the simplest structural element, the linear elastic spring, to introduce the energy methods we will use in the balance of the book. The aim here is to introduce these energy methods in as simple a way as possible by using them to rederive and extend much of what we already know about elastic springs. Thus, as you read what follows, try not to view our discussion as a rehash of everything you know about springs. Focus instead on how some new approaches and techniques can be used to derive what we know in ways that are elegant, succinct, and quite powerful.

3.1 Minimum Potential Energy and Equilibrium for a Simple Spring

Consider a simple elastic spring (Fig. 3.1). It obeys a *constitutive law* that is quite familiar and expressible as

$$F_s = k\xi \tag{3.1}$$

where we use the symbol ξ to define the relative extension of the spring, that is, the difference in the displacement between the spring's two ends, and F_s and k represent, respectively, the force in the spring and the spring stiffness. Note that we are not using the more traditional form of the equation in which x is used to stand for the spring's relative extension because throughout the book we will use x to represent a spatial coordinate that is typically an independent variable, and we do not wish to confuse the meaning.

Now if we attach one end of the spring to a fixed point and pull the other with an external force P, we know from Newton's second law that the internal spring force must be equal to the applied external force, that is,

$$F_s = P \tag{3.2}$$

We can now calculate the spring extension as a function of the applied force, either by combining Eqs. (3.1) and (3.2) or by doing the equivalent from the free-body diagram of the loaded spring shown in Fig. 3.1. This is all quite familiar, so far, and it represents what might be termed an *axiomatic approach* to deriving the equation of equilibrium of the spring expressed in terms of the spring extension and the applied load:

$$P = k\xi \tag{3.3}$$

But we now wonder whether there is another way to derive these results – and, in fact, there is.

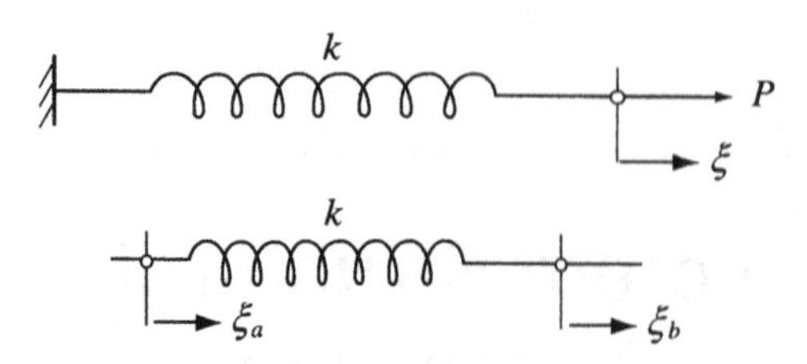

Figure 3.1. A simple linear elastic spring.

Consider the following. If we first define the *strain energy U* of the spring as the energy stored in it as it is extended or compressed, we can calculate it as

$$U = \int_0^{\xi} F_s d\eta = \int_0^{\xi} k\eta d\eta = \frac{1}{2}k\xi^2 \tag{3.4}$$

where we have used η as a dummy variable of integration in the two integrals.

We also want to account for the work done by the externally applied force P as it pulls or pushes on the free end of the spring (cf. Fig. 3.1). However, in our accounting, we actually want to calculate the potential of the applied force, in parallel with what we do in particle mechanics when calculating work and energy for rigid bodies. First, the applied load is assumed to be a *dead load* because it is constant and its magnitude and direction are unaffected by – or are independent of – the displacement at the loading point during this calculation, quite unlike the displacement produced by the internal spring force. Second, as we noted, we are really interested in expressing the work done by the applied load in terms of its *potential*. Thus, we now calculate the *potential of the applied load V* as the negative of the product of a conservative force with the displacement through which the force moves, that is, as the negative of the work done:

$$V = -P\xi \tag{3.5}$$

Now we add the strain energy U and the potential of the applied loads V to define the *total potential energy* Π for this externally loaded spring as

$$\begin{aligned}\Pi &= U + V \\ &= \frac{1}{2}k\xi^2 - P\xi\end{aligned} \tag{3.6}$$

Note that the total potential energy is a quadratic form in ξ; that is, it is a polynomial in ξ, whose exponents are of power two and less. A sketch of Eq. (3.6) is shown in Fig. 3.2, and we see that it is shaped like a well whose minimum is located at the equilibrium point $\xi = P/k$. Let us now seek the extremum of the potential energy by considering small variations in the displacement of the free end, ξ. We do this by considering *small variations* in ξ called $\delta\xi$, and we look at the variation in the total potential energy by calculating the change in the total potential energy as follows:

$$\begin{aligned}\Pi(\xi + \delta\xi) - \Pi(\xi) &= \frac{1}{2}k(\xi + \delta\xi)^2 - P(\xi + \delta\xi) - \frac{1}{2}k\xi^2 + P\xi \\ &= \frac{1}{2}k(2\xi\delta\xi) - P\delta\xi + \frac{1}{2}k(\delta\xi)^2 \\ &= (k\xi - P)\delta\xi + \frac{1}{2}k(\delta\xi)^2\end{aligned} \tag{3.7}$$

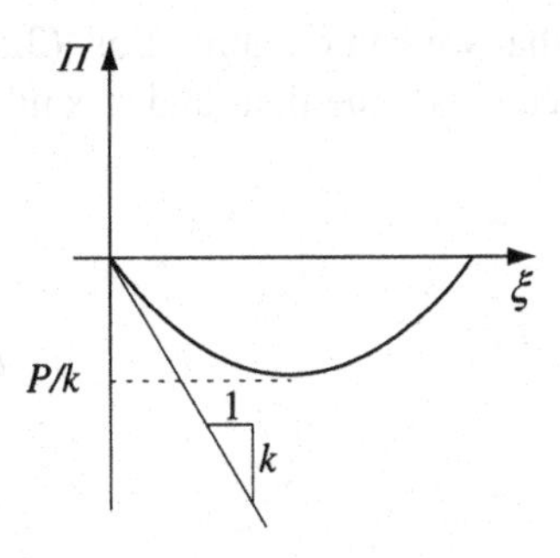

Figure 3.2. The total potential energy of a simple linear elastic spring.

Thus, the variation of the total potential energy has two terms, one of which is linear in the variation $\delta\xi$, the other of which is of order $(\delta\xi)^2$. The vanishing of the first-order term is called the *vanishing of the first variation of the total potential energy* and is written in the form

$$\delta^{(1)}\Pi = (k\xi - P)\delta\xi = 0 \tag{3.8}$$

In view of the spring's constitutive relation (Eq. (3.1)) we see that Eq. (3.8) actually expresses equilibrium in terms of ξ, the displacement variable, as we demonstrated axiomatically in Eq. (3.3).

Now, the *second variation of the total potential energy* can be written as

$$\delta^{(2)}\Pi = \frac{1}{2}k(\delta\xi)^2 \geq 0 \tag{3.9}$$

This equation shows that the second variation is positive – regardless of whether the variation $\delta\xi$ is positive or negative – and vanishes only when $\delta\xi$ is identically zero. Thus, in fact, we have found that the extreme value produced by this variational process represents a *minimum* of the total potential energy, thus confirming the sketch in Fig. 3.2. Stated as a general principle, then, we have found that by varying the total potential energy Π with respect to the displacement field, we have derived the equation of equilibrium expressed in terms of the displacement (or deflection) variable. Further, we have also shown that satisfaction of equilibrium corresponds to a minimum of the total potential energy. These results are embodied in the *principle of minimum potential energy*, which in turn can be identified as the single most important concept that underlies the approach to structural analysis presented in this book.

We need to make one more important point here. The foregoing analysis of the total potential energy of the discrete spring could have been done using elementary calculus. That is, we could have simply asked for the conditions under which the total potential (3.6) would reach an extreme value and found that the condition is given by

$$\frac{d\Pi}{d\xi} = (k\xi - P) = 0 \tag{3.10}$$

while the condition that defines the nature of this extremum is given by

$$\frac{d^2\Pi}{d\xi^2} = k > 0 \tag{3.11}$$

In fact, for this elementary discrete system, we see that we can combine Eqs. (3.8) and (3.10) and Eqs. (3.9) and (3.11) to show that we can calculate the first- and second-order variations as, respectively,

$$\begin{aligned} \delta^{(1)}\Pi &= \frac{d\Pi}{d\xi}\delta\xi \\ \delta^{(2)}\Pi &= \frac{1}{2}\frac{d^2\Pi}{d\xi^2}(\delta\xi)^2 \end{aligned} \tag{3.12}$$

So why all the fuss about small variations $\delta\xi$?

The answer is rather simple. The single-variable calculus approach will work for every model examined in *this* chapter. Remember, however, that we are intent on building models of three-dimensional, continuum structures, in which case we need some way of accounting for small variations in the displacement variables, which are themselves continuous functions of the coordinates of the structure being analyzed. The total potential energy statements for these problems will often involve integrals of continuous functions and their derivatives, as we will see in Chapter 4 and beyond. The simple rules for calculating extrema used in single-variable (or multivariate, for that matter) calculus are inadequate for this task. We are using the discrete models as more familiar examples with which we can demonstrate some of the mathematics that derives from the very elegant *calculus of variations*.

3.2 A Discrete Model of a Beam Using Minimum Potential Energy

To extend what we have just learned let us consider a structure such as that shown in Fig. 3.3. This is a *discrete model* of a beam, by which we mean that we are modeling the bending action of a continuous beam in terms of the rotational spring of stiffness k_r that is connected by two rigid, massless links to the supports. The entire assembly supports a vertically applied force P at the center. As we will see later (cf. Chapter 6), this is a very good model for the bending behavior of a centrally loaded, simply supported elastic beam, and we will be able to specifically relate k_r to the material and geometrical properties of a real beam. For now we will focus first on the moment that the rotational spring produces, without which the links cannot sustain the vertically applied force by transmitting vertical forces to the supports.

From geometry, we can identify the displacement or deflection of the beam at its center as w and we can relate this to the angles θ that the rigid links make with the horizontal, unbent configuration of the beam,

$$w = \frac{L}{2}\sin\theta \cong \frac{L}{2}\theta \tag{3.13}$$

where in the last part of this result we have assumed that the angle θ is small, which physically means that we are assuming that the slope of the bent beam is itself small. We should recall this in Chapter 6 when we derive and apply our more precise continuum model of beam deformation.

When the beam deflects by an amount w at its center, the rotational spring deforms through a total angle of $\theta - (-\theta) = 2\theta$, so that the moment M_r produced by the rotational

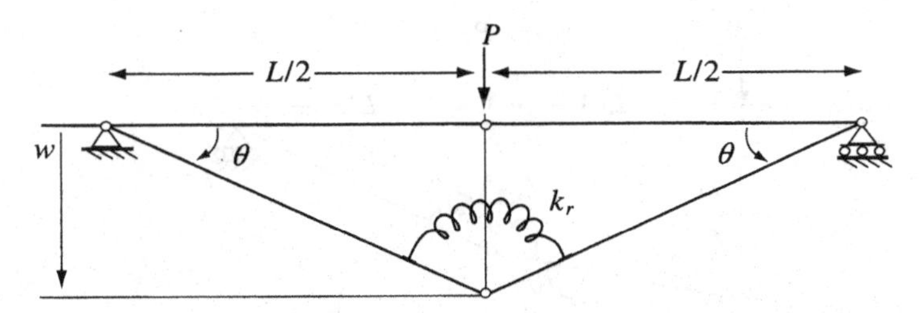

Figure 3.3. A discrete model of an elastic beam.

spring is

$$M_r = k_r(2\theta) = 2k_r\theta = 4k_r\frac{w}{L} \tag{3.14}$$

Equation (3.14) represents the constitutive law for the discrete beam model. Taking static equilibrium as axiomatic, we can calculate the vertical reactions at each end as $P/2$ acting up, so that moment equilibrium requires

$$M_r = \frac{PL}{4} \tag{3.15}$$

Now we can eliminate the internal spring moment M_r between Eqs. (3.14) and (3.15) to find the equation of equilibrium for the beam made up of a rotational spring and two massless links in terms of the displacement at the center, that is,

$$P = (16k_r/L^2)w \tag{3.16}$$

We can check the physical dimensions of this expression for consistency. Since Eq. (3.14) indicates that the rotational spring has physical dimensions of moment, that is, [force][distance], the term in parentheses in Eq. (3.14) must have those of a linear spring, [force]/[distance]. Hence, our result (3.14) is dimensionally correct.

Having identified the spring stiffness for the beam, we can calculate the strain energy stored in this beam, just as we did for the simple spring. In particular,

$$U = \int_0^\theta M_r(2d\theta) = 2k_r\theta^2 = \frac{1}{2}k_r\left(\frac{4w}{L}\right)^2 \tag{3.17}$$

The potential of the applied load for this problem is quite analogous to that used in the linear spring, that is, with the magnitude and direction of the load P being unaffected by the deflection at that point,

$$V = -Pw \tag{3.18}$$

so that the total potential energy for this problem is easily found to be

$$\begin{aligned} \Pi &= U + V \\ &= \frac{1}{2}k_r\left(\frac{4w}{L}\right)^2 - Pw \end{aligned} \tag{3.19}$$

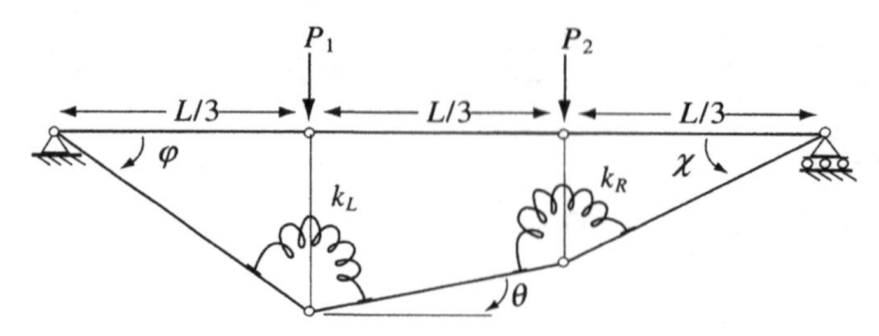

Figure 3.4. A more complicated discrete model of an elastic beam.

We can now calculate the first variation of the total potential energy for *admissible variations* in the beam center displacement, that is, for displacements δw which are arbitrary, small, and do not violate any of the boundary conditions. Then the first variation of Π can then be written as

$$\delta^{(1)}\Pi = [16k_r(w/L^2) - P]\delta w \tag{3.20}$$

The necessary and sufficient condition for equilibrium then results from the vanishing of this first variation, which is to say that we recover again the same equilibrium result that we identified in Eq. (3.16) having already incorporated both the constitutive law for the discrete beam and the equation of static equilibrium expressed in terms of moments.

Example 3.1. Combine the equilibrium equations for the two nodes of the discrete beam model shown in Fig. 3.4 with the constitutive laws of the given springs to formulate a complete model for the beam's behavior.

Inasmuch as there are two nodal torsional springs, this problem has two degrees of freedom, which means that we can express the displacements of each node from the beam's initial position as

$$\begin{aligned} w_L &= \frac{L}{3}\sin\varphi \cong \frac{L\varphi}{3} \\ w_R &= \frac{L}{3}\sin\chi \cong \frac{L\chi}{3} \end{aligned} \tag{3.21}$$

The orientation of the middle link with respect to the horizontal is constrained by the geometry to be, in both exact and linearized forms,

$$\begin{aligned} \sin\theta &= \sin\chi - \sin\varphi \\ \theta &\cong \chi - \varphi \end{aligned} \tag{3.22}$$

The constitutive equations for the two torsional springs would show the restoring moments or torques to be

$$\begin{aligned} M_L &= k_L(\varphi - \theta) = k_L(2\varphi - \chi) \\ M_R &= k_R(\chi + \theta) = k_R(2\chi - \varphi) \end{aligned} \tag{3.23}$$

We now apply Newton's law to the nodes by summing moments over, respectively, the left-hand link and the right-hand link,

$$\begin{aligned} M_L &= \frac{2P_1L}{9} + \frac{P_2L}{9} \\ M_R &= \frac{P_1L}{9} + \frac{2P_2L}{9} \end{aligned} \tag{3.24}$$

The restoring moments can be eliminated between Eqs. (3.23) and (3.24) to yield the following pair of eqautions for the two independent rotation angles:

$$\begin{aligned} -\chi + 2\varphi &= \frac{2P_1L}{9k_L} + \frac{P_2L}{9k_L} \\ 2\chi - \varphi &= \frac{P_1L}{9k_R} + \frac{2P_2L}{9k_R} \end{aligned} \tag{3.25}$$

This symmetric pair of algebraic equations can be solved for the rotation angles and, through Eqs. (3.21), the displacements of the two nodes

$$\begin{aligned} w_L &= \frac{L\varphi}{3} = \left(\frac{L^2}{81}\right)\left(\frac{4P_1}{k_L} + \frac{P_1}{k_R} + \frac{2P_2}{k_L} + \frac{2P_2}{k_R}\right) \\ w_L &= \frac{L\chi}{3} = \left(\frac{L^2}{81}\right)\left(\frac{2P_1}{k_L} + \frac{2P_1}{k_R} + \frac{P_2}{k_L} + \frac{4P_2}{k_R}\right) \end{aligned} \tag{3.26}$$

■

Example 3.2. Use the principle of minimum total potential energy to formulate a model for the discrete beam model shown in Fig. 3.4 and compare the results with those of Example 3.1.

The strain energy for the beam model under consideration is the energy stored in the two rotational springs defined by their constitutive laws (Eqs. (3.23))

$$U = U_L + U_R = \frac{1}{2}k_L(2\varphi - \chi)^2 + \frac{1}{2}k_R(2\chi - \varphi)^2 \tag{3.27}$$

while the potential of the two applied loads is

$$V = -P_1 w_L - P_2 w_R = -\frac{P_1L}{3}\varphi - \frac{P_2L}{3}\chi \tag{3.28}$$

so that the total potential for this problem is

$$\Pi = \frac{1}{2}k_L(2\varphi - \chi)^2 + \frac{1}{2}k_R(2\chi - \varphi)^2 - \frac{P_1L}{3}\varphi - \frac{P_2L}{3}\chi \tag{3.29}$$

The first variation of the potential (3.29) is, then,

$$\delta^{(1)}\Pi = k_L(2\varphi - \chi)(2\delta\varphi - \delta\chi) + k_R(2\chi - \varphi)(2\delta\chi - \delta\varphi) - \frac{P_1L}{3}\delta\varphi - \frac{P_2L}{3}\delta\chi$$

which can be regrouped and written as

$$\begin{aligned} \delta^{(1)}\Pi = &\left[(4k_L + k_R)\varphi - 2(k_L + k_R)\chi - \frac{P_1L}{3}\right]\delta\varphi \\ &+ \left[-2(k_L + k_R)\varphi + (k_L + 4k_R)\chi - \frac{P_2L}{3}\right]\delta\chi \end{aligned} \tag{3.30}$$

If we set the first variation (3.30) to zero we will obtain a pair of symmetric algebraic equations that produce exactly the same solution as given in Eqs. (3.26). It is worth noting that we have found the same result by both the (axiomatic) equilibrium method and by applying the principle of minimum total potential energy. And while the result in Eq. (3.30) does not look as clean as the pair of equations given in Eqs. (3.26), in general, we get the same or cleaner results variationally. Further, there are many problems (e.g., all indeterminate structures) where we cannot obtain a solution by writing Newton's second law. ■

3.3 Compatibility and Total Potential Energy Modeling of Structures

We observed in Chapter 1 that there are three basic concepts or principles that we apply in structural mechanics, namely, equilibrium, constitutive laws, and compatibility. It is clear from what we have already done with the principle of minimum total potential energy that we are using constitutive laws to formulate the stored (strain) energy, and the result of the application of the principle is a statement of equilibrium. What is less evident is the role of compatibility, and so it is worth discussing that before we go any further.

First, note that we have cast all of the problems we have analyzed in terms of displacements or deflections, and we have noted that all their variations were assumed to be consistent with the boundary conditions of the problem. Thus, implicitly we have chosen to base our models on displacement or deflection assumptions that were consistent or compatible. Now we take that idea as an explicit guiding principle, although it becomes significantly more involved in the case of continuous – rather than discrete – models of structures.

Second, compatibility issues become more evident in discrete modeling of more complex problems. We can see this in the following examples.

Example 3.3. Consider the pair of springs connected in series, as shown in Fig. 3.5(a). Formulate the equation(s) of equilibrium using the minimum total potential principle. How is geometric compatibility ensured in this case?

For springs in series, we must provide a sufficient number of coordinates to be able to measure the relative extension of each spring, so for the two springs shown we will need two coordinates. Then, building on the basic spring constitutive law of Eq. (3.1), we find the two internal spring forces to be

$$F_{s1} = k_1\xi_1, \qquad F_{s2} = k_2(\xi_2 - \xi_1) \tag{3.31}$$

where we take $\xi_0 = 0$. The energy stored in the ith spring is (after Eq. (3.4))

$$U_i = \int_0^{\xi_i-\xi_{i-1}} F_{si}d\eta = \int_0^{\xi_i-\xi_{i-1}} k_i\eta d\eta = \frac{1}{2}k_i(\xi_i - \xi_{i-1})^2 \tag{3.32}$$

The potential of the applied load applied at the free end of the chain of the two springs is

$$V = -P\xi_2 \tag{3.33}$$

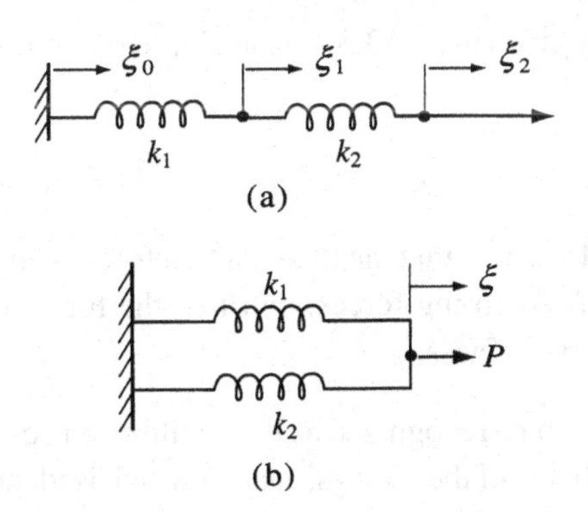

Figure 3.5. Combinations of elastic springs; (a) in series and (b) in parallel.

And the combined total potential energy is, thus,

$$\Pi = \sum_{i=1}^{2} U_i - V = \frac{1}{2}\sum_{i=1}^{2} k_i(\xi_i - \xi_{i-1})^2 - P\xi_2 \tag{3.34}$$

Now in minimizing the total potential energy represented in Eq. (3.34) we have to recognize explicitly a kinematic compatibility condition we have used in developing our model of the chain: in particular, that each spring's relative extension could only be calculated from a difference of two independent coordinates, and that we need two such coordinates – ξ_1 and ξ_2 – for the chain of two springs. Thus, in order to find the minimum of the total potential energy given in Eq. (3.34), we have to vary each coordinate independently, after which we can identify the first variation of that total potential (see also the problems at the close of the chapter) as

$$\begin{aligned}\delta^{(1)}\Pi = {} & [k_1\xi_1 - k_2(\xi_2 - \xi_1)]\delta\xi_1 \\ & + [k_2(\xi_2 - \xi_1) - P]\delta\xi_2\end{aligned} \tag{3.35}$$

The vanishing of the first variation (3.35) yields the pair of equilibrium equations that we were looking for:

$$\begin{aligned}(k_1 + k_2)\xi_1 - k_2\xi_2 &= 0 \\ -k_2\xi_1 + k_2\xi_2 &= P\end{aligned} \tag{3.36}$$

■

Example 3.4. Consider the pair of springs connected in parallel, as shown in Fig. 3.5(b). Formulate the equation(s) of equilibrium using the minimum total potential principle. How is geometric compatibility ensured in this case?

Here we recognize from the outset that, as drawn, the relative extension of each spring must be the same. Hence, we need only one independent coordinate to describe in a compatible way the (identical) movement of the free end of each of the springs. The total potential energy for this system is

$$\Pi = \sum_{i=1}^{2} U_i - V = \frac{1}{2}(k_1 + k_2)\xi^2 - P\xi \tag{3.37}$$

The first variation of the total potential energy given in Eq. (3.37) is (again, see the problems at the end of this chapter)

$$\delta^{(1)}\Pi = [(k_1 + k_2)\xi - P]\delta\xi \tag{3.38}$$

This result simply states another recognizable result: that the total applied force is equaled by the simple sum of the two parallel (internal) spring forces. Further, the force in any particular spring is directly proportional to its stiffness. ■

Each of the bracketed terms in Eq. (3.35) can be recognized as the equilibrium equation corresponding to the nodes at the right end of each of the springs, each of which is identified by the corresponding ξ_i. The physical meaning of these equations is that the force exerted at the free end of the chain is transmitted unchanged to each of the springs in the series, exactly as we would expect from free bodies drawn at each node. Further, we have applied the principle of compatibility by recognizing that the presence of each spring requires an identifiable, independent coordinate, notwithstanding the fact that the strain energy stored in each spring is expressed as the square of the difference between the coordinates at each end of the spring.

Now, although it certainly is true that we model elastic structures as assemblies of elastic springs of various kinds and of various stiffnesses, the point of these last two examples is less in anticipation of modeling real structures. Rather, the point is that compatibility (or consistency or continuity) conditions have a role to play in the variational modeling of elastic structures, especially when we use the kinematics (or geometry of deformation) as a prime building block in the variational approach.

3.4 Virtual Work in Discrete Modeling of Structures

There is one more result that is worth a brief discussion here, and that is the idea of *virtual work* (and this is related neither to current computer jargon nor to the effort needed to master the material we present in this book). The concept of virtual work is worth noting because it is an important idea, and it can be seen as a logical derivative of what we have just done.

Consider, first, the elastic spring, and in particular the statement of the first variation of its total potential energy (from Eq. (3.8)). With the aid of the spring constitutive law given in Eq. (3.1) we can write

$$\delta W_{virt} = F_s\delta\xi - P\delta\xi = (F_s - P)\delta\xi = 0 \tag{3.39}$$

Now this result properly has the physical dimensions of work, and as written we see that it represents the difference between the work done by the (elastic) spring force $F_s = k\xi$ acting through a small but arbitrary displacement $\delta\xi$ and the work done by the applied external force P acting through the same arbitrary displacement $\delta\xi$. This pair of terms represents the total *virtual work* done by all of the forces acting on the system being modeled, both internal and external, through the small yet arbitrary *virtual displacements* $\delta\xi$. Further, setting the virtual work (3.39) to zero yields a restatement of the equation of equilibrium. Note, however, that the equilibrium equation resulting from applying virtual work is cast in terms of the forces acting (recall our comments about Eq. (3.8)), and that the internal forces are functions of the actual displacements, *not* of the virtual displacements. That is, $F_s = k\xi \neq k\delta\xi$.

For the discrete model of the beam (as two links connected by a rotational spring), the corresponding virtual work statement is

$$\delta W_{virt} = M_r 2\delta\theta - P\delta w = M_r\left(\frac{4\delta w}{L}\right) - P\delta w = \left(M_r - \frac{PL}{4}\right)\left(\frac{4\delta w}{L}\right) \tag{3.40}$$

which when set to zero produces the same equilibrium equation derived before as Eq. (3.20). The term in the parentheses is the difference between the work done by the moment M_r acting through a virtual slope $4\delta w/L$ and that done by the external moment $PL/4$ acting through the same virtual slope. Thus, by formulating the virtual work done by the actual forces present, both internal and external, when acting through virtual displacements, and setting that virtual work to zero, we are in fact deriving once again the equation(s) of equilibrium.

Problems

3.1 Write linearized equation(s) of equilibrium to find the centerline deflection for the discrete beam model shown in Fig. 3.P1. Confirm your results by applying the principle of minimum total potential energy. Compare your answer to that of the discrete model having only a single spring.

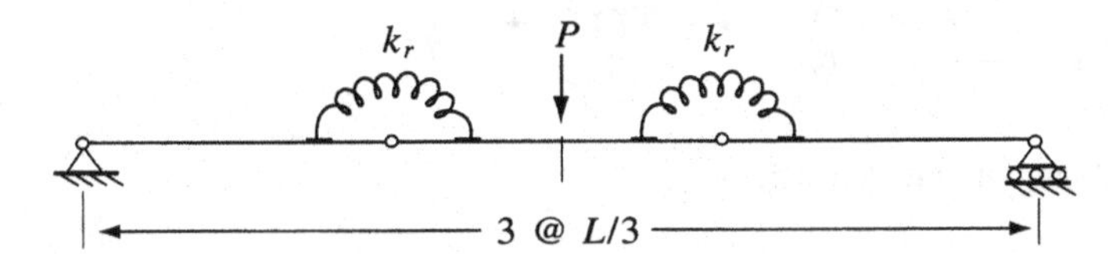

Figure 3.P1. Figure for Problem 3.1.

3.2 Write linearized equation(s) of equilibrium to find the centerline deflection for the discrete beam model shown in Fig. 3.P2. Confirm your results by applying the principle of minimum total potential energy. Compare your answer to that of the discrete model having only a single spring and the results of Problem 3.1.

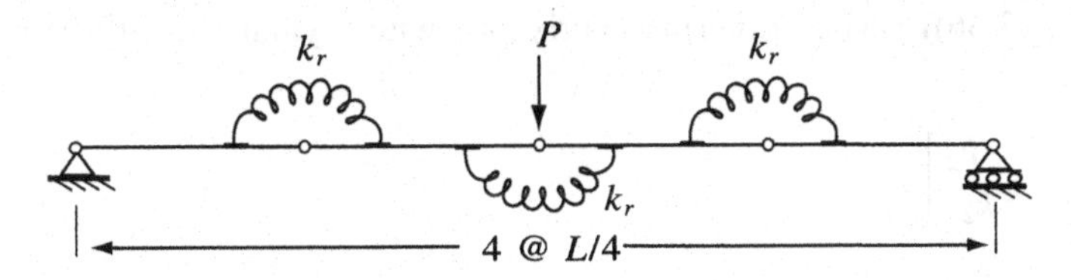

Figure 3.P2. Figure for Problem 3.2.

3.3 Write linearized equation(s) of equilibrium to find the centerline deflection for the discrete beam model shown in Fig. 3.P3. Confirm your results by applying the principle of minimum total potential energy. Compare your answer to that of the discrete model having only a single spring.

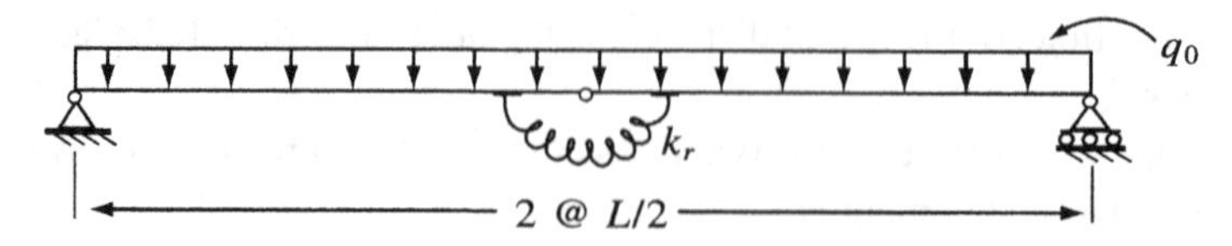

Figure 3.P3. Figure for Problem 3.3.

3.4 Write linearized equation(s) of equilibrium to find the centerline deflection for the discrete beam model shown in Fig. 3.P4. Confirm your results by applying the principle of minimum total potential energy. Compare your answer to that of the discrete model having only a single spring.

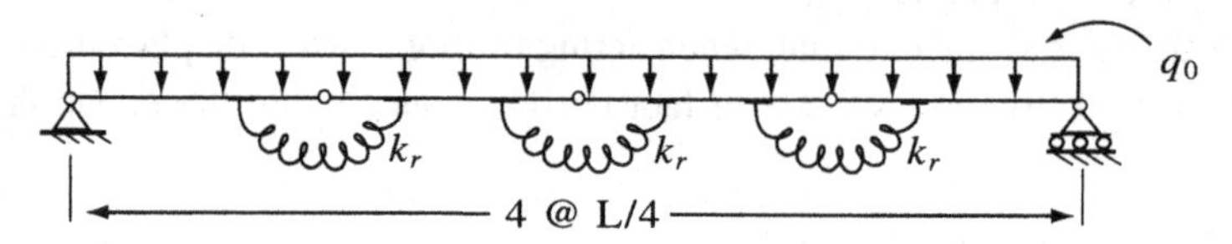

Figure 3.P4. Figure for Problem 3.4.

3.5 Formulate the total potential energy for the collection of spring shown in Fig. 3.P5. How many degrees of freedom are there?

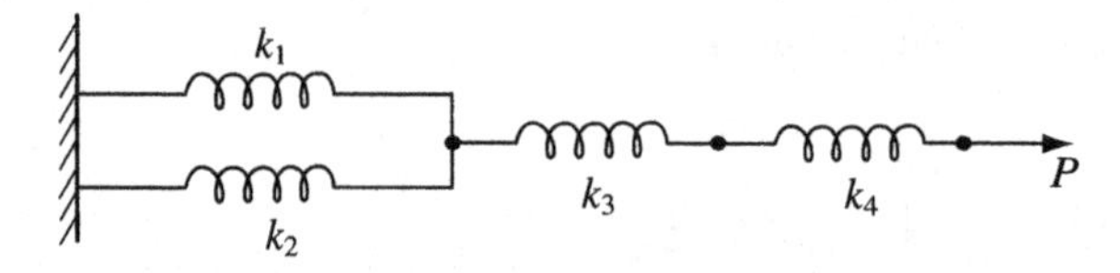

Figure 3.P5. Figure for Problem 3.5.

3.6 Derive the equilibrium equation(s) for the system of Problem 3.5.

3.7 Develop the second variation of the total potential energy of the system analyzed in Problems 3.5 and 3.6.

3.8 Calculate the total variation of the total potential energy of Eq. (3.34).

3.9 Calculate the total variation of the total potential energy of Eq. (3.37).

3.10 Show that the vanishing of the first variation of the total potential for the two springs in series (Eq. (3.36)) can be written in a matrix form with a symmetric stiffness matrix equal to

$$\begin{bmatrix} k_1 + k_2 & -k_2 \\ -k_2 & k_2 \end{bmatrix}$$

This is a view looking upward along one of the four legs of the Eiffel Tower (cf. Fig. 1.3b). Note the profusion of simple elements that are not loaded except at their endpoints or connections to other tower members. These elements evoke the one-dimensional bars whose modeling is described in this chapter. (Photo by Clive L. Dym.)

4

Bars: Axially Loaded Members

We now turn to *continuous* models of structures in which we treat structures as three-dimensional objects that must be modeled in terms of functions whose independent variables are spatial coordinates. In this chapter we will focus on long, slender elements in (and on) which all of the action takes place in the long direction. Thus, if we define a spatial coordinate, say, the x-coordinate, to run along the length of such a member, the external forces are asssumed to be applied along that axis, and the important stresses, strains, and displacements will be directed along the x-axis and vary with x. In order to do this variationally, we will have to extend our definition of the variation of a function.

The slender, axially loaded members we focus on here are called bars, rods, or struts (or "two-force" members in dated terminology). Bars play a key rule in structural engineering as the elements that make up trusses, both two dimensional (planar) and three dimensional. As we will see in Chapter 5, as aggregates of bars, trusses perform as extremely efficient beams that can carry significant loads normal to the principal axis of the truss.

We will also explicitly see here for the first time the *discretization* part of the two-stage modeling process, that is, the part of the modeling process wherein we take an idealized continuous model and find solutions, both exact and approximate, for a problem (or a class of problems) by developing a discrete solution for the continuous, idealized model of a structural problem.

4.1 Stress, Strain and Hooke's Law for Axially Loaded Members

Let us begin by reviewing some basic results from elementary elasticity theory (as presented in "Strength of Materials"). To begin with, we consider solid elastic bars (see Fig. 4.1) of cross-sectional area A and length L for which we assume that

$$\sqrt{A}/L \ll 1 \tag{4.1}$$

Thus we are assuming that the bar is slender by taking the square root of the cross-sectional area both to be representative of the thickness and width of the member and stating that it is small in comparison with the member length. We further assume that *all external loading is directed axially*, that is, directed along the x-axis, whether the load is tensile ($P \geq 0$) or compressive ($P \leq 0$). Finally, we will assume that the only stress, strain, and displacement components of interest are directed in the x-direction.

The displacement of any cross section at coordinate x along the bar's axis is taken as $u(x)$. The strain calculation (see Fig. 4.2) follows from the definition of strain as the

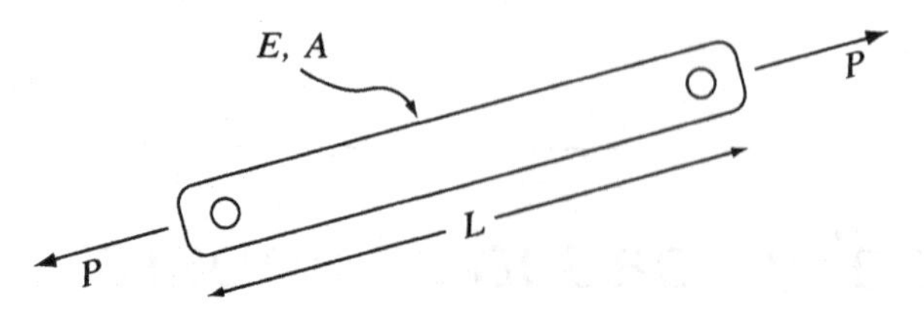

Figure 4.1. A one-dimensional model of an axially loaded member.

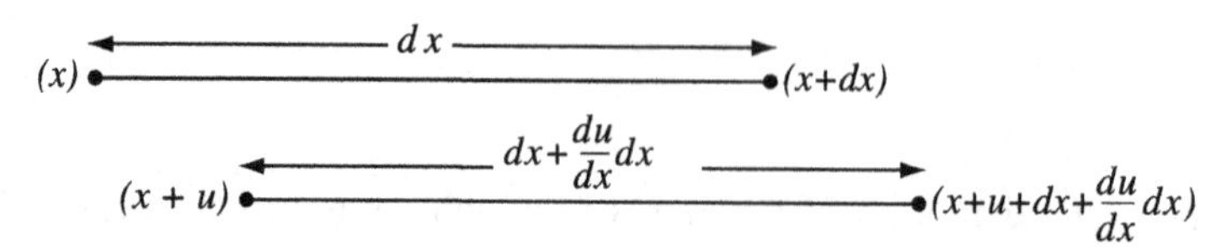

Figure 4.2. Displacement and strain in an axially loaded member.

(dimensionless) change in length of a strained element divided by the element's original length:

$$\varepsilon \equiv \varepsilon_{xx} = \frac{\left(dx + \frac{du}{dx}dx\right) - dx}{dx} = \frac{du}{dx} \tag{4.2}$$

The two other components of displacement, $v(x, y, z)$ in the y-direction and $w(x, y, z)$ in the z-direction, and the five other components of strain are here assumed to be zero.

For the stresses, since the objects are small in the y- and z-directions and there are no forces applied in these directions, we can assume that the normal stress components, σ_{yy} and σ_{zz}, are zero. Further, all of the shear stress components are zero because the corresponding shear strains vanish because of the assumed displacement field. For this one-dimensional model, then, the constitutive law is simply a one-dimensional statement of Hooke's law, that is,

$$\sigma \equiv \sigma_{xx} = E\varepsilon = E\frac{du}{dx} \tag{4.3}$$

4.2 Strain Energy and Work–Energy for Axially Loaded Members

Now, how do we calculate the stored or strain energy for this bar? In elasticity theory there is a formal mathematical result that defines strain energy in the general case of deformation of a three-dimensional solid. Here we will approach this result in two ways. In the first we will simply do for one dimension what we would do for the general case, that is, compute the elastic work done by the stresses acting on an element acting through the displacements produced by the stresses. Thus, we show in Fig. 4.3 an element with the displacements of Fig. 4.2 and the stresses we expect to find in the stressed element. First, let us perform an equilibrium calculation for the element by summing forces acting in the x-direction. Thus, seeing the change in stress between the two faces we can simply sum forces according to Newton's second law, that is,

$$\left(\sigma_{xx} + \frac{d\sigma_{xx}}{dx}dx\right)A(x) - \sigma_{xx}A(x) = \frac{d\sigma_{xx}}{dx}(dx)A(x) = 0$$

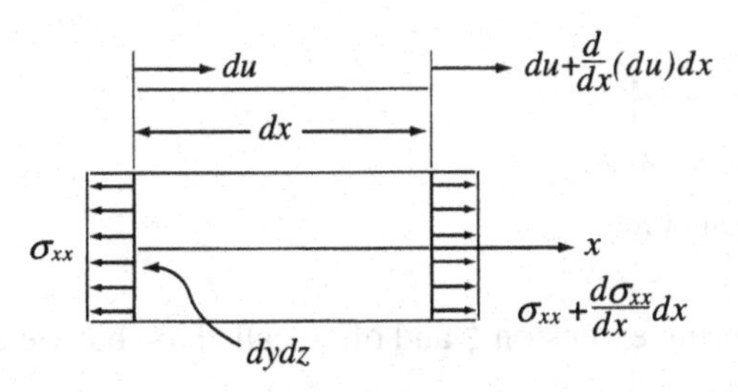

Figure 4.3. Displacement and stress in an axially loaded member.

from which it follows that

$$\frac{d\sigma_{xx}}{dx} = 0 \tag{4.4}$$

We now calculate the work done during increments of the displacement and du and $du + (d(du)/dx)dx$ on the left- and right-hand faces of the element. In so doing, we will discard higher-order terms where appropriate. We will also make use of (1) the equation of equilibrium just derived and (2) the one-dimensional strain-displacement relationship given in Eq. (4.2). After some manipulation we find the following expression for the incremental stored or strain energy:

$$\begin{aligned} dU &= \int_{Vol} dU_0 = \int_{Vol} \left[\left(\sigma_{xx} + \frac{d\sigma_{xx}}{dx}dx\right)\left(du + \frac{d(du)}{dx}dx\right) - \sigma_{xx}du(x)\right] dydz \\ &= \int_{Vol} \left(du\frac{d\sigma_{xx}}{dx} + \sigma_{xx}\frac{d(du)}{dx}dx\right) dydz \\ &= \int_{Vol} \sigma_{xx}\frac{d(du)}{dx}dxdydz \\ &= \int_{Vol} \sigma_{xx}d\varepsilon_{xx}dxdydz \end{aligned} \tag{4.5}$$

With Hooke's law, here in the form of Eq. (4.3), the last expression can be integrated over the strain produced to yield the strain energy in the form

$$U = \int_0^{\varepsilon_{xx}} \int_{Vol} \sigma_{xx}d\varepsilon_{xx}dxdydz = \int_{Vol} \frac{1}{2}E\varepsilon_{xx}^2 dxdydz \tag{4.6}$$

Note that the integrands in Eqs. (4.5) and (4.6) represent the *strain energy density*, or strain energy per unit volume U_0, and in view of Hooke's law it can be written in the following different forms:

$$U_0 = \frac{1}{2}E\varepsilon_{xx}^2 = \frac{1}{2}\sigma_{xx}\varepsilon_{xx} = \frac{1}{2E}\sigma_{xx}^2 \tag{4.7}$$

Equations (4.7) are clearly analogous to the different forms we use to write the stored energy of linear elastic springs. In fact, the second derivation of Eqs. (4.6) and (4.7) is based on the classic "Strength of Materials" results for the extension of a rod or bar. For a bar of length L, area A, and modulus E, under an axial load P, the classic formula for

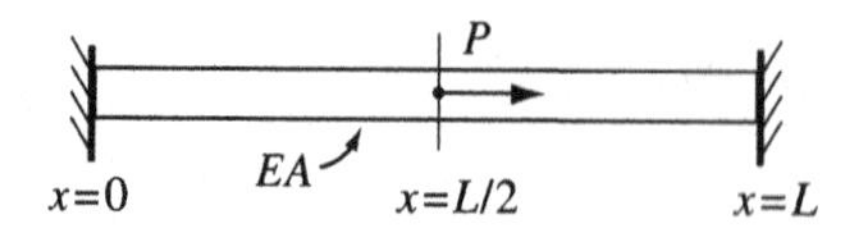

Figure 4.4. A centrally loaded bar or rod.

its extension Δ (analogous to the spring extension ξ and often called δ – but we use δ as the variational operator) is

$$\Delta = \frac{PL}{AE}$$

and the corresponding energy stored in this version of a spring is

$$U = \frac{1}{2} P\Delta = \frac{P^2 L}{2AE} = \frac{1}{2}\left(\frac{P}{A}\right)\left(\frac{P}{AE}\right) AL = \frac{1}{2}\sigma\varepsilon(Vol) \tag{4.8}$$

The equality of Eqs. (4.6), (4.7), and (4.8) speaks for itself.

Now that we have the strain energy, let us formulate the potential of the applied loading for the particular case shown in Fig. 4.4. Here we have a concentrated axial load P applied at the center of a bar that is restrained at both ends. The potential of this external load would be

$$\begin{aligned} V &= -Pu(L/2) \\ &= -P\int_0^L u(x)\delta_D(x - L/2)dx \end{aligned} \tag{4.9}$$

In Eq. (4.9) we have introduced the spatial version of the *Dirac delta function*, $\delta_D(x-L/2)$. The Dirac delta function is a *spike*, that is, it is a spatial impulse function that has an infinite amplitude but has zero width along the x-axis. When multipled by the function $u(x)$ and integrated over the length of the bar, it effectively filters out all values of $u(x)$ save that at $x = L/2$.

We can combine the potential of the external load with the previously calculated strain energy to form the total potential energy. As we do this we will cast the strain in terms of the displacement gradient, so that

$$\Pi = \frac{1}{2} E\int_{Vol}\left(\frac{du}{dx}\right)^2 A(x)dx - P\int_0^L u(x)\delta_D(x - L/2)dx \tag{4.10}$$

4.3 Total Potential Energy Principle for Axially Loaded Members

We now concern ourselves with developing the principle of minimum potential energy for the continuous model of the axially loaded bar or rod, as we did in Chapter 3 for the discrete spring-and-links models. However, in order to effect the transition to variations within continuous models, we will have to broaden our view of the meaning of variations of displacements or other quantities in such cases.

Consider the curve labeled as $u(x)$ in Fig. 4.5. We suppose it to be the solution we seek, say, to an equilibrium equation or, equivalently, as the function that minimizes the

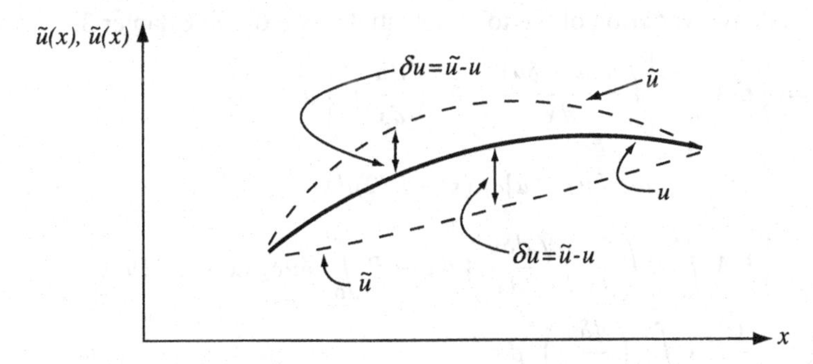

Figure 4.5. An optimum solution $u(x)$ and variations from that solution.

total potential energy. The family of curves marked $\tilde{u}(x)$ are variations of (or from) the path $u(x)$ that we seek. We then define the *delta operator* as

$$\delta[u(x)] \equiv \tilde{u}(x) - u(x) \tag{4.11}$$

Thus, the delta operator marks a variation or departure from the path $u(x)$. We assume that variation to be small in magnitude, and we further assume it to be a variation in the amplitude of $u(x)$ that is independent of – or not associated with – the independent variable x. This is clearly consistent with the way we defined small variations from discrete solutions in Chapter 3. However, it stands in marked contrast with the way we define small changes as derivatives in the calculus, wherein with each change $du(x)$ is associated with a corresponding change dx of the independent variable. Another view of what we are saying is that the variations are simply changes in amplitude or magnitude or the vertical distance between $u(x)$ and $\tilde{u}(x)$ for the same value of x.

One consequence of this assumption is that calculating the variation of a function can be viewed an application of an operator, the delta operator. A second consequence is that the variation and derivative operators commute, that is, they are invariant with respect to the order in which they are applied,

$$\begin{aligned} \frac{d}{dx}(\delta u(x)) &= \frac{d(\tilde{u}(x) - u(x))}{dx} \\ &= \frac{d\tilde{u}(x)}{dx} - \frac{du(x)}{dx} = \delta\left(\frac{du(x)}{dx}\right) \end{aligned} \tag{4.12}$$

In order to indicate how we apply the delta operator, let us find the minimum of the total potential energy for the centrally loaded bar, although for ease of arithmetic we take the bar cross section to be a constant, that is, $A(x) = A$. Thus, the total potential is here

$$\Pi = \frac{1}{2}EA\int_0^L \left(\frac{du}{dx}\right)^2 dx - P\int_0^L u(x)\delta_D(x - L/2)dx \tag{4.13}$$

Note, by the way, how clever we were to attach the subscript D to the notation we use for the Dirac delta function. In other contexts, where there is no danger of confusing the Dirac spike with a variational operation, there is no need for the subscript.

First we calculate the variation of the total potential as we did in Chapter 3,

$$\delta^{(T)}\Pi = \frac{1}{2}EA\int_0^L \left[\left(\frac{d(u+\delta u)}{dx}\right)^2 - \left(\frac{du}{dx}\right)^2\right]dx$$
$$- P\int_0^L [(u+\delta u) - u]\delta_D(x - L/2)dx$$
$$= \frac{1}{2}EA\int_0^L 2\left(\frac{du}{dx}\right)\left(\frac{d\delta u}{dx}\right)dx - P\int_0^L \delta u\delta_D(x - L/2)dx$$
$$+ \frac{1}{2}EA\int_0^L \left(\frac{d\delta u}{dx}\right)^2 dx$$

or

$$\delta^{(T)}\Pi = \int_0^L \left[EA\frac{du}{dx}\frac{d\delta u}{dx} - P\delta u\delta_D(x - L/2)\right]dx$$
$$+ \frac{1}{2}EA\int_0^L \left(\frac{d\delta u}{dx}\right)^2 dx \tag{4.14}$$

Before evaluating the total variation – and especially the first variation – given in Eq. (4.14), we show that we can also calculate the first variation of Π by applying the delta operator (Eq. (4.11)). For clarity, let us apply the delta operator to the relevant integrands of the total potential (4.13), leaving any needed integrations for just a moment because they will be the same as in Eq. (4.14). Thus,

$$\delta^{(1)}\left(\frac{du}{dx}\right)^2 = 2\left(\frac{du}{dx}\right)\delta\left(\frac{du}{dx}\right) = 2\left(\frac{du}{dx}\right)\left(\frac{d\delta u}{dx}\right); \quad \delta^{(1)}u(x) = \delta u \tag{4.15}$$

from which it follows that

$$\delta^{(1)}\Pi = \frac{1}{2}EA\int_0^L 2\frac{du}{dx}\frac{d\delta u}{dx}dx - P\int_0^L \delta u(x)\delta_D(x - L/2)dx \tag{4.16}$$

which is clearly the same as the first variation of Π contained in Eq. (4.14). Applying the operator this way is a useful shorthand, and it will also work well for us when we turn to obtaining discrete solutions directly (Section 4.4). However, for calculating higher-order variations, it is generally best to go through the algebraic steps outlined in our calculations in Chapter 3, as well as in the unnumbered equation preceding Eq. (4.14). On the other hand, for the linear elastic systems we are dealing with, it is also the case that: (1) we are rarely interested in a symbolic or numerical evaluation of the second variation because it is sufficient just to know that the second variation is normally positive-definite so equilibrium is always stable, and (2) the second variation of a total potential functional Π that is a quadratic form will always look just like the functional itself with δu substituted for u.

To return to our main point, before we set the first variation of Eqs. (4.14) or (4.16) to zero, we must integrate the first term by parts so that we can eventually write the integrand of the entire expression as [something]$\times\delta u$. Thus,

$$\delta^{(1)}\Pi = \left[EA\frac{du}{dx}\delta u\right]_0^L - EA\int_0^L \frac{d^2u}{dx^2}\delta u dx - P\int_0^L \delta u(x)\delta_D(x - L/2)dx$$
$$= \left[EA\frac{du}{dx}\delta u\right]_0^L - \int_0^L \left[EA\frac{d^2u}{dx^2} + P\delta_D(x - L/2)\right]\delta u dx \tag{4.17}$$

We see in Eq. (4.17) something more complicated than we found in Chapter 3 for our discrete models. To make this expression vanish, it is clear that we will have to apply conditions within the interval $0 < x < L$ as well as at the boundaries, $x = 0, L$. Within the interval that defines the length of the bar, in order to make the integral vanish, the fundamental theorem of the calculus tells us that for arbitrary and continuous variations δu the integral will be zero if and only if the bracketed term is itself zero. Thus, the equation of equilibrium for the centrally loaded bar is a differential equation with the displacement $u(x)$ being the dependent variable that we want to determine:

$$EA\frac{d^2u}{dx^2} + P\delta_D(x - L/2) = 0 \tag{4.18}$$

This is a second-order, ordinary differential equation, which means that we need two boundary conditions to ascertain completely its solution. Where do the appropriate boundary conditions come from? Not surprisingly, they come from the boundary terms in the first variation of the total potential, Eq. (4.17),

$$\delta^{(1)}\Pi \Rightarrow \left[EA\frac{du}{dx}\delta u\right]_0^L$$

so, at each end of the beam we must have

$$EA\frac{du}{dx}\delta u = 0 \quad \Rightarrow \quad \text{Either } EA\frac{du}{dx} = 0 \quad \text{or} \quad \delta u = 0 \tag{4.19}$$

That is, at each end of the bar, one of two conditions must be true, which means that we have two boundary conditions for the bar taken as a whole. The choice amounts to a force–displacement duality. Recognize that

$$EA\frac{du}{dx} = EA\varepsilon_{xx} = A\sigma_{xx} \equiv F_x \tag{4.20}$$

which is the axial force in the bar, F_x. Thus, Eq. (4.19) states that at each end of the bar we can prescribe a value for the force at that point (in this case, zero) or we can prescribe a particular value of the displacement, which in turn means that the variation of the displacement is zero at that point. Thus, we can view the mathematical statement of duality of Eq. (4.19) as the equivalent statement:

$$\begin{aligned} EA\frac{du}{dx}\delta u = 0 \quad \Rightarrow \quad & \text{Either } F_x = F_x^* = \text{prescribed} \\ & \text{or} \quad u = u^* = \text{prescribed} \end{aligned} \tag{4.21}$$

A final note on nomenclature. The boundary conditions on force are often called the *natural* (or *force*) boundary conditions, while those on displacement terms resulting from displacement variations are called the *geometric* boundary conditions.

Now, with Eqs. (4.18) and (4.21) we have a complete mathematical model for our idealization of slender, axially loaded members. Thus, it's now time to think about solutions.

4.4 Exact and Approximate Solutions for Bars and Rods

We want to solve our model's governing equation in two ways. In the first, which we call the *indirect* method, we will apply standard mathematics to solve the differential equation that we derived variationally for our continuous idealization or model of the bar. The second method, the *direct* method, emerges when we substitute a discrete representation into the idealized variational model – the step preceding the application of the principle of minimum potential energy – and perform the minimization of the energy with respect to parameters in the discrete representation of the solution. What this really means will become much clearer in this section.

4.4.1 An Exact Solution for the Centrally Loaded Bar

To begin with, we can integrate Eq. (4.18) rather straightforwardly, and the solution for the displacement can be written as

$$u(x) = \begin{cases} \dfrac{Px}{2EA} & \text{for } 0 \leq x \leq L/2 \\ \dfrac{P(L-x)}{2EA} & \text{for } L/2 \leq x \leq L \end{cases}$$
$$= (P/EA)\,[x/2 - (x - L/2)H(x - L/2)] \tag{4.22}$$

where in the second form of Eq. (4.22) we have introduced the Heaviside unit step function that results from the integration of the Dirac delta function:

$$\int_{-\infty}^{x} \delta_D(x - x_0)dx = 0 \qquad x < x_0$$
$$= H(x - x_0) = 1 \quad x \geq x_0 \tag{4.23}$$

Note that the boundary conditions we impose are as in our picture, where walls at each end restrain the bar from moving, so that $u(0) = u(L) = 0$. We show this result for the displacement in Fig. 4.6, along with the stress distribution along the bar length. We obtain the stress distribution simply by differentiating the solution just found to find the strain, and then multiplying that strain by Young's modulus. The net result is

$$\sigma_{xx}(x) = \begin{cases} \dfrac{P}{2A} & \text{for } 0 \leq x \leq L/2 \\ -\dfrac{P}{2A} & \text{for } L/2 \leq x \leq L \end{cases}$$
$$= (P/A)\,[1/2 - H(x - L/2)] \tag{4.24}$$

The displacement and stress fields given in Eqs. (4.22) and (4.24) and shown in Fig. 4.6 are consistent with our intuitive expectations. The displacement is both symmetric about the center of the bar, and continuous, while the stress (and therefore the strain) exhibit a jump or discontinuity at the point where the load is being applied. It also feels appropriate that the left side of the bar should be in tension, with $\sigma_{xx} > 0$, and the right-hand side should be in compression, $\sigma_{xx} < 0$. It is easy to show that the external reactions that occur at the ends of the bar are consistent with this, that is,

$$R_{left} = P/2 \Leftarrow; \qquad R_{right} = P/2 \Leftarrow \tag{4.25}$$

The arrows indicate the directions of the two forces exerted on the ends of the bar by the confining walls, and these results confirm our intuition about the distribution of tensile and compressive stresses in the bar. They also show that the bar and its supports are in

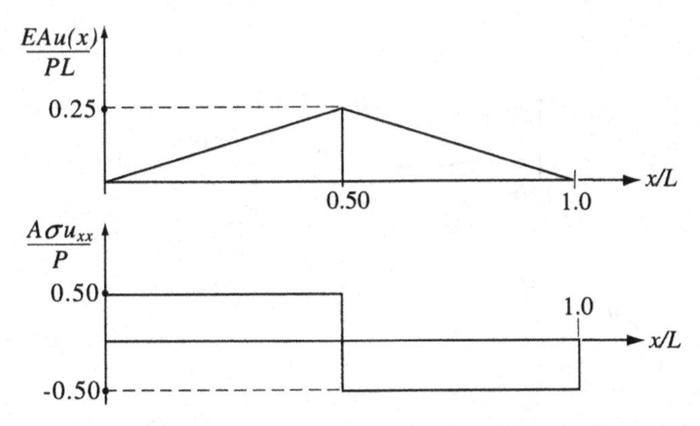

Figure 4.6. The displacement and stress fields of a centrally loaded bar.

equilibrium. So we have derived equilibrium variationally (confirming Newton's second law) and satisfied it by applying the mathematics of differential equations. It is in this sense we say we are applying the minimum potential energy principle *indirectly*, because we are using it "only" to derive the correct equilibrium equation and boundary conditions.

4.4.2 Trigonometric Series Approximations for the Centrally Loaded Bar

The other method of applying the minimum potential energy principle consists of assuming a *shape* or *trial function* as a candidate solution to the equilibrium condition and an *arbitrary multiplier* of the shape, with respect to which we minimize the total potential energy. This is known as the *Rayleigh–Ritz method*, or the *direct displacement* method. Since we have an exact solution available for the centrally loaded bar, let us find some direct solutions this problem. These direct solutions are called *discretized* approximations because we are seeking an approximate solution to the differential equation of equilibrium by expressing the total potential energy in terms of the discretized amplitudes of the candidate trial functions.

Example 4.1. Find the displacement and stresses for the centrally loaded bar for the assumed displacement field:

$$u(x) = u_1(x) = u_1 \sin\frac{\pi x}{L} \tag{4.26}$$

Where did this come from? Well, imagine a sine curve in the interval $(0, \pi)$, which clearly corresponds to the range of the sinusoid in Eq. (4.26) over the domain $0 \le x \le L$. Notice that Eq. (4.26) satisfies the boundary conditions on displacement at both ends, and it is symmetric about the center of the bar. Thus, it appears to have a shape that is roughly similar to the exact solution, but how do we determine the magnitude u_1? The answer is that we will turn again to our variational principle. Let us substitute Eq. (4.26) into the energy functional of Eq. (4.13):

$$\begin{aligned}\Pi_1 &= \frac{1}{2}EA\int_0^L\left(u_1\frac{\pi}{L}\cos\frac{\pi x}{L}\right)^2 dx - Pu_1\sin\frac{\pi}{2}\\ &= \left(\frac{\pi}{2}\right)^2\left(\frac{EA}{L}\right)u_1^2 - Pu_1\end{aligned} \tag{4.27}$$

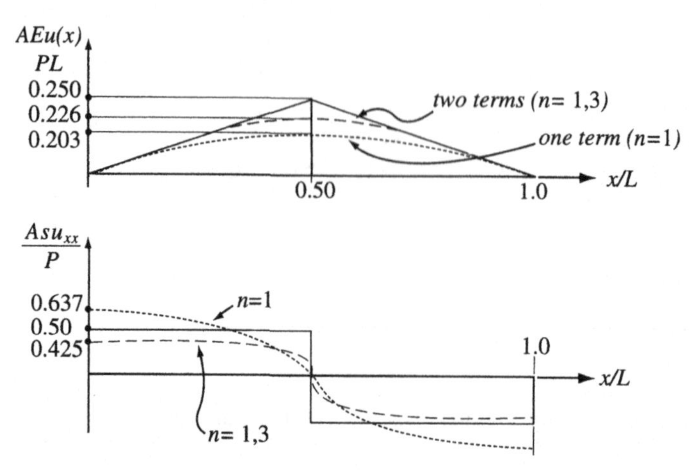

Figure 4.7. Exact and approximate displacement and stress fields of a centrally loaded bar.

Thus, we have converted our total potential energy functional – an expression involving the function $u(x)$ – into the algebraic polynomial in u_1. In order to determine the minimum value of the total potential polynomial, we simply set its derivative to zero:

$$\delta^{(1)}\Pi_1 \equiv \frac{d\Pi_1}{du_1}\delta u_1 = \left(\frac{\pi^2}{2}\left(\frac{EA}{L}\right)u_1 - P\right)\delta u_1 = 0 \tag{4.28}$$

Equation (4.28) is a simple equation for the amplitude of the trial shape of Eq. (4.26), so we can write our complete approximation solution as

$$u_1(x) = \left(\frac{2}{\pi^2}\right)\left(\frac{PL}{AE}\right)\sin\frac{\pi x}{L} \cong (0.203)\left(\frac{PL}{AE}\right)\sin\frac{\pi x}{L} \tag{4.29}$$

to which there corresponds a stress distribution

$$\sigma_{xx}^1(x) = E\frac{du_1(x)}{dx} = \left(\frac{2}{\pi}\right)\left(\frac{P}{A}\right)\cos\frac{\pi x}{L} \cong (0.637)\left(\frac{P}{A}\right)\cos\frac{\pi x}{L} \tag{4.30}$$

In Fig. 4.7 we show plots of the dimensionless displacement and the dimensionless stress for the approximate solution we have just derived and the exact results found just before. We can see that we haven't done too badly on the displacement, although we don't reflect the slope discontinuity at the center and the peak displacement differs from the exact result by about 19%. The stress results are even further off, however, as we miss the peak value by 27% and, more disturbing, we don't see the stress discontinuity at all. ■

What can we learn from Example 4.1, and how can we do a better job of getting approximate solutions? And why are we interested in approximate solutions to an elementary problem for which we already have the exact solution? The answer to the latter question is the reminder that what we are doing for this simple problem is a conceptual model for what we can do – and in fact, we *do* do in engineering practice and research – for problems that are far more complicated and for which we do not have neat closed-form solutions.

As to what we learned from this first approximate result, the most evident lesson is that we did far worse with our stress approximation than with the displacement approximation. The reason for this is rooted in basic calculus: Integration is a *smoothing* process that takes results and makes them less "ugly," while differentiation does the opposite by making results more ragged, and thus uglier. We found the stress by differentiating the displacement. Consequently, we should expect it to be harder to get good approximations to stress distributions. Stated another way, in order to assess the "goodness" of an assumed displacement approximation, we are likely to impose a stricter metric based on the accuracy of the resulting stress distribution.

Example 4.2. Try to improve the result found in Example 4.1 by generalizing Eq. (4.26) with the following assumed displacement:

$$u(x) = u_2(x) = u_1 \sin\frac{\pi x}{L} + u_3 \sin\frac{3\pi x}{L} \tag{4.31}$$

Here we are adding a second term to what we could call a *Fourier series* representation for the displacement. However, we are choosing our terms with care. Note that what might be called the second term, corresponding to $n = 2$, we have taken to be zero. Why? Because this term is *antisymmetric* (or an *odd* function of x) about the centerline and we know the displacement should be *symmetric* (or an *even* function of x) about that point. (In fact, we can show that this term and all even-integer terms are zero for this case. See Problem 4.3.)

Substituting the approximation (4.31) into the total potential (4.13) yields

$$\begin{aligned}
\Pi_2 &= \frac{1}{2}EA\int_0^L \left(u_1\frac{\pi}{L}\cos\frac{\pi x}{L} + u_3\frac{3\pi}{L}\cos\frac{3\pi x}{L}\right)^2 dx \\
&\quad - P\left(u_1 \sin\frac{\pi}{2} + u_3\sin\frac{3\pi}{2}\right) \\
&= \left(\frac{\pi}{2}\right)^2\left(\frac{EA}{L}\right)\left[u_1^2 + 9u_3^2\right] - P(u_1 - u_3)
\end{aligned} \tag{4.32}$$

The minimization of this potential energy is done by extending what we did for the one-term approximation, that is,

$$\begin{aligned}
\delta^{(1)}\Pi_2 &= \frac{d\Pi_2}{du_1}\delta u_1 + \frac{d\Pi_2}{du_3}\delta u_3 \\
&= \left(\frac{\pi^2}{2}\left(\frac{EA}{L}\right)u_1 - P\right)\delta u_1 \\
&\quad + \left(\frac{(3\pi)^2}{2}\left(\frac{EA}{L}\right)u_3 + P\right)\delta u_3 = 0
\end{aligned} \tag{4.33}$$

In order for this first variation to vanish, we have two separate equations for the two independent variables, u_1 and u_3, that we can write in matrix form

$$\begin{bmatrix} \dfrac{\pi^2}{2}\left(\dfrac{EA}{L}\right) & 0 \\ 0 & \dfrac{(3\pi)^2}{2}\left(\dfrac{EA}{L}\right) \end{bmatrix} \begin{Bmatrix} u_1 \\ u_3 \end{Bmatrix} = P\begin{Bmatrix} 1 \\ -1 \end{Bmatrix} \tag{4.34}$$

The matrix Eq. (4.34) can be solved for the amplitudes with which we can calculate the displacement field achieved with this approximation:

$$u_2(x) = \left(\frac{2}{\pi^2}\right)\left(\frac{PL}{AE}\right)\left[\sin\frac{\pi x}{L} - \left(\frac{1}{3^2}\right)\sin\frac{3\pi x}{L}\right]$$
$$\cong (0.203)\left(\frac{PL}{AE}\right)\left[\sin\frac{\pi x}{L} - (0.111)\sin\frac{3\pi x}{L}\right] \tag{4.35}$$

and its corresponding stress distribution:

$$\sigma_{xx}^2(x) = E\frac{du_2(x)}{dx} = \left(\frac{2}{\pi}\right)\left(\frac{P}{A}\right)\left[\cos\frac{\pi x}{L} - \left(\frac{1}{3}\right)\cos\frac{3\pi x}{L}\right]$$
$$\cong (0.637)\left(\frac{P}{A}\right)\left[\cos\frac{\pi x}{L} - (0.333)\cos\frac{3\pi x}{L}\right] \tag{4.36}$$

Numerical results for the results (4.35) and (4.36) have been shown in Fig. 4.7, where we can compare them to the exact results as well as those of our first approximation. We see that the error in the maximum displacement has been reduced to 9.6%, while the error in the peak stress has been reduced to 15%, so our situation has been improved. Note, however, we have not reproduced the stress discontinuity at $x = L/2$, although the slope in the stress curve there is much steeper than it is for the one-term approximation. ■

Given the above results, it should be no surprise to suggest that we can further improve our results by adding more terms to our series, that is, by assuming that

$$u(x) = \sum_{n=1}^{\infty} u_n \sin\frac{n\pi x}{L} \tag{4.37}$$

Problems 4.5 and 4.6 ask for a demonstration that the approximate solution (4.37) converges to the exact result. That is, if we take the entire Fourier series embodied in Eq. (4.37), it will reproduce the discontinuities in the slope of the displacement and in the stress that are found in the exact solution.

There are other kinds of series that can be used to develop trial solutions in the direct method. One of these is a *generalized Fourier series* wherein

$$u(x) = \sum_{n=0}^{\infty} u_n \phi_n(x) \tag{4.38}$$

We require that the individual $\phi_n(x)$ satisfy *at least* all the geometric boundary conditions, that is, the conditions on $u(x)$. Further, it is highly desirable that the $\phi_n(x)$ satisfy as many of the natural or force conditions as possible (on $\sigma_{xx}(x)A = EAu'(x)$). This will become more evident in Chapter 7 when we model beams. Finally, it is also nice – although not essential – if the trial functions form an *orthogonal set*, that is,

$$\int_0^L \phi_n(x)\phi_m(x)dx = 0 \quad \text{for } n \neq m \tag{4.39}$$

This property is very attractive. Why? Recall that the matrix equation (4.34), turned out to be a *diagonal matrix* where all terms off the main diagonal are zero because the mixed-product integral in Eq. (4.32) vanishes:

$$\int_0^L \cos\frac{\pi x}{L}\cos\frac{3\pi x}{L}dx = 0$$

because the trigonometric functions typically form an orthogonal set. In fact, it is worth noting that (see Problem 4.4)

$$\int_0^L \sin\frac{n\pi x}{L}\sin\frac{m\pi x}{L}dx = \int_0^L \cos\frac{n\pi x}{L}\cos\frac{m\pi x}{L}dx$$
$$= \begin{cases} 0 & \text{for } n \neq m \\ L/2 & \text{for } n = m \end{cases} \tag{4.40}$$

Thus, if the $\phi_n(x)$ form an orthogonal set, the resulting linear equations for the coefficients u_n are *uncoupled*, and the matrix equation for the u_n is diagonal.

4.4.3 Polynomial Approximations for the Centrally Loaded Bar

There are other kinds of trial solutions that can be used to obtain direct, discretized solutions. One of these is a polynomial in the coordinates.

Example 4.3. Find displacement and stress approximations for the centrally loaded bar that result from the polynomial trial function

$$u_3(x) = u_0 x(L - x) \tag{4.41}$$

We can also write our trial function in dimensionless terms,

$$u_3(x) = u_0 L^2 (x/L)(1 - (x/L)) \tag{4.42}$$

This polynomial trial function satisfies the geometric boundary conditions. It has one independent parameter to be used as the basis for minimizing the total potential. In fact, the total potential in this case takes the form

$$\Pi_3 = \frac{1}{2}EA\int_0^L (u_0(L-2x))^2 dx - Pu_0\left(\frac{L}{2}\right)\left(\frac{L}{2}\right)$$
$$= \frac{1}{6}EAL^3u_0^2 - \frac{1}{4}PL^2u_0 \tag{4.43}$$

The first variation of Eq. (4.43) is then

$$\delta^{(1)}\Pi_3 = \left(\frac{1}{3}EAL^3u_0 - \frac{1}{4}PL\right)\delta u_0 \tag{4.44}$$

The vanishing of this variation produces an amplitude, which in turn can be used to obtain the following displacement and stress results:

$$u_3(x) = \left(\frac{3}{4}\right)\left(\frac{PL}{AE}\right)\left(\frac{x}{L}\right)\left(1 - \frac{x}{L}\right)$$
$$\sigma_{xx}^3(x) = E\frac{du_3(x)}{dx} = \left(\frac{3}{4}\right)\left(\frac{P}{A}\right)\left(1 - 2\left(\frac{x}{L}\right)\right) \tag{4.45}$$

We have summarized these results and compared them to the exact solution in Table 4.1. And we can quickly see that this approximation isn't very good, when measured in terms of values or of the variation of the stress or the displacement. ■

Table 4.1. Comparing results for the approximation (4.41) with those for its exact counterpart.

	$x = 0$	$x = L/2$	$x = L$	*Variation in x*
$\dfrac{AEu_3(x)}{PL}$	0	0.1875	0	quadratic
$\dfrac{AEu_{exact}(x)}{PL}$	0	0.250	0	bi-linear
$\dfrac{A\sigma_{xx}^3(x)}{P}$	0.75	0	−0.75	linear
$\dfrac{A\sigma_{xx}^{exact}(x)}{P}$	0.50	jump	−0.50	"square wave"

It is worth noting, however, that as bad as the approximation (4.41) turned out to be, it is a special case of the more general polynomial approximation

$$u_{poly}(x) = a + bx + cx^2 \tag{4.46}$$

The general result does, in fact, reduce to Eq. (4.41) if the two geometric boundary conditions are satisfied. However, unlike the Fourier or trigonometric series, it is less than obvious that extending the polynomials to include higher powers would be especially useful, especially given the shape of the exact solution, that is, the slope discontinuity at the center of the bar.

What the exact solution does suggest, however, is that perhaps different regimes ought to established, each having its own trial function. For example,

$$u_4(x) = \begin{cases} a + bx & \text{for } 0 \le x \le L/2 \\ c + dx & \text{for } L/2 \le x \le L \end{cases} \tag{4.47}$$

This might be called a *linear interpolation* over the interval $0 \le x \le L$. In fact, it is easily verified (cf. Problem 4.7) that after the geometric boundary conditions and an appropriate compatibility condition are satisfied, this interpolated form produces the exact result we had found in Section 4.4.1.

What is perhaps most interesting about this interpolation is the suggestion that one way to approximate the behavior of a function over a domain is to break up the domain into smaller subdomains or elements, within each of which we can use some simple algebraic forms to represent the function's variation within that element. This is very much like saying that if we take short enough intervals along the abscissa of a plot of a function, we can represent a very complicated curve by simple functions within each interval, as long as we remember to satisfy boundary conditions at the ends of our domain and compatibility conditions at the intersections of the various elements or sub-domains. In fact, combining the interpolation process for large numbers of small elements with the variational approach, which is used to directly derive the discretized equilibrium matrices (cf. Eq. (4.34)), is what the well-known finite element method is all about. Clearly, for complex problems, there are many more complicated issues and opportunities that must be addressed. But this method and some of its variants are increasingly becoming the standard means by which structural engineering calculations are done.

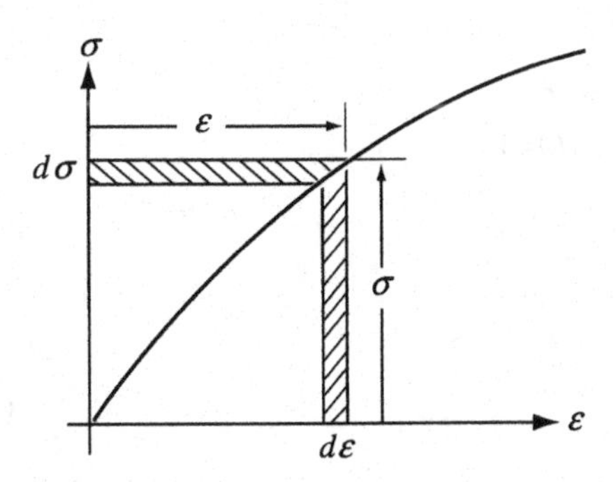

Figure 4.8. Strain and complementary energy densities for elastic materials.

4.5 Castigliano's Theorems for Bars and Rods

This is a good place to introduce some additional calculations that are based on the energy stored in an element or a structure. In demonstrating these ideas, we will extend our formulation to introduce a complement to the strain energy where we cast the energy in terms of forces, rather than displacements. Recall that in our discussions of the principle of minimum potential energy we cast the stored energy in terms of kinematic variables, that is, of strain and displacement. We did this for both the discrete models (Chapter 3) and the continuous models introduced earlier in this chapter.

Recall the strain energy density U_0 from Eqs. (4.5) and (4.6), which is nothing more than the strain energy per unit volume. We have sketched a general stress–strain curve in Fig. 4.8, and the corresponding strain energy density is easily seen to be the area under the curve

$$U_0 = \int_0^{\varepsilon_{xx}} \sigma_{xx} d\varepsilon_{xx} \tag{4.48}$$

For a linearly elastic body, of course, the strain energy can be put in several equivalent forms, as shown in Eq. (4.7), the last of which is written entirely in terms of stresses. So we might ask whether this is just happenstance, or is there something noteworthy and useful about it? Well, if we apply the principles of calculus to the curve shown in Fig. 4.8, we can calculate the complement to the strain energy area by finding the area above the curve, the area between the curve and the ordinate. We define this area to be the *complementary energy density*:

$$U_0^* = \int_0^{\sigma_{xx}} \varepsilon_{xx} d\sigma_{xx} \tag{4.49}$$

And for a linearly elastic material, in parallel with Eq. (4.7),

$$U_0^* = \frac{1}{2}\frac{\sigma_{xx}^2}{E} = \frac{1}{2}\sigma_{xx}\varepsilon_{xx} = \frac{1}{2}E\varepsilon_{xx}^2 \tag{4.50}$$

Thus, for a *linearly elastic solid*, it is clearly the case that

$$U_0 = U_0^* = \frac{1}{2}\sigma_{xx}\varepsilon_{xx} \tag{4.51}$$

Consider now the simple problem of a bar fixed at the left end and being pulled on by the external force P at the right end (Fig. 4.9). The well-known solution for this problem

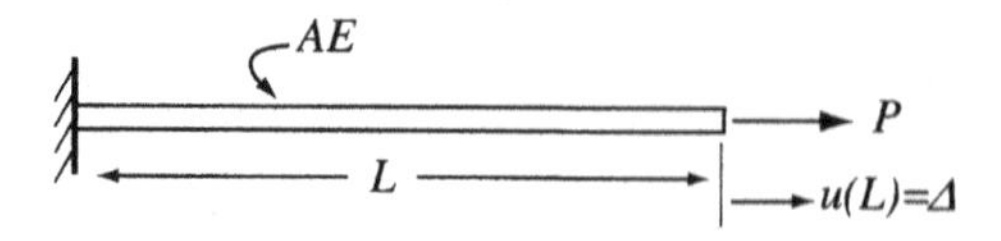

Figure 4.9. A simply extended bar.

is

$$u(x) = \frac{Px}{AE} \equiv \Delta\left(\frac{x}{L}\right) \tag{4.52}$$

where we have cast this result in terms of the displacement at the loaded end, $\Delta \equiv u(L) = PL/AE$. With the aid of Eq. (4.52) we can calculate the stress, the strain, and the strain energy of the loaded bar. In particular,

$$U = \int_0^L U_0 A dx = \frac{1}{2} AE \int_0^L \left(\frac{\Delta}{L}\right)^2 dx = \frac{1}{2}\left(\frac{AE}{L}\right)\Delta^2 \tag{4.53}$$

This looks exactly as it should for a spring of stiffness $k = AE/L$, but of greater interest is that we can now calculate the rate of change of U with respect to Δ:

$$\frac{\partial U}{\partial \Delta} = \left(\frac{AE}{L}\right)\Delta = \left(\frac{AE}{L}\right)\left(\frac{PL}{AE}\right) \equiv P \tag{4.54}$$

Thus, by differentiating the strain energy with respect to the displacement or deflection at a point, we find the value of the external load applied at that point.

It also turns out that we can do a similar calculation with the complementary energy. Since we know the stress everywhere along the length of the bar is

$$\sigma_{xx}(x) = \frac{P}{A} \tag{4.55}$$

we can calculate the complementary energy as

$$U^* = \int_0^L U_0^* A dx = \frac{1}{2}\left(\frac{A}{E}\right)\int_0^L \left(\frac{P}{A}\right)^2 dx = \frac{1}{2}\left(\frac{L}{AE}\right)P^2 \tag{4.56}$$

The complementary energy is cast here in terms of a *flexibility coefficient* that is the reciprocal of the spring stiffness, that is, $f = 1/k = L/AE$. In the same way that k represents how stiff a spring the structure appears to be in this configuration and under this particular load, the coefficient f is a mark of how flexible it is, of how easily this structural configuration responds to the load in question. It is intuitively evident that f is a measure of flexibility because a bar is bound to be *more* flexible if we increase its length or decrease its area or its modulus.

Further, once again there is one more step that is of great interest. We now calculate the rate of change of U^* with respect to P:

$$\frac{\partial U^*}{\partial P} = \left(\frac{L}{AE}\right)P \equiv \Delta \tag{4.57}$$

Thus, by differentiating the strain energy with respect to the external load applied at a point, we find the value of the deflection at that point.

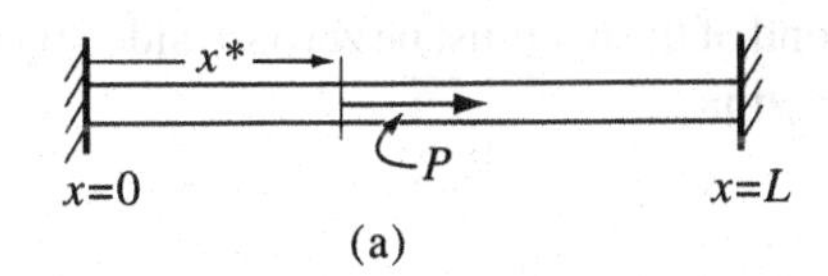

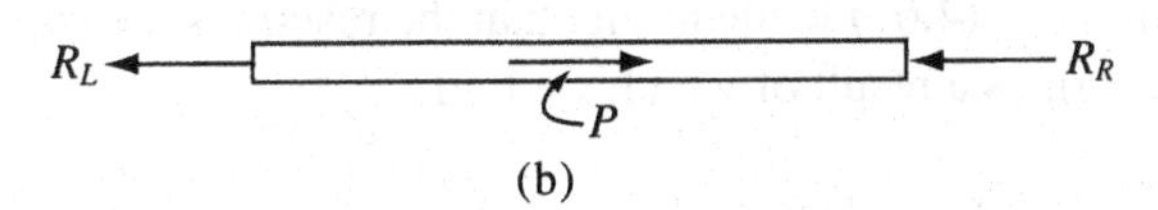

Figure 4.10. (a) An axially loaded bar and (b) its free-body diagram.

Equations (4.54) and (4.57) are demonstrations of two very famous theorems attributed to a well-known Italian engineer. In Eq. (4.54) we have shown *Castigliano's first theorem*, while in Eq. (4.57) we have shown *Castigliano's second theorem.* We will derive them in more general terms in Chapter 6, but we will continue to apply them to one-dimensional elements in this chapter.

The two bar problems we have considered so far are very similar, but there is one important difference we have not discussed in any depth. In particular, the centrally loaded bar, fixed as it is between two rigid supports, is statically indeterminate, which is to say, we cannot determine the two reaction forces (at either end) from the equation of static equilibrium taken by itself. (Of course, at the close of Section 4.4.1 we did find that we could calculate the reactions once we had determined the internal displacement and stress fields.) It also turns out that Castigliano's second theorem can be very helpful to us in such cases.

Consider the bar pictured in Fig. 4.10(a), which is a variation of the centrally-loaded bar wherein the axial load P is applied at an arbitrary location x^*. From the free-body diagram shown in Fig. 4.10(b), we can confirm Newton's second law along the bar axis to be

$$R_L + R_R = P \tag{4.58}$$

We cannot determine both reactions from this one equation. However, we can calculate the stress field in the bar in terms of the reactions, that is,

$$\begin{aligned} \sigma &= \frac{R_L}{A} = \frac{P - R_R}{A} \qquad 0 \le x \le x^* \\ &= \frac{-R_R}{A} \qquad\qquad\quad x^* \le x \le L \end{aligned} \tag{4.59}$$

and with the aid of Eqs. (4.59) we can calculate the complementary energy in terms of the reactions

$$\begin{aligned} U^* &= \frac{1}{2AE}\left[\int_0^{x^*} R_L^2 dx + \int_{x^*}^{L} R_R^2 dx\right] \\ &= \frac{1}{2AE}\left[R_L^2 x^* + R_R^2 (L - x^*)\right] \end{aligned} \tag{4.60}$$

We know that the deflection at each end of the bar must be zero, so aided by Castigliano's second theorem (Eq. (4.57)) we can say that

$$\frac{\partial U^*}{\partial R_R} = \Delta_R = 0 \tag{4.61}$$

We now substitute Eq. (4.60) into Eq. (4.61) while noting that the reactions are explicitly related by equilibrium (Eq. (4.58)), as a result of which we find

$$\begin{aligned}\frac{\partial U^*}{\partial R_R} &= \frac{1}{AE}\left[x^* R_L \frac{\partial R_L}{\partial R_R} + (L - x^*)R_R\right] \\ &= \frac{1}{AE}\left[x^*(P - R_R)(-1) + (L - x^*)R_R\right] = 0\end{aligned} \tag{4.62}$$

Equation (4.62) thus turns out to be the second equation we need to calculate the reactions. Combining Eqs. (4.58) and (4.62) produces the evident results:

$$R_R = \frac{x^*}{L}P \quad \text{and} \quad R_L = \frac{L - x^*}{L}P \tag{4.63}$$

These results satisfy equilibrium and are intuitively pleasing with respect to the relation between the magnitude of a reaction and the distance between that reaction and the applied load. Equations (4.63) also produce the same results for the centrally loaded bar that we have already obtained. However, the most important point is that we have used Castigliano's second theorem to find the redundant reaction of a statically indeterminate structure.

4.6 The Force Approach and the Displacement Approach

This is a good moment to summarize what we have found altogether from our examples of both link-and-spring models and of axially loaded bars. We have used two basic approaches (or methods, or paradigms) to formulate and analyze models of structural bars. In one we worked with displacements or deflections as the lead variables, while in the other we used forces as the lead variables. In the *force method*, often also called the *flexibility method*, we used the principle of minimum complementary energy to derive equations that can be seen as compatibility conditions, and the key structural parameters were expressed as flexibility coefficients that depend on the type of material and the specific geometry underlying the idealization. For example, at the very end of the last section we used the force method as expressed in Castigliano's second theorem to solve for the reactions of the indeterminate bar. Equation (4.62) can be read as a compatibility condition: The displacement at the coordinate x^* must be the same whether measured for the left side of the bar ($x \leq x^*$) or the right side ($x \geq x^*$):

$$(P - R_R)\frac{x^*}{AE} = R_R\frac{(L - x^*)}{AE} \tag{4.64}$$

where we recognize in this result the flexibilities of the left and right sides of the bar, (x^*/AE) and $((L - x^*)/AE)$, respectively.

In the *displacement* or *stiffness method*, we used the principle of minimum potential energy to derive equilibrium equations, and the key structural parameters were expressed as stiffnesses that depend on the type of material and the specific geometry underlying the idealization. We used the displacement approach in the derivations and examples of Sections 4.2 and 4.4.

We note again that the ideas reflected above form the underpinnings of the finite element method as applied to structural analysis and design. Moreover, the displacement method is used far more often in constructing FEM programs. Thus, let us apply the displacement method to the indeterminate problem of Section 4.5 because that problem's simplicity of kinematics and geometry allows us to cast it into two essentially equivalent statements.

Example 4.4. Formulate an energy statement for the loaded bar of Fig. 4.10 by extrapolating the solution for the centrally applied load (cf. Eq. (4.22)).

We start by assuming the following displacement field:

$$u(x) = u^* \left(\frac{L}{x^*}\right)\left[\frac{x}{L} - \left(\frac{x - x^*}{L - x^*}\right) H(x - x^*)\right] \tag{4.65}$$

where u^* represents the displacement at the point x^* where the load is applied and the displacement's slope is discontinuous. Then the strains in each bar segment can be computed, along with the corresponding strain energy terms, that is,

$$\begin{aligned} U &= \frac{AE}{2}\left[\int_0^{x^*} \left(\frac{u^*}{x^*}\right)^2 dx + \int_{x^*}^{L} \left(\frac{-u^*}{(L - x^*)}\right)^2 dx\right] \\ &= \frac{AE}{2}\left[\frac{L(u^*)^2}{x^*(L - x^*)}\right] \end{aligned} \tag{4.66}$$

Thus, we can then calculate the total potential energy for this problem as

$$\Pi = \frac{AE}{2}\left[\frac{L(u^*)^2}{x^*(L - x^*)}\right] - Pu^* \tag{4.67}$$

It is then easy to see (cf. Problem 4.8) that minimization of the potential (4.67) yields values of u^* that can be substituted into the displacement assumption (4.65), after which we can find the strains and the corresponding stresses that are exactly the same as implied by the combination of Eqs. (4.63) and (4.59). ■

Example 4.5. Formulate an equivalent discrete statement for the loaded bar of Fig. 4.10.

We model this problem as a pair of springs in parallel (cf. Example 3.4 with $\xi = u^*$). With spring stiffnesses for the two bar segments being modeled as $k_1 = AE/x^*$ and $k_2 = AE/(L - x^*)$, respectively, the total potential energy for this problem is (from either Example 3.4 or Eq. (4.67))

$$\Pi = \frac{1}{2}(k_1 + k_2)(u^*)^2 - Pu^* \tag{4.68}$$

The first variation of the total potential energy (4.68) is, then,

$$\delta^{(1)}\Pi = [k_1 u^* + k_2 u^* - P]\delta u^* \tag{4.69}$$

And it is easily verified that the vanishing of the first variation (4.69) produces the same result as the vanishing of the first variation of Eq. (4.67). ■

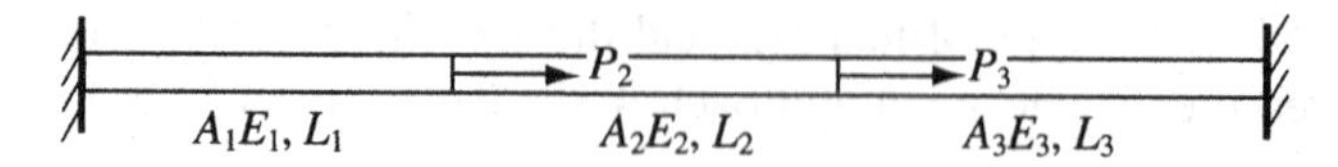

Figure 4.11. A three-segment, axially loaded bar.

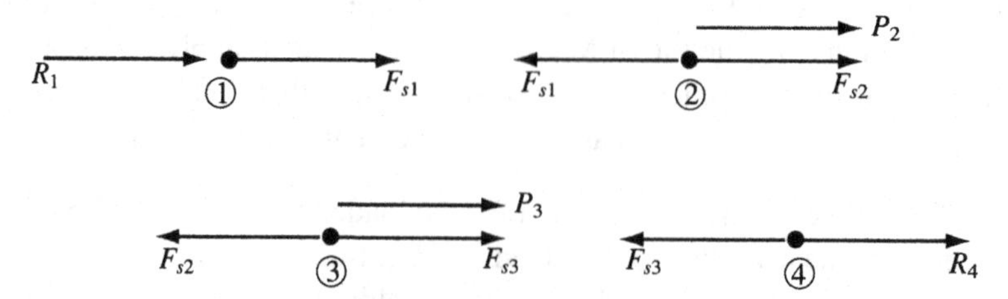

Figure 4.12. Free-body diagrams for the discrete three-segment model.

Example 4.6. Find the reactions and the displacements at the points where the loads are applied for the three-segment bar shown in Fig. 4.11. Each of the segments are assumed to have different material and geometric properties.

We can model this structure as three discrete springs in series, or we can develop a continuous model based on a linear interpolation for the displacement in each segment. Since both of these models produce identical results for this type of problem, we will use the springs-in-series model here.

There are four spring nodes whose free-body diagrams are shown in Fig. 4.12. We model the stiffness of each spring as $k_i = A_i E_i / L_i$ and we denote the displacement at the ith node by u_i. The constitutive law for the ith spring is

$$F_{si} = \frac{A_i E_i}{L_i}(u_{i+1} - u_i) = k_i(u_{i+1} - u_i) \tag{4.70}$$

In addition, we will assume initially that we don't know the values at the two ends, that is, we will take u_1 and u_4 as unspecified.

We can write the discrete total potential energy for this case as

$$\begin{aligned} \Pi &= \sum_{i=1}^{3} U_i - V \\ &= \frac{1}{2}\sum_{i=1}^{3} k_i(u_{i+1} - u_i)^2 - P_2 u_2 - P_3 u_3 - R_1 u_1 - R_4 u_4 \end{aligned} \tag{4.71}$$

Note that in writing Eq. (4.71) we have also included the potential of the reactions R_1 and R_4 because we have not yet specified values for the displacements at the ends of the bar. The first variation of Eq. (4.71) is

$$\begin{aligned} \delta^{(1)}\Pi = {} & [-k_1(u_2 - u_1) - R_1]\delta u_1 \\ & + [k_1(u_2 - u_1) - k_2(u_3 - u_2) - P_2]\delta u_2 \\ & + [k_2(u_3 - u_2) - k_3(u_4 - u_3) - P_3]\delta u_3 \\ & + [k_3(u_4 - u_3) - R_4]\delta u_4 \end{aligned} \tag{4.72}$$

The vanishing of the first variation (4.72) produces four linear equations for the four unknown nodal displacements, which we write in matrix form:

$$\begin{bmatrix} k_1 & -k_1 & 0 & 0 \\ -k_1 & k_1+k_2 & -k_2 & 0 \\ 0 & -k_2 & k_2+k_3 & -k_3 \\ 0 & 0 & -k_3 & k_3 \end{bmatrix} \begin{Bmatrix} u_1 \\ u_2 \\ u_3 \\ u_4 \end{Bmatrix} = \begin{Bmatrix} R_1 \\ P_2 \\ P_3 \\ R_4 \end{Bmatrix} \tag{4.73}$$

This is a *symmetric matrix*, that is, its elements are arrayed symmetrically with respect to the matrix's diagonal. Further, note that all of the elements in this matrix are expressed in terms of the stiffness coefficients, as a result of which we can identify a *stiffness matrix* for the structural system of Figs. 4.11 and 4.12:

$$[K] = \begin{bmatrix} k_1 & -k_1 & 0 & 0 \\ -k_1 & k_1+k_2 & -k_2 & 0 \\ 0 & -k_2 & k_2+k_3 & -k_3 \\ 0 & 0 & -k_3 & k_3 \end{bmatrix} \tag{4.74}$$

We can then write Eq. (4.73) in the form

$$[K] \begin{Bmatrix} u_1 \\ u_2 \\ u_3 \\ u_4 \end{Bmatrix} = \begin{Bmatrix} R_1 \\ P_2 \\ P_3 \\ R_4 \end{Bmatrix} \tag{4.75}$$

In principle, the matrix equation represented in Eqs. (4.73) and (4.75) says that we can determine the four nodal displacements as a function of the four loads that make up the *column matrix* (or *vector*) that is the right-hand side of both forms of the matrix equation. The theory of linear equations also allows us to recast this set of algebraic equations simply by listing them in a different order, that is, by shifting the ordering of the rows in Eq. (4.73). If we move the fourth row up to be the second row, we can segregate or partition the forces so that the reactions appear together and the applied loads also appear together. Then our matrix equation looks like the following:

$$\begin{bmatrix} k_1 & 0 & -k_1 & 0 \\ 0 & k_3 & 0 & -k_3 \\ -k_1 & 0 & k_1+k_2 & -k_2 \\ 0 & -k_3 & -k_2 & k_2+k_3 \end{bmatrix} \begin{Bmatrix} u_1 \\ u_4 \\ u_2 \\ u_3 \end{Bmatrix} = \begin{Bmatrix} R_1 \\ R_4 \\ P_2 \\ P_3 \end{Bmatrix} \tag{4.76}$$

At this point we choose to recognize that the physical configuration shown in Fig. 4.11 does impose a constraint on two of our nodal displacements, that is, $u_1 = u_4 = 0$. Having made this recognition, we can reduce the four-by-four matrix equation to the last two equations for the two remaining unknowns:

$$\begin{bmatrix} k_1+k_2 & -k_2 \\ -k_2 & k_2+k_3 \end{bmatrix} \begin{Bmatrix} u_2 \\ u_3 \end{Bmatrix} = \begin{Bmatrix} P_2 \\ P_3 \end{Bmatrix} \tag{4.77}$$

while the unknown (redundant) reactions can be expressed in terms of these nodal displacements u_2 and u_3 by virtue of the first two rows of Eq. (4.76)

$$R_1 = -k_1 u_2 \quad \text{and} \quad R_4 = -k_3 u_3 \tag{4.78}$$

This last result makes perfect sense in terms of the free-body diagrams of the elements (Fig. 4.12) because it says that the reactions at the walls must be balanced by the forces produced by the relative extensions of the springs adjacent to each wall. Of course, calculating these reactions depends on obtaining the solution to the matrix equation (4.77). ■

We achieved several things in Example 4.6. First, we have shown that we can apply the displacement or stiffness method to an indeterminate structure (even though we have not explicitly solved the matrix equation (4.77) for the unknown nodal displacements). We did see that we can obtain a complete solution, including explicit results for both support reactions (in Eqs. (4.78)). Second, in Eqs. (4.73) and (4.75) we put forward a *matrix representation* of the equilibrium equations that result from the principle of minimum potential energy. Although we make only limited use of matrix representation as convenient notation, it is also symbolic of a third point, that this example is an abstraction of the finite element method.

Why is this true? First, after choosing an *idealization* – the long, slender, axially loaded bar – we *discretized* that model by representing the bar as a set of three linear springs connected in series at four discrete nodes, including both supports. Second, we used the displacement approach to write the total potential energy in terms of the four nodal displacements, and we then minimized that potential energy with respect to the nodal displacements to find the four equilibrium equations (cf. the matrix forms (4.73) and (4.75)). Inherent in these expressions is a stiffness matrix (4.74), which represents the stiffness of our ideal, discrete model. Finally, we showed that by carefully manipulating the matrix equilibrium equation we could modify and partition the stiffness matrix in order to solve for the interior nodal displacements in one step, and for the unknown, indeterminate support reactions in a second step. This form of matrix manipulation is easily automated, and it too is a part of the FEM process.

For more complex structures, loadings, and geometry, the solution process will clearly be considerably more complicated than what we have done. The physical models could include other complicating effects, including nonlinear materials and geometry, vibration and dynamics, and so on. There are also very serious issues resulting from attempts to achieve accuracy for models that have a very large number of unknowns – in the thousands and even tens of thousands – and these issues involve both matrix mathematics and high-end computing. Nonetheless, it is fair to say that the process exemplified in Example 4.6 is an abstract, but nontheless real, model of FEM structural modeling.

Problems

4.1 Obtain the exact solution to the equilibrium equation for a bar fixed at the left end ($x = 0$) and subjected to an external tensile force P applied at $x = L$. *Hints*: Assume that the proper differential equation is

$$EA\frac{d^2u}{dx^2} = 0$$

and that the appropriate boundary condition at $x = L$ is $u'(L) = P/EA$.

4.2 Obtain a second solution for the bar fixed at the left end ($x = 0$) and subjected to an external tensile force P applied at $x = L$, subject to the boundary condition

$u'(L) = 0$, and where the equilibrium equation is

$$EA\frac{d^2u}{dx^2} + P\delta_D(x - L) = 0$$

Compare the answer with that of Problem 4.1 and explain any differences. *Hints*: Are the problems the same? Are their mathematical representations the same? Can we use a Dirac spike to put a load at a boundary point?

4.3 For the bar having an axial load P at its centerline, with both ends restrained from moving, use the direct method to obtain an approximate solution in the form:

$$u(x) = u_{approx}(x) = u_1 \sin\frac{\pi x}{L} + u_2 \sin\frac{2\pi x}{L}$$

4.4 Verify Eqs. (4.40) by performing the required integrations.

4.5 For the bar having an axial load P at its centerline, with both ends restrained from moving, use the direct method to obtain a solution in the form:

$$u(x) = \sum_{n=1}^{\infty} u_n \sin\frac{n\pi x}{L}$$

Hint: For a complete solution, or even for a viable approximation, do we need all terms in the series or would a subset ($n = 1, 3, 5, \ldots$ or $n = 2, 4, 6, \ldots$) suffice?

4.6 Obtain a Fourier series expansion of the function

$$\begin{aligned} u(x) &= \frac{Px}{2AE} \qquad 0 \le x \le L/2 \\ &= \frac{P(L-x)}{2AE} \qquad L/2 \le x \le L \end{aligned}$$

and compare your results to those of Problem 4.5.

4.7 Starting with the interpolated approximate solution (4.47), satisfy the geometric boundary conditions and satisfy continuity of the displacement at the midpoint of the bar to find

$$u_4(x) = \begin{cases} bx & \text{for} \quad 0 \le x \le L/2 \\ b(L-x) & \text{for} \quad L/2 \le x \le L \end{cases}$$

Then find the minimum of the corresponding total potential energy and show that the resulting solution is equivalent to the exact solution (4.22).

4.8 Minimize the total potential formulation of Eq. (4.67) with respect to u^* and calculate the resulting strains and stresses in the bar segments. Compare these to the results of substituting the reactions (4.63) into the stress Eqs. (4.59).

4.9 For the axial bar and loads shown in Fig. 4.11, use the direct method of minimizing the total potential energy to obtain two different approximate solutions for the reactions R_1, R_4 and the axial displacements u_2, u_3. For each solution, show the equilibrium equations resulting from the direct variational process and compare them (conceptually) to the results derived in class for the linear spring, $k\xi = P$. For this problem, $A_i E_i = AE = 1.8(10^3)$ kN; $L_1 = 1.5$ m; $L_2 = 1.0$ m; $L_3 = 1.2$ m; $P_2 = 10$ kN, $P_3 = 20$ kN.

The bridge shown here is not sufficiently noteworthy to warrant a name. The set of five trusses carry the weight of trains across a small tributary somewhere near Salinas, California. One sees many such undistinguished bridges around the world, and they serve as simple yet eloquent testament to the functional merit and efficiency of the basic truss. (Photo by Clive L. Dym.)

5

Trusses: Assemblages of Bars

Chapter 4 contained an extensive discussion of a structural element whose range of application seems quite limited. After all, just how often can we use a bar built into to adjacent walls and restricted to carry only loads directed along its axis? Well, remember that in Chapter 2 we talked about two basic one-dimensional elements, one being the axially loaded bar, the other being the beam in bending. Further, and of direct consequence here, we noted that axially loaded bars can be combined to produce a structure that, in aggregate, behaves like a very efficient bent beam. Thus, for example, trusses are used in bridges and in towers. So now we turn to such assemblies of axially loaded bars with an eye toward applying both energy methods and classical methods of analysis.

We begin by reviewing some basic ideas about trusses that are likely already familiar, after which we will describe the classical methods of truss analysis, the methods of joints and of sections. We complete the chapter with a discussion of energy methods.

5.1 Traditional Truss Analysis; the Methods of Joints and of Sections

There are two ways of calculating the bar forces in a statically determinate truss, both based on an *axiomatic approach* to solving mechanics problems wherein we apply Newton's second law to trusses taken in their entirety and to their respective parts. These traditional ways date from precomputer times when it was better to manually decompose problems into smaller, more easily solved subproblems, especially when the calculations were to be done by hand or with the aid of ancient tools such as the slide rule. In these traditional methods we draw free-body diagrams and write equilibrium equations for them, in one case for each pin of interest, in the second case for sections of the entire truss. First, however, we review the basic assumptions made in classical truss analysis.

5.1.1 On Trusses (Again)

Consider the plane trusses shown in Fig. 5.1. The structural analysis of trusses consists of calculating the support reactions and bar forces under a specified set of externally applied loads. The rules for analyzing these structural devices are as follows. Each and every element is an axially loaded bar, which means that we cannot apply transverse loads at any points along the interior of a bar. Indeed, we assume that all loads acting on a truss, both applied loads and reactions, are applied at the connections or joints. (Recall the discussion in Section 2.4.1.) We also assume that the bar elements that make up the truss are pinned together, that is, the bars are connected with *pins* or *pin joints*, and that these pins are *ideal* or *frictionless*. Further, the moments of all forces about each bar end must be zero because the net load on each bar is directed along its axis, loads are applied only at the ends of the bars, and all pins are frictionless.

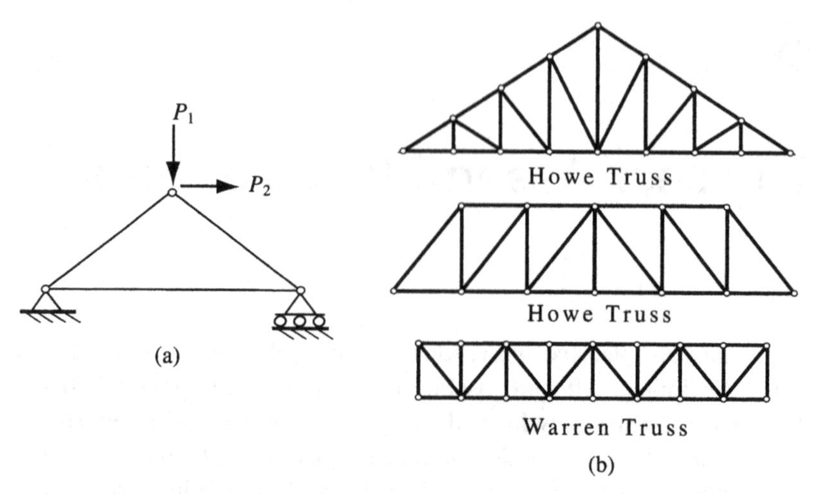

Figure 5.1. Trusses: (a) a very elementary truss and (b) more useful trusses.

As a practical, real-world matter, truss connections are not idealized pins. Rather, truss connections consist of gussett plates and rivets or bolts. However, the connections are designed so that the centroidal axes of the bars at a pin intersect at the center of the pin. This means that although some bending stresses, called *secondary stresses*, may be introduced on top of the axial tensile or compressive bar stresses, they will be sufficiently small that they are not normally analyzed for practical truss design.

Another consequence of the pin-joint assumption, as we noted in Section 2.3, is that at each pin (or node) where members are connected or where there is an external support, there are two equations of equilibrium that can be applied to account for all of the forces acting there. We normally take these directions along the conventional x- and y-axes, but any pair of orthogonal directions will work. This suggests that for a truss with j pins, there are $2j$ equilibrium equations that can be applied. How does this reflect on the issue of statical determinacy?

We have already provided an answer to this question in Section 2.3.1, although the reasoning depends in part on viewing the triangle as the basic truss configuration element. The total number of unknowns to be determined for a given truss is the sum of the number of reactions r plus the number of bars b, the latter because each bar has just a single axial force as its "unknown." Then, if the structure is statically determinate, the number of equations available must equal the number of unknowns (cf. Section 2.3). Hence,

$$r + b = 2j \tag{5.1}$$

It turns out that Eq. (5.1) is a necessary – but not sufficient – condition for the truss to be statically determinate and stable. (Remember that we assemble bars in triangular configurations in order to ensure internal stability.)

Recall also from Section 2.3.1 that while there are three equations for planar equilibrium of a planar truss taken in its entirety, and there are two available at each pin, this doesn't mean that there are $2j + 3$ independent equilibrium equations for the truss. When we work out the $2j$ equilibrium equations for all of the joints of a truss, we must account for all of the forces acting on each pin, including all externally applied loads, all reactions, and the bar forces of each bar attached to the pin in question. If we properly add up these $2j$ equations for all j pins, paying attention to both directions and sign, the bar forces will

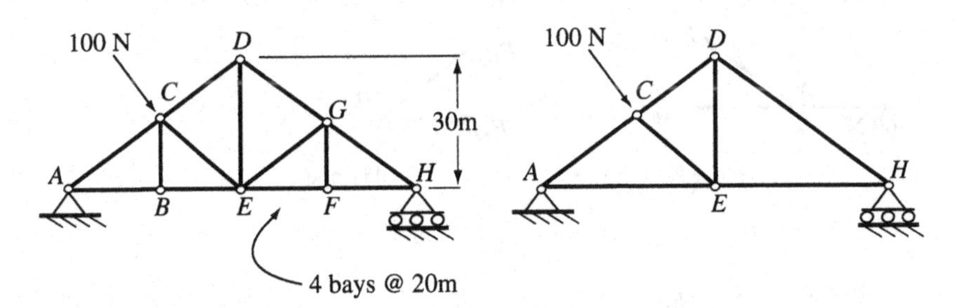

Figure 5.2. A truss to analyze.

cancel out in that sum. In addition, the reactions had better balance the external forces acting on the truss, in terms of both force and moment equilibrium. Thus, the equilibrium of the truss taken on a pin-by-pin basis must subsume or include the equilibrium of the entire truss, satisfying $\sum F_x = 0$, $\sum F_y = 0$, and $\sum M_x = 0$ (see Problems 5.1 and 5.2). Thus, there are only $2j$ *independent* equilibrium equations for the entire problem posed for a truss with j joints or pins.

5.1.2 The Method of Joints

In the *method of joints* we work through a truss on a pin-by-pin basis, paying early attention to joints where we have no more than two unknown bar forces per pin. Then we work through the truss as methodically as we can.

Example 5.1. Apply the method of joints to find the bar forces in the truss shown in Fig. 5.2(a).

We first calculate the reactions from static equilibrium of the entire truss:

$$R_{Ax} = 60\text{ N} \leftarrow; \qquad R_{Ay} = 48.8\text{ N} \uparrow; \qquad R_{Hy} = 31.2\text{ N} \uparrow$$

Now consider the joints labeled B and F. Note that in both instances there are no applied loads shown at these two pins. It follows from vertical equilibrium at each of these pins that the bar forces in bars BC and FG must be zero because they are the *only* terms that appear in these vertical equilibrium equations. Thus, these members are called *zero-force members*. It is a general principle that if all members but one are in the same direction at a joint and if no external forces are applied at that joint, the "out-of-line" member is a zero-force member. Indeed, for the truss of Fig. 5.2(a), having made the determination that

$$F_{BC} = F_{FG} = 0$$

it follows from the consideration of pin G that member GE is a zero-force member, that is, $F_{GE} = 0$. It also follows from these three joint analyses that we can immediately identify the following results:

$$F_{AB} = F_{BE}; \qquad F_{EF} = F_{FH}; \qquad F_{DG} = F_{GH}$$

To return to the problem of determining the bar forces for the "reduced" truss of Fig. 5.2(b), we will apply the method of joints to pins in the following order: A, H, C, D, and E. We have drawn the free-body diagrams for pins A, H, C, and D in Fig. 5.3 using the following sign conventions. Bar forces are considered as *positive in tension* and negative in compression, and we reflect this in the diagrams by drawing the forces as vectors acting along the lines of the bars and directed or pointing away from the pins.

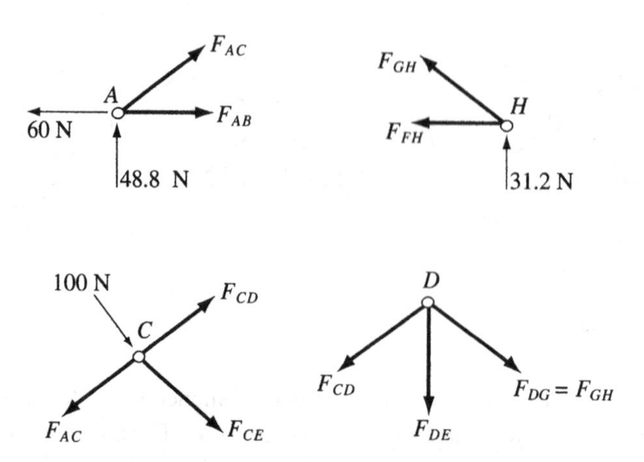

Figure 5.3. Free-body diagrams for the truss of Fig. 5.2.

Why did we specify a particular order for analyzing the pins? The reason is that at two joints there are at least three bars connected whose forces are as yet unknown. At joint D we have unknowns for bars CD, DE, and DG, while at joint E we have unknowns for bars BE, CE, DE, EG, and EF, if we have not analyzed the other joints first. Pins A and H each have only two unknowns, and we can use the two equilibrium equations at each pin to determine them. Further, at each of these pins, the vertical equilibrium equation yields a bar force without having to solve even two simultaneous equations. That is, for pins A and H, respectively,

$$\sum F_{Ay} = \frac{3}{5}F_{AC} + 48.8 = 0$$

$$\sum F_{Hy} = \frac{3}{5}F_{GH} + 31.2 = 0$$

The calculation of the remaining bar forces is straightforward (cf. Problem 5.3), so we leave this truss and the subject of the method of joints for now. Suffice it to say that it is useful to work out a few examples just to have some facility at recognizing both the simplifications and the complexities that arise due to the geometry (e.g., the triangles in the example truss just completed were "3–4–5 triangles") or some aspect of the loading. ■

Note that with its zero-force members removed, the truss of Example 5.1 looks like the structure shown in Fig. 5.2(b), so it is perhaps worth asking why one would design a truss to include members that carry no force. The answer is, in this case, relatively simple. It is rather unlikely that the external loading shown in Figs. 5.2 is a realistic loading situation. In general, loads are certainly distributed over all of the top joints in a roof truss and over all of the bottom pins in a bridge truss. In addition, even for a roof truss it is likely there will be loads along the bottom chord (that is, the bottom line of bars and pins) because of interior loads, such as ceilings and lighting, interior travel cranes in industrial buildings, and so on. Indeed, one American structure, the Kansas City Hyatt Regency Hotel, became famous because of the collapse that resulted from failed connections of a walkway that was suspended from the roof truss over a very large and high atrium space. The point here is that trusses typically do carry loads along both top and bottom chords, such as the three suspended walkways in the Hyatt Regency, and thus a designer must be careful if a loading pattern produces a configuration as unrealistic as Fig. 5.2(b). However, sometimes we do "insert" zero-force members in order to maintain triangular shapes in a large truss, so zero-force bars do appear in the real world.

The method of joints can be formulated systematically as a computational algorithm. This means we translate the topology or geometric configuration of the truss, together with its support reactions and applied loads, into a form that the computer can "recognize" and use to perform the calculations we want. We identify each member and incorporate its orientation, count the number of joints, and account for all reactions and loads. Then we solve the $2j$ equations of equilibrium of the joints in the truss. Since there is commercially available software that casts the method of joints in matrix format, we leave its application as an exercise in programming that is not part of our agenda.

5.1.3 The Method of Sections

Now we will apply Newton's second law to sections of the entire truss, that is, we will illustrate the *method of sections*. We typically apply the method of sections to situations in which we want to find the forces in some particular bars or in which we can exploit some particular aspect of the geometry.

Example 5.2. Apply the method of sections to find the bar forces in the section shown for the truss pictured in Fig. 5.4(a).

The geometry in this truss looks more complicated because there are triangles of several different proportions, and they are not all familiar. Suppose we wanted to know the forces in bars *DG*, *EG*, and *EF* for this so-called Parker truss. Note that these members all cross the third panel of the truss, counting from the left support. Thus, after determining the two symmetric reaction forces at joints *A* and *L*, it makes sense to consider a section of the truss defined as drawn in Fig. 5.4(b), in which we have drawn a section line through the third panel and drawn a free-body of the entire truss acting to the left of the section.

One way of analyzing this section would be to write down the equilibrium equations corresponding to summing forces in the horizontal and vertical directions, as well as for the moments about some point. However, it is even easier if we note the following. If we projected the line of member *EF* back down to where it intersects a projection of the horizontal chord, we would find a point about which we could take moments and have an equation having only one unknown, F_{EG} (and be careful when interpreting this result!):

$$\sum M_0 = 48\left(\frac{36}{60}F_{EG}\right) + 8(36)\left(\frac{48}{60}F_{EG}\right) - 6(36)(125)$$
$$+ 7(36)(50) + 8(36)(50)$$

Similarly, we could sum moments about the point corresponding to the location of joint *G* on the complete truss, in which case we would find an equation involving only F_{EF}:

$$\sum M_G = 48\left(\frac{6}{\sqrt{37}}F_{EF}\right) + 36\left(\frac{1}{\sqrt{37}}F_{EF}\right) + 3(36)(125)$$
$$- 2(36)(50) - 1(36)(50)$$

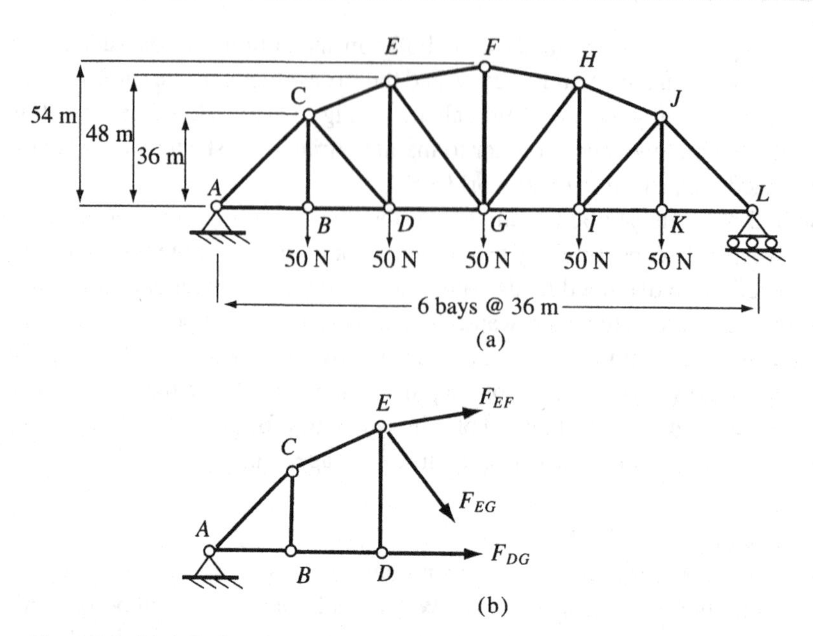

Figure 5.4. (a) A Parker truss and (b) a section of the Parker truss.

Note that in both of these calculations we have broken the unknown bar forces into their horizontal and vertical components, respectively, to make it easier to calculate the moments, and we have arbitrarily chosen to take clockwise moments as positive. Further, having done either of these calculations, we are now in a position to easily find the rest of the bar forces for the chosen panel. ∎

We could clearly further apply the method of sections to this truss and others just like it (cf. Problems 5.4 and 5.5). It is a skill that can be honed with exercise, especially since the method of sections depends particularly on using engineering insight to choose sections and points about which to sum moments.

5.2 Castigliano's Second Theorem and Trusses as Assemblages of Bars

In this section we describe applications of Castigliano's second theorem to the analysis of trusses, including the calculation of joint deflections for determinate trusses, the determination of redundant bar forces for indeterminate trusses, and the matrix formulation of equations determining deflections of indeterminate trusses.

5.2.1 Castigliano's Second Theorem, Unit Loads, and Virtual Work

We now wish to return to our consideration of energy methods with an eye toward calculating the deflections of specific pins of a truss. In so doing we will apply Castigliano's second theorem to trusses and, as a consequence, produce the principle of virtual work for trusses (see also Section 3.4).

Example 5.3. Use Castigliano's second theorem to calculate the movement of pin C in both the horizontal and vertical directions for the very simple two-bar, trusslike assembly shown in Fig. 5.5. There are two external loads applied at the pin C, one horizontal

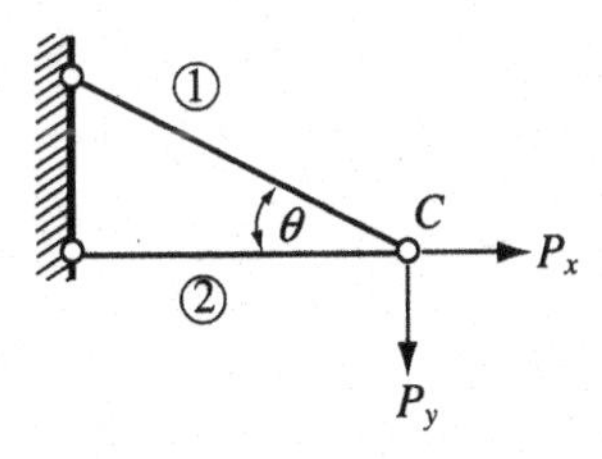

Figure 5.5. A simple two-bar, trusslike assembly.

and one vertical, and the bars have different properties, that is, $(A_1, E_1, L_1 = L)$ and $(A_2, E_2, L_2 = L\cos\theta)$.

The complementary energy is

$$U^* = \sum_{i=1}^{2} U_i^* = \frac{1}{2}\sum_{i=1}^{2}\frac{F_i^2 L_i}{A_i E_i} \tag{5.2}$$

Applying equilibrium to joint C allows us to express the bar forces in terms of the externally applied loads:

$$\begin{aligned} F_1 &= P_y/\sin\theta \\ F_2 &= P_x - P_y\cot\theta \end{aligned} \tag{5.3}$$

With the aid of these results we can write the complementary energy in terms of the applied loads:

$$U^* = \frac{1}{2}\left(\frac{L_1}{A_1E_1}\right)\left(\frac{P_y}{\sin\theta}\right)^2 + \frac{1}{2}\left(\frac{L_2}{A_2E_2}\right)(P_x - P_y\cot\theta)^2 \tag{5.4}$$

We apply Castigliano's second theorem to calculate the horizontal and vertical movement of pin C by differentiating Eq. (5.4):

$$\begin{aligned} \delta_{Ch} &= \frac{\partial U^*}{\partial P_x} = \left(\frac{L_2}{A_2E_2}\right)(P_x - P_y\cot\theta) \\ \delta_{Cv} &= \frac{\partial U^*}{\partial P_y} = \left(\frac{L_1}{A_1E_1}\right)\left(\frac{P_y}{\sin^2\theta}\right) + \left(\frac{L_2}{A_2E_2}\right)(P_y\cot^2\theta - P_x\cot\theta) \end{aligned} \tag{5.5}$$

Equations (5.5) are a complete solution to the problem posed, and for particular cases we have only to plug in parameter values to calculate numerical values for the deflections sought. ■

The solution (5.5) displays another interesting facet of applying Castigliano's second theorem to calculate deflections. In particular, note that joint C will move horizontally even if P_x vanishes, and there will be movement in the vertical direction even if P_y is zero. This certainly reflects the fact that motion of a particular pin reflects the extensions of all of the bars in a truss, not just the efffects of loads applied at that pin. However, it also shows that we can apply Castigliano's second theorem to calculate the movement of a pin by placing a phantom load at the pin, in the direction of the movement we want to find, and then letting that phantom load vanish after we apply Castigliano's theorem. Thus, the

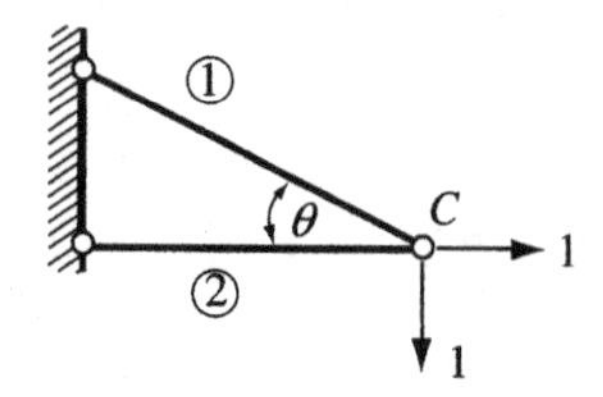

Figure 5.6. Unit loads applied to the assembly of Fig. 5.5.

horizontal movement of pin C due to a vertical load P_y acting alone is

$$\delta_{Ch}\Big|_{\substack{P_y \\ P_x=0}} = \frac{\partial U^*}{\partial P_x}\Big|_{P_x=0} = \left(\frac{L_2}{A_2 E_2}\right)(-P_y \cot\theta) \tag{5.6}$$

and the vertical movement of pin C due to a horizontal load P_x acting alone is

$$\delta_{Cv}\Big|_{\substack{P_x \\ P_y=0}} = \frac{\partial U^*}{\partial P_y}\Big|_{P_y=0} = \left(\frac{L_2}{A_2 E_2}\right)(-P_x \cot\theta) \tag{5.7}$$

Equations (5.5) are also interesting in still another way. If we write out the sum in Eq. (5.2), the complementary energy is

$$U^* = \frac{1}{2}\left(\frac{F_1^2 L_1}{A_1 E_1}\right) + \frac{1}{2}\left(\frac{F_2^2 L_2}{A_2 E_2}\right) \tag{5.8}$$

Then the components of the movement of pin C are given by

$$\begin{aligned} \delta_{Ch} &= \frac{\partial U^*}{\partial P_x} = \left(\frac{F_1 L_1}{A_1 E_1}\right)\frac{\partial F_1}{\partial P_x} + \left(\frac{F_2 L_2}{A_2 E_2}\right)\left(\frac{\partial F_2}{\partial P_x}\right) \\ \delta_{Cv} &= \frac{\partial U^*}{\partial P_y} = \left(\frac{F_1 L_1}{A_1 E_1}\right)\frac{\partial F_1}{\partial P_y} + \left(\frac{F_2 L_2}{A_2 E_2}\right)\left(\frac{\partial F_2}{\partial P_y}\right) \end{aligned} \tag{5.9}$$

We see that each term in Eq. (5.9) contains a "gradient" of a bar force that we can determine from joint equilibrium (Eqs. (5.2)):

$$\begin{aligned} \frac{\partial F_1}{\partial P_x} &= 0, & \frac{\partial F_2}{\partial P_x} &= 1 \\ \frac{\partial F_1}{\partial P_y} &= \frac{1}{\sin\theta}, & \frac{\partial F_2}{\partial P_y} &= -(1)\cot\theta \end{aligned} \tag{5.10}$$

And a simple examination (see Fig. 5.6) of these formulas tells us that these bar force gradients actually represent the forces in each bar that result from placing horizontal and vertical *unit loads* at pin C (in place of P_x and P_y). We introduce the following notation for the bar forces generated by the unit loads:

$$\begin{aligned} (f_{Ch})_1 &= \frac{\partial F_1}{\partial P_x} = 0, & (f_{Ch})_2 &= \frac{\partial F_2}{\partial P_x} = 1, \\ (f_{Cv})_1 &= \frac{\partial F_1}{\partial P_y} = \frac{1}{\sin\theta}, & (f_{Cv})_2 &= \frac{\partial F_2}{\partial P_y} = -(1)\cot\theta \end{aligned} \tag{5.11}$$

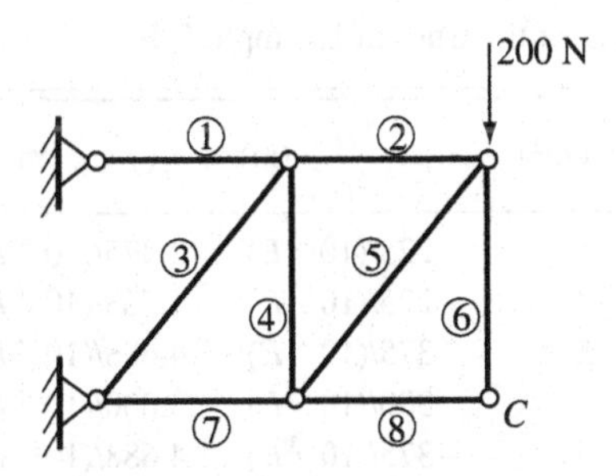

Figure 5.7. A loaded truss; we want the vertical deflection of pin C.

The notation is a bit cumbersome, but it is consistent with that used in much of the civil engineering literature. To represent the bar forces resulting from unit loads, we use the first inner subscript to denote the pin whose deflection we seek, the second inner subscript to denote the direction of the deflection sought (and toward which we direct the unit load), and the outer subscript to denote the particular bar whose response is being calculated.

With this notation, we can rewrite Eqs. (5.9) for the deflections of pin C as

$$\begin{aligned}\delta_{Ch} &= \frac{\partial U^*}{\partial P_x} = \left(\frac{F_1 L_1}{A_1 E_1}\right)(f_{Ch})_1 + \left(\frac{F_2 L_2}{A_2 E_2}\right)(f_{Ch})_2 \\ \delta_{Cv} &= \frac{\partial U^*}{\partial P_y} = \left(\frac{F_1 L_1}{A_1 E_1}\right)(f_{Cv})_1 + \left(\frac{F_2 L_2}{A_2 E_2}\right)(f_{Cv})_2\end{aligned} \tag{5.12}$$

or, in a compact form more amenable to generalization to an n-bar truss

$$\begin{aligned}\delta_{Ch} &= \sum_{i=1}^{2}\left(\frac{F_i L_i}{A_i E_i}\right)(f_{Ch})_i \\ \delta_{Cv} &= \sum_{i=1}^{2}\left(\frac{F_i L_i}{A_i E_i}\right)(f_{Cv})_i\end{aligned} \tag{5.13}$$

For an n-bar truss, of course, the upper limit in these sums would be n.

It is worth noting that because we are able to identify the bar force gradients of Eqs. (5.13) as the result of applying unit loads to a truss, the method we have just outlined for calculating truss deflections is frequently called the *unit load method* or the *dummy load method.* However, it clearly can be derived from Castigliano's second theorem. Apart from how we choose to view their origins, Eqs. (5.13) are very useful for calculating truss deflections.

Example 5.4. Calculate the vertical deflection at pin C of the truss shown in Fig. 5.7.

In Table 5.1 we show the computation of the desired vertical deflection using the formulation (5.13) in a standard spreadsheet. Note that we have identified bars by their own "bar numbers" rather than by letters or numbers of the joints.

It is also evident from the equilibrium of pin C that the bar forces F_6 and F_8 are both zero. Recognizing that fact can save us some computational time, but it does *not* mean that pin C doesn't move at all. Remember that the other bars in this truss shorten or lengthen because of the forces caused by the external loads, so we should expect a nontrivial result for the deflection of pin C.

Table 5.1. Calculating the deflection of pin C of the truss of Example 5.4.

Member	F_i (kN)	$(f_{Cv})_i$	L_i (m)	A_i (cm^2)	$\delta_i \equiv \dfrac{F_i L_i}{A_i E_i}$ (m)	$\delta_i (f_{Cv})_i$ (m)
1	300	3/2	4.5	6	$225/(10^{-3}E)$	$3.375/(10^{-6}E)$
2	150	3/4	4.5	3	$225/(10^{-3}E)$	$1.688/(10^{-6}E)$
3	−250	−5/4	7.5	5	$-375/(10^{-3}E)$	$4.688/(10^{-6}E)$
4	200	1	6.0	4	$300/(10^{-3}E)$	$3.000/(10^{-6}E)$
5	−250	−5/4	7.5	5	$-375/(10^{-3}E)$	$4.688/(10^{-6}E)$
6	0	1	6.0	4	0	0
7	−150	−3/4	4.5	3	$-225/(10^{-3}E)$	$1.688/(10^{-6}E)$
8	0	0	4.5	3	0	0
Total						$1.913/(10^{-7}E)$

It is also interesting to note that the bar forces along the top chord are in tension, while the bottom chord and the diagonals are in compression, thus it behaves just like a tip-loaded cantilever beam. Thus, we confirm again that trusses are often behave in aggregate like beams in bending. ■

Another interesting feature is that we can also identify the final result, Eq. (5.13), with the concept of virtual work that we introduced in Section 3.4. This is true because we can rewrite Eqs. (5.13) in the following form:

$$\begin{aligned} 1 \bullet \delta_{Ch} &= \sum_{i=1}^{2} (f_{Ch})_i \left(\frac{F_i L_i}{A_i E_i} \right) \\ 1 \bullet \delta_{Cv} &= \sum_{i=1}^{2} (f_{Cv})_i \left(\frac{F_i L_i}{A_i E_i} \right) \end{aligned} \tag{5.14}$$

What is so significant about this form? First, note that both sides of these equations represent products of forces multiplying deflections, that is, work. Second, the work done on both sides of the equations involves unit forces (on the left-hand sides) or *their* consequent bar forces (e.g., the $(f_{Ch})_i$ on the right-hand sides) multiplying the actual deflections produced at the points where they are sought (on the left-hand sides) or the bar deflections produced by the applied loads (the $(F_i L_i / A_i E_i)$ on the right-hand sides). Thus, the result (5.14) can be seen as another statement of the principle of virtual work in which, corresponding to Eqs. (3.39) and (3.40), we can write

$$\begin{aligned} \delta W_{virt/h} &= \sum_{i=1}^{2} (f_{Ch})_i \left(\frac{F_i L_i}{A_i E_i} \right) - 1 \bullet \delta_{Ch} \\ \delta W_{virt/v} &= \sum_{i=1}^{2} (f_{Cv})_i \left(\frac{F_i L_i}{A_i E_i} \right) - 1 \bullet \delta_{Cv} \end{aligned} \tag{5.15}$$

Here, of course, the "actual work" is that done by the unit loads and its consequent bar forces, while the "virtual displacements" are the actual bar elongations and joint deflections.

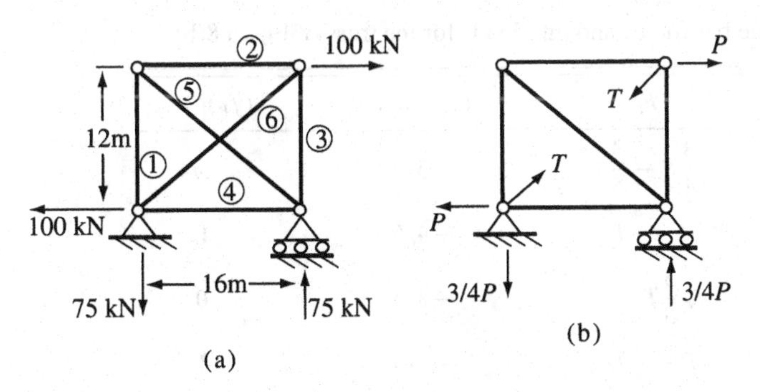

Figure 5.8. (a) An internally indeterminate truss and (b) a determinate version.

5.2.2 Castigliano's Second Theorem for Internally Indeterminate Trusses

We now extend Castigliano's second theorem to calculating bar forces in trusses that are internally indeterminate by considering the truss shown in Fig. 5.8(a). It is clear that the truss fails the test of the necessary condition for determinacy given in Eq. (5.1) because we have only eight equations to use to solve for the three reactions and six unknown bar forces. So, how do we deal with this internally indeterminate truss?

The answer is fairly straightforward. First we presume that we know the redundant bar force, say, in bar 6, and we call it T. The truss can then be taken as determinate and we can calculate its bar forces under the combined loading of P and T (cf. Fig. 5.8(b)). The bar forces F_i are listed in the second column of Table 5.2, and we see that they can be written in the form

$$F_i \equiv p_i P + t_i T \tag{5.16}$$

where the $p_i P$ are the contributions of the external load P to each bar force and the $t_i T$ are the contributions of the force T, which is here the total force in bar 6, so that $t_6 = 1$ and $p_6 = 0$.

Now, in the same way that we defined bar force gradients in Eqs. (5.11), we can define bar force gradients or unit loads generated by a unit load acting in place of the redundant T in bar 6,

$$(f_T)_i \equiv \frac{\partial F_i}{\partial T} = t_i \tag{5.17}$$

and unit loads generated by a unit load acting in place of the applied load P,

$$(f_P)_i \equiv \frac{\partial F_i}{\partial P} = p_i \tag{5.18}$$

so that the total bar forces of Eq. (5.16) can be written as

$$F_i = (f_P)_i P + (f_T)_i T \tag{5.19}$$

The second step in analyzing our indeterminate truss is to imagine that bar 6 has been cut so that it can't carry the load T, or any load for that matter. The faces of the cut separate by a distance δ_T – but that is a separation or cut that cannot happen in the real truss because it would violate compatibility. Since Castigliano's second theorem tells us how to calculate

Table 5.2. The bar forces and unit loads for the truss of Fig. 5.8(b).

Member	F_i	$(f_T)_i = t_i$	$(f_P)_i = p_i$
1	$\frac{3}{4}P - \frac{3}{5}T$	−3/5	3/4
2	$P - \frac{4}{5}T$	−4/5	1
3	$-\frac{3}{5}T$	−3/5	0
4	$P - \frac{4}{5}T$	−4/5	1
5	$-\frac{5}{4}P + T$	1	−5/4
6	T	1	0

compatible deformations produced by given loads on a structure in equilibrium, we can use that theorem here to find the value of T that keeps the truss deflections compatible. The complementary energy for our six-bar truss is

$$U^* = \sum_{i=1}^{6} U_i^* = \frac{1}{2}\sum_{i=1}^{6} \frac{F_i^2 L_i}{A_i E_i} \tag{5.20}$$

Since compatibility dictates that the separation δ_T between the cut ends of bar 6 must vanish, Castigliano's second theorem (Eqs. (5.9)) suggests that

$$\delta_T = \frac{\partial U^*}{\partial T} = \sum_{i=1}^{6} \frac{F_i L_i}{A_i E_i}\frac{\partial F_i}{\partial T} = 0 \tag{5.21}$$

Thus, with the results (5.17–5.19) we can recast Eq. (5.21) in the form

$$\begin{aligned}\delta_T &= \sum_{i=1}^{6} \frac{L_i}{A_i E_i}\big((f_P)_i P + (f_T)_i T\big)(f_T)_i \\ &= P\sum_{i=1}^{6} \frac{L_i}{A_i E_i}(f_P)_i (f_T)_i + T\sum_{i=1}^{6} \frac{L_i}{A_i E_i}(f_T)_i^2 = 0\end{aligned} \tag{5.22}$$

and then the redundant can be found by dividing the two sums of Eq. (5.22):

$$T = -\frac{\displaystyle\sum_{i=1}^{6} \frac{L_i}{A_i E_i}(f_P)_i (f_T)_i}{\displaystyle\sum_{i=1}^{6} \frac{L_i}{A_i E_i}(f_T)_i^2} P \tag{5.23}$$

The result (5.23) is very much like what we found in Section 5.2.1, although now we are able to handle internally indeterminate trusses. Thus, a calculation based on Eq. (5.23), like the results shown in Table 5.1, is easily done in a standard spreadsheet using the unit load results shown in Table 5.2. But there is just one important point that must be remembered as we apply these results to particular problems (e.g., Problems 5.7–5.9), that is, the contributions of the external loads are calculated after the redundant bar force has been replaced by the presumed-known force T, so that both the externally applied loads and the redundant replacement(s) are included in equilibrium at the truss joints.

5.2.3 Matrix Formulation of Castigliano's Second Theorem

As with other techniques developed earlier, it is clear that matrix representations and related computational algorithms make it easy to apply Castigliano's second theorem to real engineering structures. To illustrate the matrix formulation of Castigliano's second theorem, let us again consider the truss shown in Fig. 5.8(a), but now we add that we want to find the horizontal movement of the joint at which the load P is applied. Then, in a straightforward application of Castigliano's second theorem, we want to calculate the following two deflections by differentiating the complementary energy for this truss:

$$\begin{aligned} \delta_P &= \frac{\partial U^*}{\partial P} = \sum_{i=1}^{6} \frac{F_i L_i}{A_i E_i} \frac{\partial F_i}{\partial P} \\ \delta_T &= \frac{\partial U^*}{\partial T} = \sum_{i=1}^{6} \frac{F_i L_i}{A_i E_i} \frac{\partial F_i}{\partial T} \end{aligned} \tag{5.24}$$

and where we will later set $\delta_T = 0$ so that we can determine the redundant T.

Now, given the definitions of the unit loads (Eqs. (5.17) and (5.18)) and the bar forces (Eqs. (5.19)), the truss deflections (Eqs. (5.24)) can be written as

$$\delta_P = \sum_{i=1}^{6} \frac{L_i}{A_i E_i} \big((f_P)_i P + (f_T)_i T\big)(f_P)_i$$

$$\delta_T = \sum_{i=1}^{6} \frac{L_i}{A_i E_i} \big((f_P)_i P + (f_T)_i T\big)(f_T)_i$$

which can be further manipulated into the form

$$\begin{aligned} \delta_P &= P \sum_{i=1}^{6} \frac{L_i}{A_i E_i} (f_P)_i^2 + T \sum_{i=1}^{6} \frac{L_i}{A_i E_i} (f_T)_i (f_P)_i \\ \delta_T &= P \sum_{i=1}^{6} \frac{L_i}{A_i E_i} (f_P)_i (f_T)_i + T \sum_{i=1}^{6} \frac{L_i}{A_i E_i} (f_T)_i^2 \end{aligned} \tag{5.25}$$

Now, Eq. (5.25) can clearly be represented as a matrix equation:

$$\begin{Bmatrix} \delta_P \\ \delta_t \end{Bmatrix} = \begin{bmatrix} f_{11} & f_{12} \\ f_{21} & f_{22} \end{bmatrix} \begin{Bmatrix} P \\ T \end{Bmatrix} \tag{5.26}$$

where the elements of the (symmetric) *extended flexibility matrix* $[f]$ are

$$\begin{aligned} f_{11} &= \sum_{i=1}^{6} \frac{L_i}{A_i E_i} (f_P)_i^2 \\ f_{12} &= \sum_{i=1}^{6} \frac{L_i}{A_i E_i} (f_T)_i (f_P)_i = f_{21} \\ f_{22} &= \sum_{i=1}^{6} \frac{L_i}{A_i E_i} (f_T)_i^2 \end{aligned} \tag{5.27}$$

Since we have only one redundant in this case, we can determine its value by setting $\delta_T = f_{21}P + f_{22}T = 0$, which yields

$$T = -f_{22}^{-1} f_{21} P \tag{5.28}$$

Then the deflection of the truss joint at which the load P is applied can be found from the first of Eqs. (5.26) with the redundant T determined by Eq. (5.26):

$$\begin{aligned}\delta_P &= f_{11}P + f_{12}T \\ &= \left(f_{11} - f_{12}f_{22}^{-1}f_{21}\right)P\end{aligned} \tag{5.29}$$

Note that the flexibility coefficients f_{ij} defined in Eqs. (5.27) include two factors. One is the flexibility of each bar, L_i/E_iA_i, whereas the second is a multiplier that reflects components of the given load and/or the redundant that appear in that bar. This multiplier thus effectively reflects the orientation of each bar in the truss as "measured" through the joint equilibrium equations.

Equation (5.29) is also an interesting result because it mirrors the general force (flexibility) method of structures that we will develop in greater detail in Section 10.2. In fact, we have written the result in a way that evokes the appearance of a matrix formulation in which the displacement and force vectors could have many more elements, both as given forces and redundants, and the flexibility elements can be interpreted as partitioned matrices of the significantly more extended flexibility matrix that would occur in more complex problems. (For more immediate gratification, compare the form of Eq. (5.27) with Eq. (10.20).)

Problems

5.1 For the very elementary truss shown in Fig. 5.1(a), apply the method of joints to all three pins and then subsequently combine the resulting equations of equilibrium to show that the set of the pin equilibrium equations is equivalent to equilibrium taken for the truss as a whole.

5.2 What happens to the determinacy of the elementary triangular truss if we add two more bars connected by one additional pin? And we add still another pair of bars joined by a pin? And so on?

5.3 Complete the calculation of the bar forces for the truss shown in Fig. 5.2 while using the free-body diagrams shown in Fig. 5.3.

5.4 Using the method of sections on the truss shown in Fig. 5.4, complete the calculations begun in Section 5.1.3.

5.5 Using the method of sections on the truss shown in Fig. 5.4, calculate the forces in members *FG*, *HJ*, *IJ*, and *IK*.

5.6 Calculate the bar forces for all of the members of the truss of Fig. 5.4.

5.7 Use the second theorem of Castigliano to find the bar forces for the truss of Fig. 5.8(a), assuming it is 12 m high and 16 m long and that $P = 100$ kN. Express the answers in terms of common values of A and E.

5.8 Use the second theorem of Castigliano to find the bar forces for the truss of Fig. 5.P8, assuming each panel is 20 m × 20 m and that $P = 150$ kN. Express the answers in terms of common values of A and E.

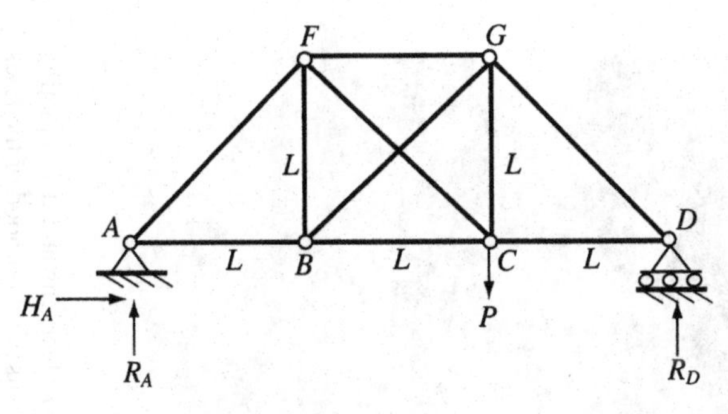

Figure 5.P8. Figure for Problem 5.8.

5.9 Apply the second theorem of Castigliano to find the bar forces for the truss of Fig. 5.P9 with the dimensions shown and with $P = 100$ kN. Express the answers in terms of common values of A and E.

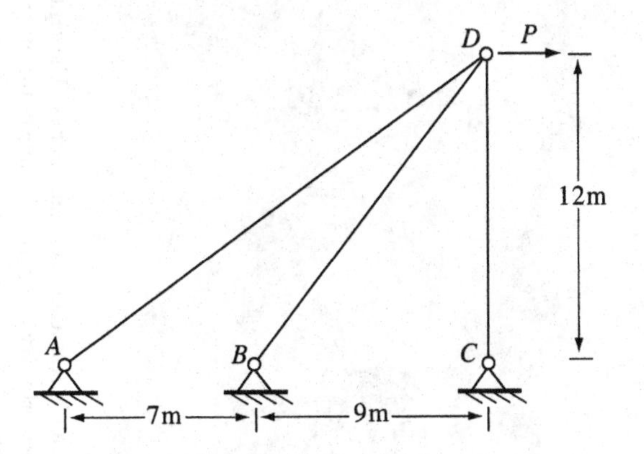

Figure 5.P9. Figure for Problem 5.9.

5.10 Calculate the extended flexibility matrix for the truss of Fig. 5.8(a), assuming it is 12 m high and 16 m long and that all bars have common values of A and E.

5.11 Apply phantom forces to the pin C of the truss in Fig. 5.7 to find the horizontal and vertical movement of the pin using Castigliano's second theorem. Compare the results to those given in Example 5.4.

James B. Eads designed the bridge named after him as a set of elegant arches. The central spans are some 510 feet long and at completion in 1874 were the longest of their kind in the world. By comparison, some of the other bridges in the St. Louis neighborhood of the Eads Bridge appear to cross the Mississippi River far less gracefully. The Gateway Arch, on the other hand, confirms Eads' view of the arch as simply elegant art. (Photo by J. Wayman Williams.)

6

Energy Principles for Calculating Displacements and Forces

In this chapter we present in more general terms the energy methods that we introduced for discrete systems (Chapter 3) and for axially loaded bars (Chapter 4) and trusses (Chapter 5). We do this to lay a foundation for applying energy methods to more complicated, more interesting structural engineering problems. As a starting point we will review the two paradigms we have discussed, with an eye toward asking how we can generalize them for more effective use.

6.1 Recapitulation and Review

We have illustrated two paradigms, the first in which we cast our idealizations in terms of displacements and stiffnesses, the second in which we wrote our idealizations in terms of forces and flexibility parameters. In the stiffness methods our discretizations were expressed in terms of approximations for the displacement fields, while for the flexibility methods we discretized the forces.

Our energy methods involved two kinds of calculations. The first was styled in terms of minimum energy principles, as a result of which we developed a second calculation as reflected in the two Castigliano theorems. Let us take the simple elastic bar as our illustrative model (cf. Fig. 4.9). Thus, an axial force P produces a deflection Δ. In the displacement approach we could write the principle of minimum potential energy in terms of the stored strain energy and the potential of the applied loads as follows:

$$\Pi = \frac{1}{2}k\Delta^2 - P\Delta \tag{6.1}$$

whose first variation produces

$$\delta^{(1)}\Pi = (k\Delta - P)\delta\Delta = 0 \tag{6.2}$$

We recognize here that the vanishing of the first variation produces the equation of equilibrium for the discretized rod in terms of the displacement variable Δ and the bar stiffness k.

The second energy-based calculation in the displacement paradigm is the first theorem of Castigliano in which we can calculate the force required to produce the deflection Δ as (recall Eq. (4.56))

$$\frac{\partial U}{\partial \Delta} = k\Delta = \left(\frac{AE}{L}\right)\Delta \equiv P \tag{6.3}$$

There are calculations in the force paradigm that parallel these results. The complementary potential energy (which we have not actually introduced before) is written as the sum of the complementary energy and the potential of the load, although we would here regard the load as the system variable:

$$\Pi^* = \frac{1}{2} f P^2 - P\Delta \tag{6.4}$$

whose first variation produces

$$\delta^{(1)}\Pi^* = (fP - \Delta)\delta P = 0 \tag{6.5}$$

We recognize here that the vanishing of the first variation produces the equation that ensures compatibility for the discretized rod in terms of the force P and the flexibility of the bar, f.

The second energy-based calculation in the force paradigm is the second theorem of Castigliano in which we can calculate deflection Δ produced by the force (recall Eq. (4.57)):

$$\frac{\partial U^*}{\partial P} = fP = \left(\frac{L}{AE}\right) P \equiv \Delta \tag{6.6}$$

The principal question before us now is: How do we generalize these results so that we can use them in more complex problems? A related issue is concerned with developing different ways of representing both the continuous elastic media of which our elastic structures are made and the relevant kinds of forces and displacements at the various points on a structure's surface on which external forces and geometric constraints (for support purposes and to prevent unwanted motion) are applied.

6.2 Work–Energy Relationships and Methods

We start by writing the energy *stored* in an elastic structure in terms of the actual work done by the applied external loads. Note that in examining this work we do not need to include terms for work done by immovable reactions because they can do no work, as a result of which there is no movement of the structure at those points that produces any stored energy. Further, we will limit our present considerations to the static response of linear elastic structures, by which we mean that we will not include inertia or related gravitational acceleration terms, nor will we worry about nonlinear behavior resulting from nonlinear stress–strain laws or unusually large structural deformation. The extension to dynamic behavior is consistent with the results presented here, while the problems posed by the modeling of nonlinear behavior are altogether another set of subjects – even though the ideas we focus on in this book and their modern computational implementations have produced enormous gains in our ability to analyze and understand a great deal of nonlinear behavior.

Initially, we focus our discussion on the strain energy as the form of stored energy, remembering that for a linear elastic system the strain energy has a magnitude equal to the complementary energy for a given state of stress and strain. We will calculate the strain energy by calculating the work done by the applied loads acting through the displacements produced where those external forces are applied. When we speak of such applied loads, we mean *generalized forces*, by which we intend to include external loads such as moments and torques. Consequently, when we speak of the displacements produced by the generalized loads, we mean *generalized displacements*, which can include rotations or angle changes. The important point is that we are calculating work, which in dimensional terms means [force]×[distance], and expressing it in terms of the resulting energy stored. Thus, for an elastic spring

$$W_{ext/spr} = \int_0^{\xi} F_s(\eta)d\eta = \int_0^{\xi} k\eta d\eta = \frac{1}{2}k\xi^2 \tag{6.7a}$$

while for the axially loaded bar the work done by the external load is written as

$$W_{ext/bar} = \frac{1}{2}k\Delta^2 = \frac{1}{2}\left(\frac{AE}{L}\right)\Delta^2 \tag{6.7b}$$

The displacements produced by the applied forces are in the classical linear form. For the spring,

$$\xi = k^{-1}P = fP \tag{6.8a}$$

while for the axially loaded bar,

$$\Delta = k^{-1}P = fP = \left(\frac{AE}{L}\right)^{-1}P \tag{6.8b}$$

Clearly, the work done by the external forces on an elastic system must produce an equal amount of stored energy because we are talking about a conservative system, so that the work done is conserved in the form of stored energy. Thus, for the spring,

$$W_{ext/spr} \equiv U_{spr} = \frac{1}{2}k\xi^2 \tag{6.9a}$$

while for the bar,

$$W_{ext/bar} \equiv U_{bar} = \frac{1}{2}k\Delta^2 = \frac{1}{2}\left(\frac{AE}{L}\right)\Delta^2 \tag{6.9b}$$

The resemblances between the alphabetized pairs of Eqs. (6.7), (6.8), and (6.9) speak for themselves. It clearly would be worth generalizing these results so as to incorporate multiple inputs, in fact, an arbitrary number of inputs.

We generalize in the tradition of adding or superposing the responses for linear systems. Thus, building on the results of Eqs. (6.8), we would say that for two forces applied to an elastic system, P_1 and P_2, the total displacement u_i at point i is

$$u_i = f_{i1}P_1 + f_{i2}P_2 \tag{6.10}$$

Extending this result still further, for n loads P_j, the deflection or displacement at the ith point is

$$u_i = f_{i1}P_1 + f_{i2}P_2 + \cdots + f_{in}P_n$$

$$= \sum_{j=1}^{n} f_{ij}P_j \tag{6.11}$$

where the f_{ij} represent *flexibility coefficients* or *influence coefficients* that indicate the proportion of the movement at point i we should attribute to a load P_j. Thus, Eq. (6.11) tells us that the deflection at the ith point is a *linear sum* of the deflections at that point produced by the n loads, each multiplied by a suitably identified flexibility coefficient. The calculation of the specifics of the f_{ij} is clearly a central concern of this book, because if we can do this in general, we can analyze the response of any linear structure to any collection of loads.

Now, in order to calculate the work done by a given load while acting on and deforming the body, for example, at point 1, we would simply write that

$$W_1 = \frac{1}{2} P_1 u_1 \tag{6.12}$$

The factor of one-half arises because the displacement is linearly related to the deflection, as in the spring and the bar. In fact, there are some thermodynamics considerations that underlie the use of the one-half factor.

Consider that over some time period the structure is either thermally isolated (*adiabatic*) or that we can neglect any temperature expansions (*isothermal*) that occur during this time. The first law of thermodynamics states that the work done on the body during that time is equal to the sum of the energy stored and the change in the kinetic energy. We can ignore the kinetic energy if the loads are applied "slowly" – our choice here – or we could add in the proper terms to account for the kinetic energy. In either case, the displacements reach their final values slowly and in proportion to the loads. By the first law of thermodynamics, we can identify the stored energy as energy that can be converted into mechanical work. Thus, as we will shortly do, we can equate the external work to the stored strain energy. However, as we will write the stored energy in terms of flexibility coefficients and applied loads, we will consider that the stored energy is the (equal) complementary energy.

If we now calculate the work done by all of the forces acting at their various points of applications – assuming that all of the "corresponding" displacements of those points are in the same directions as the forces – we would have

$$W_{ext/total} = \sum_{i=1}^{n} W_i = \sum_{i=1}^{n} \frac{1}{2} P_i u_i \tag{6.13}$$

Since we can write the displacements in terms of the loads by virtue of Eq. (6.11), we can write the total work done by all of the external loads as

$$W = W_{ext/total} = \frac{1}{2} \sum_{i=1}^{n} \sum_{j=1}^{n} f_{ij} P_i P_j \tag{6.14}$$

which by our thermodynamic argument means that

$$U \equiv W = \frac{1}{2} \sum_{i=1}^{n} \sum_{j=1}^{n} f_{ij} P_i P_j = U^* \tag{6.15}$$

We can write this result in extended form as

$$\begin{aligned} U^* = W = & \frac{1}{2} f_{11} P_1^2 + \frac{1}{2} f_{12} P_1 P_2 \cdots \frac{1}{2} f_{1n} P_1 P_n \\ & + \frac{1}{2} f_{21} P_2 P_1 + \frac{1}{2} f_{22} P_2^2 \cdots \frac{1}{2} f_{2n} P_2 P_n \\ & \vdots \\ & + \frac{1}{2} f_{n1} P_n P_1 + \frac{1}{2} f_{n2} P_n P_2 \cdots \frac{1}{2} f_{nn} P_n^2 \end{aligned} \tag{6.16}$$

Now that we have this seemingly elegant result, we will inquire into its meaning in somewhat more detail.

Consider the following. Equation (6.11) clearly represents the total displacement at each of i points as the loads P_j are applied. Should we worry about the order in which the loads are applied? Common sense would seem to suggest that the order in which the loads are applied should not matter for a truly linear system. In an easy example, suppose we applied a load P_1' at point 1, followed by a load P_1''. It is clear that these loads produce displacements at point i, say, whose sum we can calculate by virtue of applying Eq. (6.11) as

$$\begin{aligned} u_i &= u_i' + u_i'' = f_{i1} P_1' + f_{i1} P_1'' \\ &= f_{i1}(P_1' + P_1'') \end{aligned} \tag{6.17}$$

whose sum is clearly independent of the order in which two loads are applied at the same point.

Suppose now a second case, where we apply a load P_1 at point 1, followed by a load P_2 at point 2. Or would it matter if we apply the load P_2 first, after which we apply load P_1? No, it wouldn't matter: Equations (6.15) and (6.16) make it clear that the total work done is independent of the order in which the loads are applied. But there is an important consequence of the invariance of the work done with respect to the order in which loads are applied. When we apply load P_1 at point 1 first, it produces displacements u_1' and u_2'. Corresponding displacements u_1'' and u_2'' are produced by the second load P_2 at point 2. The work done is

$$W_{1 \text{ then } 2} = \frac{1}{2} P_1 u_1' + + P_1 u_1'' + \frac{1}{2} P_2 u_2'' = \frac{1}{2} f_{11} P_1^2 + P_1(f_{12} P_2) + \frac{1}{2} f_{22} P_2^2 \tag{6.18}$$

where in both middle terms we recognize that the already present load P_1 "rides through" at full value the displacement produced at point 1 by the load P_2, and we also use Eq. (6.11) to express the displacements in terms of corresponding influence coefficients and appropriate loads. On the other hand, for the case in which we apply load P_2 first and then load P_1, the total work done is

$$W_{2 \text{ then } 1} = \frac{1}{2} P_2 u_2' + + P_2 u_2'' + \frac{1}{2} P_1 u_1'' = \frac{1}{2} f_{22} P_2^2 + P_2(f_{21} P_1) + \frac{1}{2} f_{11} P_1^2 \tag{6.19}$$

where here as well we apply similar reasoning to the middle terms and we invoke Eq. (6.11) repeatedly. Again, as with a repeated load at the same point, since we know that the final result should produce a final result independent of the order in which the loads P_1 and P_2 are applied, we expect that

$$W_{1 \text{ then } 2} = W_{2 \text{ then } 1}$$

from which (together with Eqs. (6.18) and (6.19)) we obtain the following remarkable result:

$$f_{12} = f_{21} \tag{6.20}$$

Equation (6.20) states that because the work done by these two different loads on a linear elastic structure is independent of the order in which they were applied, the flexibility coefficients most directly involved would have to be symmetric. This result, one of our more elegant and useful results, is called the *Maxwell reciprocal principle.* In its most general form, Maxwell's principle of reciprocity is

$$f_{ij} = f_{ji} \tag{6.21}$$

There is another version of reciprocity that is worth noting, particularly because it is reminiscent of the principle of virtual work. Consider that two systems of loads (P_i and P_i') are applied, each producing corresponding displacements (u_i and u_i'). Then, by virtue of Eq. (6.11) we can relate these displacements to their corresponding loads:

$$u_i = f_{ij}P_j \qquad \text{and} \qquad u_i' = f_{ij}P_j' \tag{6.22}$$

The work done by the forces P_i' acting through the displacements u_i can then be calculated as

$$\begin{aligned} P_i'u_i &= P_i'(f_{ij}P_j) = f_{ij}P_i'P_j \\ &= f_{ji}P_j'P_i = f_{ij}P_j'P_i \end{aligned}$$

or

$$P_i'u_i = P_iu_i' \tag{6.23}$$

In other words, the work done by the forces P_i' acting through displacements u_i is equal to the work done by the forces P_i acting through displacements u_i'. Note that we have used the symmetry of the flexibility coefficients (Eq. (6.21)) to complete this calculation and obtain the result (6.23), which is often called the *Betti–Rayleigh reciprocal theorem.*

Thus, we have now identified the stored energy as being equal to the work done on an elastic structure or solid by an arbitrary number of (generalized) loads acting through corresponding (generalized) displacements. And while we have worked largely in terms of forces and flexibilities, the foregoing results are valid for all elastic structures and can be represented in either paradigm: forces and flexibilities or displacements and stiffnesses. Thus, the force–displacement relation that parallels Eq. (6.11) can be written as

$$\begin{aligned} P_i &= k_{i1}u_1 + k_{i2}u_2 + \cdots + k_{in}u_n \\ &= \sum_{j=1}^{n} k_{ij}u_j \end{aligned} \tag{6.24}$$

where the k_{ij} are the stiffness coefficients. Further, with the aid of Eqs. (6.24) we can rewrite the total external work – or the total energy stored – beginning with the form (6.13). Thus, in extended form, and in analogy with Eq. (6.16), the stored strain energy is

$$\begin{aligned} U = W = &\frac{1}{2}k_{11}u_1^2 + \frac{1}{2}k_{12}u_1u_2 + \cdots + \frac{1}{2}k_{1n}u_1u_n \\ &+ \frac{1}{2}k_{21}u_2u_1 + \frac{1}{2}k_{22}u_2^2 + \cdots + \frac{1}{2}k_{2n}u_2u_n \\ &\vdots \\ &+ \frac{1}{2}k_{n1}u_nu_1 + \frac{1}{2}k_{n2}u_nu_2 + \cdots + \frac{1}{2}k_{nn}u_n^2 \end{aligned} \tag{6.25}$$

which appears more compactly as

$$U = W = \frac{1}{2}\sum_{i=1}^{n}\sum_{j=1}^{n} k_{ij}u_i u_j \tag{6.26}$$

We are now in a position to derive some work-energy relations for use with either of our two paradigms. We have not exhausted all of the properties of the linear constructs we have just outlined – for example, that the loads and displacements found by using Eqs. (6.15) and (6.26), respectively, are unique – but we have laid the foundations for their application.

6.2.1 Work–Energy in the Force (Flexibility) Paradigm

We now present two work-energy relations in the force (flexibility) paradigm. The first is Castigliano's second theorem, which we can derive rather easily by differentiating either of Eqs. (6.15) or (6.16) with respect to the ith load component P_i. To show how this works, let us calculate the derivative of U^* from Eq. (6.15) with respect to P_2:

$$\begin{aligned}\frac{\partial U^*}{\partial P_2} &= \frac{1}{2}f_{12}P_1 + \frac{1}{2}f_{21}P_1 + f_{22}P_2 + \cdots + \frac{1}{2}f_{n2}P_n + \frac{1}{2}f_{2n}P_n \\ &= f_{22}P_2 + \sum_{n\neq 2} f_{n2}P_n \equiv u_2\end{aligned} \tag{6.27}$$

In more general terms,

$$\frac{\partial U^*}{\partial P_i} = \sum_{j=1}^{n} f_{ij}P_j \equiv u_i \tag{6.28}$$

We now define the *total complementary energy* Π^* as

$$\Pi^* \equiv U^* - \sum_{S_u} P_i u_i \tag{6.29}$$

where S_u denotes the areas or points on a structure where displacements are prescribed, such as at rigid supports or similar constraints. At such points we can admit small variations of the loads in a variational process, which we cannot do for points where the loads are prescribed.

The restriction just placed on the loading potential can be viewed in the following way. If we were to show (and list) all of the external loads and reaction forces on an elastic body or structure (cf. Fig. 6.1), they would take up the entire surface or set of points on which all forces and loads external to the body are applied, that is, $S = S_p + S_u$. We can separate these points and forces into two groups.

Some of the forces on the elastic body are *given external loads*, that is, their magnitudes and directions are fixed by our (externally made) choices. These given external loads occupy the set of points or the surface S_p. In a variational process, the magnitudes of these given loads cannot be made subject to small variations because they are prescribed loads or forces. On the other hand, the deflections that occur at these points are determined

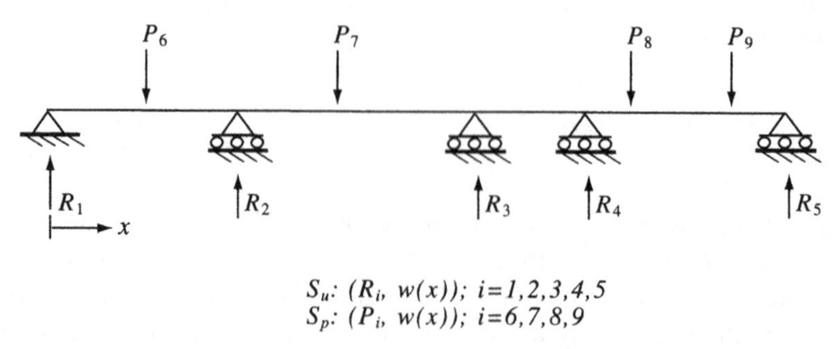

Figure 6.1. Admissible forces and displacements for an elastic beam.

as a result of our analysis of the structure, and therefore we can admit small, compatible variations of the displacements (i.e., *admissible displacements*) at these points.

Now, there are also points on a structure where we choose to *prescribe displacements*, as we do at rigid supports, for example (Fig. 6.1). These points comprise the surface S_u. In a variational process, we cannot violate geometric constraints by ignoring or overriding any prescribed displacements. On the other hand, the forces at these points are determined from our analysis of the structure, and therefore we can admit small variations of the forces at these points that satisfy equilibrium (i.e., *admissible forces*).

As a practical matter, it often turns out that the total complementary energy is just the complementary energy, that is, U^* all by itself, because the potential term is likely to be zero. For example, all of the reactions of the beam shown in Fig. 6.1 are rigid supports. Hence,

$$w(x_i) = 0 \quad \text{for } i = 1, 2, 3, 4, 5$$

This implies that the potential over S_u is zero, so that here we have $\Pi^* = U^*$.

We now return to considering the *total* complementary energy (so that we can be, in principle, reasonably general). By a straightforward (if somewhat simplified) variational process, we can find the first variation of the total complementary energy to be

$$\delta^{(1)}\Pi^* = \sum_{i=1}^{n}\sum_{j=1}^{n}(f_{ij}P_j - u_i)\delta P_i = 0 \tag{6.30}$$

The requirement that Eqs. (6.30) vanish for each and every variation of the applied loads yields the *Principle of Minimum Total Complementary Energy*:

> Among all force fields (or sets of forces) that satisfy equilibrium and all boundary conditions where forces are prescribed, the force field that minimizes the total complementary energy is the one that produces a field of compatible displacements, that is, displacements that are continuous, single valued, and do not violate any displacement boundary conditions or geometric constraints.

6.2.2 Work–Energy in the Displacement (Stiffness) Paradigm

We now use Eqs. (6.25) and (6.26) to derive the work-energy relations that parallel precisely the results achieved above, only this time for the displacement (stiffness) paradigm.

First we will derive Castigliano's first theorem. We could differentiate Eq. (6.26) with respect to the ith displacement component u_i. However, first we will differentiate the U of Eq. (6.26) with respect to a specific displacement u_2:

$$\begin{aligned}\frac{\partial U}{\partial u_2} &= \frac{1}{2}k_{12}u_1 + \frac{1}{2}k_{21}u_1 + k_{22}u_2 + \cdots + \frac{1}{2}k_{n2}u_n + \frac{1}{2}k_{2n}u_n \\ &= k_{22}u_2 + \sum_{n\neq 2} k_{n2}u_n \equiv P_2 \end{aligned} \tag{6.31}$$

In more general terms,

$$\frac{\partial U}{\partial u_i} = \sum_{j=1}^{n} k_{ij}u_j \equiv P_i \tag{6.32}$$

We now define the *total potential energy* Π as

$$\Pi \equiv U - \sum_{S_p} P_i u_i \tag{6.33}$$

where S_p denotes the areas or points on a structure where external forces or loads are prescribed. As we have already noted, at such points we can admit small variations of the displacements during a variational process. Thus, for the beam illustrated in Fig. 6.1, the potential of the applied loads will be

$$\begin{aligned}\sum_{S_p} P_i u_i &= P_6 w(x_6) + P_7 w(x_7) + P_8 w(x_8) + P_9 w(x_9) \\ &= \sum_{i=6}^{9} P_i w(x_i)\end{aligned}$$

Then, by our usual variational process, we can find the first variation of the total potential energy to be

$$\delta^{(1)}\Pi = \sum_{i=1}^{n}\sum_{j=1}^{n}(k_{ij}u_j - P_i)\delta u_i = 0 \tag{6.34}$$

For the beam example just cited, the variation of the potential of the applied loads would just be

$$-\sum_{S_p} P_i \delta u_i = -\sum_{i=6}^{9} P_i \delta w(x_i)$$

The requirement that Eqs. (6.34) vanish for each and every variation of the applied loads yields the *Principle of Minimum Total Potential Energy*:

> Among all displacement fields (or sets of displacements) that satisfy compatibility and all geometric boundary conditions (where displacements are prescribed), the displacement field that minimizes the total potential energy is the one that produces a field of forces that satisfies equilibrium and all boundary conditions on forces.

6.2.3 Matrix Formulations of Work–Energy Relations

To complete Section 6.2 we want to present the principal results in matrix notation, with a view toward encouraging familiarity with a representational style. Recall that we used matrix representation in Section 4.6 to help model the three-segment, axially loaded bar. And, as we have noted before, we are not going to use matrix techniques very much, but the general results just found are also elegantly stated in this style.

Our starting point is the displacement–force relation (6.11), which we now write as

$$\{u\} = [f]\{P\} \tag{6.35}$$

where the displacements and forces are represented as column matrices or vectors, and the flexibility coefficients appear in the $n \times n$ flexibility matrix.

The force–displacement relation (6.24) is similarly represented, only now the corresponding square matrix is a stiffness matrix, that is,

$$\{P\} = [k]\{u\} \tag{6.36}$$

Clearly, the two square matrices represent each other's inverses

$$[k] = [f]^{-1} \tag{6.37}$$

This is an important result because it serves to remind us that flexibility and stiffness coefficients are *not* inverses on a one-to-one basis, but on a systems basis.

Given Eq. (6.35), we can rewrite the complementary energy (Eq. (6.15)) as

$$W = U^* = \frac{1}{2}\{P\}^T[f]\{P\} \tag{6.38}$$

where $\{P\}^T$ is the *transpose* of the vector $\{P\}$, that is, the conversion of the column vector to a *row vector*:

$$\{P\}^T = \{\, P_1 \quad P_2 \quad \cdots \quad P_n \,\} \tag{6.39}$$

The reason that the transpose of the load vector appears is that it guarantees that the matrices are *conformable*, by which we mean that the indicated matrix multiplications can be carried out, on the usual row-by-column basis. That is, the product $[f]\{P\}$ produces a column vector, which means it can only be multiplied by a row vector. Note that when we get through with the matrix multiplications called for in Eq. (6.38), the result is the total energy stored, which is a scalar. This is another way of confirming the outcome of multiplying a conformable pair of row and column matrices.

In a similar vein, the strain energy can be written as

$$W = U = \frac{1}{2}\{u\}^T[k]\{u\} \tag{6.40}$$

And we can now write the total energy principles corresponding to our two paradigms. In the force paradigm, the total complementary energy becomes

$$\Pi^* = \frac{1}{2}\{P\}^T[f]\{P\} - \{P\}^T\{u\} \tag{6.41}$$

where our restriction of the loading potential to those forces at points where the displacements are prescribed is understood. The first variation that establishes the principle of

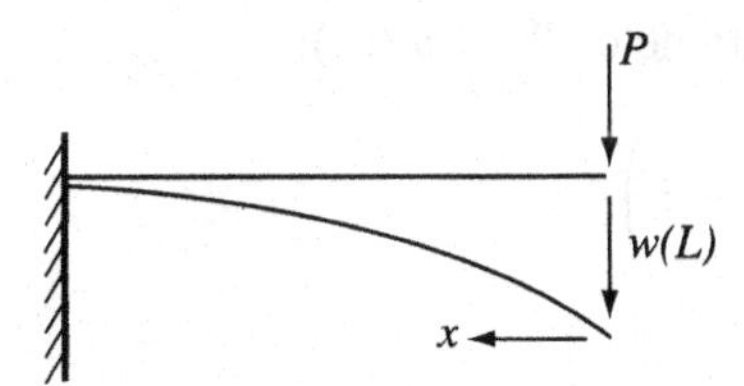

Figure 6.2. A tip-loaded cantilever beam.

minimum total complementary energy is then

$$\delta^{(1)}\Pi^* = \{\delta P\}^T[[f]\{P\} - \{u\}] \tag{6.42}$$

Remember that the transpose of the row vector appears where it does – as a pre-multiplier to the matrix compatibility equation – to maintain the conformability of the relevant matrix multiplications.

In the displacement paradigm, the total potential energy becomes

$$\Pi = \frac{1}{2}\{u\}^T[k]\{u\} - \{P\}^T\{u\} \tag{6.43}$$

where now our restriction of the loading potential to those displacements at points where the forces are prescribed is understood. The first variation that establishes the principle of minimum total potential energy is then

$$\delta^{(1)}\Pi = \{\delta u\}^T[[k]\{u\} - \{P\}] \tag{6.44}$$

And, again, the transpose of the row vector appears as a premultiplier to the matrix equilibrium equation to maintain the conformability of the relevant matrix multiplications.

6.3 Minimum Total Potential Energy Principle

The results we have developed in Section 6.2 are quite general and are applicable without exception to any linearly elastic structure. However, they are also incomplete in the following sense. We have provided no clue (beyond the bar example cited in Chapter 4) about how we would apply these results to specific elastic structures. Stated otherwise, how would we compute the stiffness or the flexibility coefficients, the k_{ij} or the f_{ij}, for a specific beam or arch or whatever. The answer is that we must resort to continuum models of the class of elastic solids and structures, from which we can derive suitable idealizations and discretizations for different physical configurations of both the geometry of a structure and the nature (magnitude, direction, type) of the externally applied loads. We will present briefly the continuum versions of the variational principles needed to develop the idealizations and discretizations that allow us to calculate the k_{ij} and the f_{ij} that were applied in Section 6.2. Our summary of the three-dimensional continuum principles will follow closely the developments of Sections 4.1 and 4.2, and we leave for futher readings the formal derivation of the total potential energy principle in three dimensions.

By way of illustration, consider the classic, tip-loaded cantilever beam (Fig. 6.2). We will find in Chapter 7 some well-known results from "Strength of Materials." Specifically, the moment in the cantilever cross section is

$$M(x) = -EIw''(x) = -Px \tag{6.45}$$

while the bending deflection is (introducing a parameter $W = w(0)$):

$$w(x) = \left(\frac{PL^3}{3EI}\right)\left(\frac{1}{2}\left(\frac{x}{L}\right)^3 - \frac{3}{2}\left(\frac{x}{L}\right) + 1\right)$$
$$\equiv W\left(\frac{1}{2}\left(\frac{x}{L}\right)^3 - \frac{3}{2}\left(\frac{x}{L}\right) + 1\right) \tag{6.46}$$

Now, if we want to write the strain or complementary energies for the cantilever beam in the forms developed in Section 6.2 (i.e., as in Eq. (6.15) or Eq. (6.26)), the only way we can get there from the preceding continuous results is to formally evaluate the continuous model for the strain energy. That is, we will find that the strain energy for a bent beam can be written as

$$U = \frac{1}{2}\int_0^L EI(w''(x))^2 dx \tag{6.47}$$

If we substitute the second of Eqs. solution (6.46) into this integral, we will find

$$U = \frac{1}{2}\frac{3EI}{L^3}W^2 = \frac{1}{2}kW^2 \tag{6.48a}$$

which parallels Eq. (6.26), or as the (equal) complementary energy,

$$U = \frac{1}{2}fP^2 = \frac{1}{2}\frac{L^3}{3EI}P^2 = U^* \tag{6.48b}$$

which parallels Eq. (6.15). In Eqs. (6.48) we have used the stiffness and flexibility coefficients for this beam, k and f. However, the important point is that we were able to achieve these results – and by extension, results for more complicated problems – only by invoking continuous models to describe the specifics of the geometry and loading of the cantilever. Unfortunately, there is no general result or unifying theorem that allows us to go directly from an idealization of a real, continuous problem into the elegant and concise formalisms developed in Section 6.2. However, energy-based principles allow us to handle the transition from continuous to discrete and, at the same time, allow us to develop consistent continuous models for specific idealizations.

6.3.1 Minimum Total Potential Energy in Three Dimensions

We now present the principle of minimum total potential energy in general terms in three dimensions. We present statements of the principle in two forms, first in extended form for Cartesian coordinates and then in matrix notation.

We start by requiring that the elastic properties of the bodies or structures we are modeling can be represented by the following *strain energy density*:

$$U_0 = \frac{\nu E}{2(1+\nu)(1-2\nu)}(\varepsilon_{xx} + \varepsilon_{yy} + \varepsilon_{zz})^2 + G\left(\varepsilon_{xx}^2 + \varepsilon_{yy}^2 + \varepsilon_{zz}^2\right)$$
$$+ \frac{G}{2}\left(\gamma_{xy}^2 + \gamma_{xz}^2 + \gamma_{yz}^2\right) \tag{6.49}$$

in which we have written the shear strains in their engineering formulation (e.g., $\gamma_{xy} = 2\varepsilon_{xy}$). To further simplify Eq. (6.49), we introduce two new symbols. One is an elastic constant, called the *Lamé constant,*

$$\lambda = \frac{\nu E}{(1+\nu)(1-2\nu)} \tag{6.50}$$

The second is the sum of the normal strains, called the *dilatation*, which is the relative change in volume of the element $dxdydz$ (see Problem 6.3),

$$e = \varepsilon_{xx} + \varepsilon_{yy} + \varepsilon_{zz} \tag{6.51}$$

Then, we can write the strain energy density in three dimensions as

$$U_0 = \frac{\lambda}{2}e^2 + G\left(\varepsilon_{xx}^2 + \varepsilon_{yy}^2 + \varepsilon_{zz}^2\right) + \frac{G}{2}\left(\gamma_{xy}^2 + \gamma_{xz}^2 + \gamma_{yz}^2\right) \tag{6.52}$$

Equation (6.52) is a *quadratic form* because all of the strain terms have exponents of 2, and it is *positive definite* because it is always positive for all nonzero values of the strains and zero when the strains are zero. Further, the stresses and strains in our elastic solid are related through the strain energy density according to

$$\begin{aligned}
\sigma_{xx} &= \frac{\partial U_0}{\partial \varepsilon_{xx}}, & \sigma_{yy} &= \frac{\partial U_0}{\partial \varepsilon_{yy}}, & \sigma_{zz} &= \frac{\partial U_0}{\partial \varepsilon_{zz}} \\
\tau_{xy} &= \frac{\partial U_0}{\partial \gamma_{xy}}, & \tau_{xz} &= \frac{\partial U_0}{\partial \gamma_{xz}}, & \tau_{yz} &= \frac{\partial U_0}{\partial \gamma_{yz}}
\end{aligned} \tag{6.53}$$

The corresponding three-dimensional stress–strain laws are found by substituting the strain energy density (6.52) into the stress definitions (6.53). Now, having an appropriate strain energy density (6.52), we can write the total strain energy in the elastic structure or body as

$$U = \int_{Vol} U_0\, dV = \int_{Vol} U_0\, dxdydz \tag{6.54}$$

We also assume that the body or structure has loads applied to its surface that are represented by applied *stress vectors* or *tractions*, $\bar{T}_x^{(n)}$, $\bar{T}_y^{(n)}$, and $\bar{T}_z^{(n)}$, where the subscripts denote the directions of the force per unit area of a surface S with a unit normal vector $\boldsymbol{n}$. Thus, the potential of the applied surface loads is

$$V_{Surface} = -\int_{S_p} \left(\bar{T}_x^{(n)} u + \bar{T}_y^{(n)} v + \bar{T}_z^{(n)} w\right) dS \tag{6.55}$$

where u, v, and w are the displacement components of the surface in the x, y, and z directions.

Now we account for forces such as gravity or other mass- or volume-centered forces, B_x, B_y, and B_z, which are termed *body forces*, and which have the dimensions of force per unit volume,

$$V_{Body} = -\int_{Vol} (B_x u + B_y v + B_z w)\, dV \tag{6.56}$$

Then the total potential energy in three dimensions is the sum of Eqs. (6.54), (6.55), and (6.56), with the strain energy density given by Eq. (6.52):

$$\Pi = \int_{Vol} \left[\frac{\lambda}{2} e^2 + G\left(\varepsilon_{xx}^2 + \varepsilon_{yy}^2 + \varepsilon_{zz}^2\right) + \frac{G}{2}\left(\gamma_{xy}^2 + \gamma_{xz}^2 + \gamma_{yz}^2\right) \right] dV$$
$$- \int_{S_p} \left(\bar{T}_x^{(n)} u + \bar{T}_y^{(n)} v + \bar{T}_z^{(n)} w\right) dS - \int_{Vol} \left(B_x u + B_y v + B_z w\right) dV \qquad (6.57)$$

In analogy with the calculations we have done for the discrete models (cf. Chapter 3) and the axially loaded bar (cf. Section 4.2), we can now examine the total variation of Eq. (6.57). We would introduce a variation of an *admissible displacement field*, that is, a set of small yet arbitrary displacements that are continuous and violate neither geometric boundary conditions nor constraints. Then, after performing the variation of the total potential energy (6.57) for the admissible field, we would find that the result contains both linear and quadratic terms in the variations of strain. The second-order terms comprise the *second variation of the total potential*, and it can be shown to be a positive-definite function of the strain variations that correspond to the displacement variations:

$$\delta^{(2)}\Pi \equiv \int_{Vol} \left[\begin{array}{l} \frac{\lambda}{2}(\delta e)^2 + G(\delta\varepsilon_{xx}\delta\varepsilon_{xx} + \delta\varepsilon_{xx}\delta\varepsilon_{xx} + \delta\varepsilon_{xx}\delta\varepsilon_{xx}) \\ + \frac{G}{2}(\delta\gamma_{xy}\delta\gamma_{xy} + \delta\gamma_{xz}\delta\gamma_{xz} + \delta\gamma_{yz}\delta\gamma_{yz}) \end{array} \right] dV \geq 0 \qquad (6.58)$$

Thus, the vanishing of the first variation will correspond to a *minimum* of the total potential energy.

The first variation can be calculated and, after invoking the stress–strain law, cast in the following form:

$$\delta^{(1)}\Pi = \int_{Vol} [\sigma_{xx}\delta\varepsilon_{xx} + \sigma_{yy}\delta\varepsilon_{yy} + \sigma_{zz}\delta\varepsilon_{zz} + \tau_{xy}\delta\gamma_{xy} + \tau_{xz}\delta\gamma_{xz} + \tau_{yz}\delta\gamma_{yz}]dV$$
$$- \int_{S} \left(\bar{T}_x^{(n)} u + \bar{T}_y^{(n)} v + \bar{T}_z^{(n)} w\right) dS - \int_{Vol} \left(B_x u + B_y v + B_z w\right) dV \quad (6.59)$$

Note that we have substituted the total area S for the area S_p in Eq. (6.59) because the displacements are prescribed on S_u, so their variations vanish there. Equation (6.59) can be integrated by parts using some well-known theorems of vector calculus:

$$\delta^{(1)}\Pi = \int_{Vol} \left\{ \begin{array}{l} \left(\frac{\partial \sigma_{xx}}{\partial x} + \frac{\partial \tau_{xy}}{\partial y} + \frac{\partial \tau_{xz}}{\partial z} + B_x\right)\delta u \\ + \left(\frac{\partial \tau_{yx}}{\partial x} + \frac{\partial \sigma_{yy}}{\partial y} + \frac{\partial \tau_{yz}}{\partial z} + B_y\right)\delta v \\ + \left(\frac{\partial \tau_{zx}}{\partial x} + \frac{\partial \tau_{zy}}{\partial y} + \frac{\partial \sigma_{zz}}{\partial z} + B_z\right)\delta w \end{array} \right\} dxdydz$$
$$- \int_S \left(\bar{T}_x^{(n)} - \sigma_{xx}n_x - \tau_{xy}n_y - \tau_{xz}n_z\right)\delta u dS$$
$$- \int_S \left(\bar{T}_y^{(n)} - \tau_{yx}n_x - \sigma_{yy}n_y - \tau_{yz}n_z\right)\delta v dS$$
$$- \int_S \left(\bar{T}_x^{(n)} - \tau_{zx}n_x - \tau_{zy}n_y - \sigma_{zz}n_z\right)\delta w dS \qquad (6.60)$$

We now make the first variation vanish by setting Eq. (6.60) to zero, which yields the equilibrium equations for the stresses in our elastic body,

$$\begin{aligned}\frac{\partial \sigma_{xx}}{\partial x}+\frac{\partial \tau_{xy}}{\partial y}+\frac{\partial \tau_{xz}}{\partial z}+B_x=0\\ \frac{\partial \tau_{yx}}{\partial x}+\frac{\partial \sigma_{yy}}{\partial y}+\frac{\partial \tau_{yz}}{\partial z}+B_y=0\\ \frac{\partial \tau_{zx}}{\partial x}+\frac{\partial \tau_{zy}}{\partial y}+\frac{\partial \sigma_{zz}}{\partial z}+B_z=0\end{aligned} \tag{6.61}$$

and the dual boundary conditions on the body's surface where we prescribe tractions or stresses (S_p) or we prescribe or constrain displacements (S_u),

$$\begin{aligned}\left(\bar{T}_x^{(n)}-\sigma_{xx}n_x-\tau_{xy}n_y-\tau_{xz}n_z\right)\bullet\delta u=0\\ \left(\bar{T}_y^{(n)}-\tau_{yx}n_x-\sigma_{yy}n_y-\tau_{yz}n_z\right)\bullet\delta v=0\\ \underbrace{\left(\bar{T}_y^{(n)}-\tau_{zx}n_x-\tau_{zy}n_y-\sigma_{zz}n_z\right)}_{\text{On } S_p}\bullet\underbrace{\delta w}_{\text{On } S_u}=0\end{aligned} \tag{6.62}$$

Thus, the extreme value of the total potential energy (6.57), sought as a function of an *admissible displacement field* (small in magnitude but compatible in shape), is in fact a minimum, and this minimum is equivalent – meaning both *necessary and sufficient* – to the statement that the corresponding stress field given in Eqs. (6.53) satisfies the *equations of equilibrium* (Eqs. (6.61)).

We can also write corresponding results in matrix notation, although it requires new notation for stress and strain vectors. The stress–strain law is written in terms of a symmetric constitutive matrix $[D]$, that is,

$$\{\sigma\}\equiv[D]\{\varepsilon\} \tag{6.63}$$

where the stress and strain components are cast as 6×1 column vectors:

$$\{\sigma\}=\begin{Bmatrix}\sigma_{xx}\\ \sigma_{yy}\\ \sigma_{zz}\\ \tau_{yz}\\ \tau_{xz}\\ \tau_{xy}\end{Bmatrix}\equiv\begin{Bmatrix}\sigma_1\\ \sigma_2\\ \sigma_3\\ \sigma_4\\ \sigma_5\\ \sigma_6\end{Bmatrix},\quad \{\varepsilon\}=\begin{Bmatrix}\varepsilon_{xx}\\ \varepsilon_{yy}\\ \varepsilon_{zz}\\ \gamma_{yz}\\ \gamma_{xz}\\ \gamma_{xy}\end{Bmatrix}\equiv\begin{Bmatrix}\varepsilon_1\\ \varepsilon_2\\ \varepsilon_3\\ \varepsilon_4\\ \varepsilon_5\\ \varepsilon_6\end{Bmatrix} \tag{6.64}$$

The strain energy that corresponds to the stress–strain law (6.63) is

$$\begin{aligned}U&=\frac{1}{2}\int_{Vol}\{\sigma\}^T\{\varepsilon\}dV\\ &=\frac{1}{2}\int_{Vol}\{\varepsilon\}^T[D]\{\varepsilon\}dV\end{aligned} \tag{6.65}$$

where the transposition of the stress vector is required in order that it can properly (i.e., conformably) multiply the strain vector, and $[D]^T=[D]$ because $[D]$ is a symmetric matrix (see the Appendix for a short introduction to matrix manipulation).

The potential of the applied surface load vector $\{\bar{T}^{(n)}\}$ and body force vector $\{B\}$ as they act through a displacement vector $\{u\}$ is a straightforward extrapolation of the extended forms (6.55) and (6.56):

$$V = -\int_S \{\bar{T}^{(n)}\}^T \{u\} dS - \int_{Vol} \{B\}^T \{u\} dV \tag{6.66}$$

so that the total potential energy in matrix form is the sum of Eqs. (6.65) and (6.66):

$$\Pi = \frac{1}{2}\int_{Vol} \{\varepsilon\}^T [D] \{\varepsilon\} dV - \int_S \{\bar{T}^{(n)}\}^T \{u\} dS - \int_{Vol} \{B\}^T \{u\} dV \tag{6.67}$$

whose first variation appears as

$$\delta^{(1)}\Pi = \int_{Vol} \{\delta\varepsilon\}^T [D] \{\varepsilon\} dV - \int_S \{\bar{T}^{(n)}\}^T \{\delta u\} dS - \int_{Vol} \{B\}^T \{\delta u\} dV \tag{6.68}$$

and whose second variation can be shown to be

$$\delta^{(2)}\Pi = \int_{Vol} \{\delta\varepsilon\}^T [D] \{\delta\varepsilon\} dV \geq 0 \tag{6.69}$$

which is clearly positive definite. Thus, the vanishing of the first variation (6.68) corresponds to a minimum of the total potential energy (6.67).

There are two ways to evaluate what happens when the matrix form of the first variation (6.68) is set to zero. One, the *continuous* approach, is the multivariable vector calculus approach just completed. This approach is hard to do in general terms because it is hard to manipulate the strain-displacement relations. Also, it wouldn't add much to our computational ability.

The second approach involves *discretizing* both the partial derivatives and the volume integrals that appear in Eq. (6.68). Discretizing this matrix form of the first variation of the total potential energy is the starting point for the finite element method, which we describe somewhat in Section 8.3.3 but which we will not pursue any further in this text.

6.3.2 Virtual Work and Reciprocity in Three Dimensions

We now want to note that the minimum total potential energy principle, as stated in Eqs. (6.59) and (6.68), can also be viewed as strongly related to the principle of virtual work. In extended nomenclature first, consider the last two integrals in Eq. (6.59) from a slightly different perspective, namely, that the body forces (B_x, B_y, B_z) and surface tractions ($\bar{T}_x^{(n)}$, $\bar{T}_y^{(n)}$, $\bar{T}_z^{(n)}$) on S_p are applied external loads that do work on an elastic body through small, arbitrary, *virtual displacements* that are consistent with (i.e., do not violate) any prescribed conditions on the displacements (prescribed on S_u). The *external virtual work* thus performed by the applied loads acting through the virtual displacements is

$$\begin{aligned} \delta W_{virtual} = &\int_S (\bar{T}_x^{(n)}\delta u + \bar{T}_y^{(n)}\delta v + \bar{T}_z^{(n)}\delta w) dS \\ &+ \int_{Vol} (B_x \delta u + B_y \delta v + B_z \delta w) dV \end{aligned} \tag{6.70}$$

The surface over which we integrate is extended to the entire surface S, that is, we include S_u because the variations of prescribed displacements are zero.

The first integrand in Eq. (6.59) can be shown to equal the *internal virtual* work of an equilibrium system of stresses acting through a admissible virtual strains derived from admissible virtual displacements:

$$\delta W_{\text{virtual}} = \int_{Vol} [\sigma_{xx}\delta\varepsilon_{xx} + \sigma_{yy}\delta\varepsilon_{yy} + \sigma_{zz}\delta\varepsilon_{zz} + \tau_{xy}\delta\gamma_{xy} + \tau_{xz}\delta\gamma_{xz} + \tau_{yz}\delta\gamma_{yz}]\, dV \tag{6.71}$$

If we now compare our two statements of virtual work, Eqs. (6.70) and (6.71), we find that we have in fact written the *principle of virtual work*:

$$\int_S (\bar{T}_x^{(n)}\delta u + \bar{T}_y^{(n)}\delta v + \bar{T}_z^{(n)}\delta w) dS + \int_{Vol} (B_x\delta u + B_y\delta v + B_z\delta w)\, dV = \int_{Vol} [\sigma_{xx}\delta\varepsilon_{xx} + \sigma_{yy}\delta\varepsilon_{yy} + \sigma_{zz}\delta\varepsilon_{zz} + \tau_{xy}\delta\gamma_{xy} + \tau_{xz}\delta\gamma_{xz} + \tau_{yz}\delta\gamma_{yz}]\, dV \tag{6.72}$$

Thus, we have found that the *external virtual work* done by applied forces that are in equilibrium while acting through admissible (arbitrary, small, and compatible) displacements is equal to the *internal virtual work* done by the equilibrium stress field acting through admissible virtual strains corresponding to the admissible virtual displacements. That is, we have found that the external virtual work of a set of forces acting through an admissible but otherwise unrelated set of displacements *is equal to* the internal virtual work of the equilibrated stresses that correspond to the forces acting through an admissible but otherwise unrelated set of strains that correspond to the displacements.

Of particular note is the remarkable fact that *we have not invoked or used a constitutive law* in deriving the principle of virtual work. That is, the stresses in Eq. (6.72) are in no way related to the strains in Eq. (6.72). However, there is a continuum counterpart to the discrete versions of the reciprocal theorems derived in Section 6.2. For a linear elastic solid, in fact,

$$\int_S (\bar{T}_x^{(n)/A}u^B + \bar{T}_y^{(n)/A}v^B + \bar{T}_z^{(n)/A}w^B) dS + \int_{Vol} (B_x^A u^B + B_y^A v^B + B_z^A w^B) dV = \int_S (\bar{T}_x^{(n)/B}u^A + \bar{T}_y^{(n)/B}v^A + \bar{T}_z^{(n)/B}w^A) dS + \int_{Vol} (B_x^B u^A + B_y^B v^A + B_z^B w^A) dV \tag{6.73}$$

while in matrix form

$$\{\bar{T}^{(n)/A}\}^T\{u^B\} + \{\bar{B}^A\}^T\{u^B\} = \{\bar{T}^{(n)/B}\}^T\{u^A\} + \{\bar{B}^B\}^T\{u^A\} \tag{6.74}$$

Equations (6.73) and (6.74) both state that the work done by a system of forces A acting through displacements caused by a force system B must be equal to the work done by a system of forces B acting through displacements caused by a force system A. This result, a clear generalization of the results derived for the discrete formulations of structures problems in Section 6.2, is often called the *Maxwell–Betti reciprocal theorem.*

6.4 Minimum Total Complementary Energy Principle

We will now outline the principles of minimum total complementary energy and of complementary virtual work as counterparts to the total potential energy results. We will provide less detail than before because the derivations are very similar. Also, the total potential energy methods – and the displacement (or stiffness) approach – most often form the basis of the approximation methods that are used, whether applied analytically or in a computer-based algorithm.

6.4.1 Complementary Virtual Work in Three Dimensions

In contrast to the work completed in Section 6.3, we now examine problems wherein we assume some displacements are prescribed at various points on an elastic body, and that the externally applied forces and the resulting internal stresses are allowed to vary. We do require that the force and stress variations satisfy equilibrium within the body and any imposed surface traction boundary conditions. Thus, variations of the stresses ($\delta\sigma_{xx}\ldots\delta\tau_{xz}$), the body forces ($\delta B_x, \delta B_y, \delta B_z$), and of the surface tractions ($\delta T_x^{(n)}, \delta T_y^{(n)}, \delta T_z^{(n)}$) are required to satisfy the equilibrium equations (6.61) within the volume V, as well as boundary conditions (6.62) on the surface S_p. We can then define the *complementary external virtual work* as:

$$\delta W^*_{virtual} = \int_S \left(u\delta T_x^{(n)} + v\delta T_y^{(n)} + w\delta T_z^{(n)}\right) dS + \int_{Vol} (u\delta B_x + v\delta B_y + w\delta B_z)\, dV \tag{6.75}$$

where the (u, v, w) form any set of admissible displacements. Further, the surface over which we integrate can be extended to the entire surface S, that is, we include S_p because the variations of any prescribed forces will automatically vanish. Then, in a process that parallels the discussion of Section 6.3.2, we can write the *principle of complementary virtual work*:

$$\int_S \left(u\delta T_x^{(n)} + v\delta v T_y^{(n)} + w\delta T_z^{(n)}\right) dS + \int_{Vol} (u\delta B_x + v\delta B_y + w\delta B_z)\, dV = \int_{Vol} [\varepsilon_{xx}\delta\sigma_{xx} + \varepsilon_{yy}\delta\sigma_{yy} + \varepsilon_{zz}\delta\sigma_{zz} + \gamma_{xy}\delta\tau_{xy} + \gamma_{xz}\delta\tau_{xz} + \gamma_{yz}\delta\tau_{yz}]\, dV \tag{6.76}$$

Thus, the *complementary external virtual work* done by admissible variations of the external forces (in equilibrium at the boundaries) while acting through arbitrary displacements is equal to the *internal complementary virtual work* done by the admissible variations of the stresses (in equilibrium within the interior) acting through arbitrary strains corresponding to arbitrary displacements. We can show that the resulting displacement and strain fields that result from applying the principle of complementary virtual work are the equations of compatibility, which the displacements and strains must satisfy.

Thus, in a manner symmetric to the principle of virtual work, which produces equilibrium equations for compatible variations in displacement, the principle of complementary virtual work will produce compatibility conditions for equilibrated variations of the stress field. Similarly, *we have not invoked or used a constitutive law* in deriving the principle of complementary virtual work. That is, the strains in Eq. (6.76) are not related to the stresses in Eq. (6.76).

6.4.2 Minimum Total Complementary Energy in Three Dimensions

The principle of complementary virtual work embodied in Eq. (6.76) can be used, with an appropriate constitutive law, to derive a principle of minimum total complementary energy. The constitutive law assumes that the complementary energy is a function of the stresses that represents the area above a stress–strain curve. It complements the strain energy, which is the area below the stress–strain curve. We thus assume a three-dimensional complementary energy density such that for all components of stress and strain,

$$\begin{aligned} \varepsilon_{xx} &= \frac{\partial U_0^*}{\partial \sigma_{xx}}, & \varepsilon_{yy} &= \frac{\partial U_0^*}{\partial \sigma_{yy}}, & \varepsilon_{zz} &= \frac{\partial U_0^*}{\partial \sigma_{zz}} \\ \gamma_{xy} &= \frac{\partial U_0^*}{\partial \tau_{xy}}, & \gamma_{xz} &= \frac{\partial U_0^*}{\partial \tau_{xz}}, & \gamma_{yz} &= \frac{\partial U_0^*}{\partial \tau_{yz}} \end{aligned} \tag{6.77}$$

which we can substitute into Eq. (6.76) to find

$$\begin{aligned} &\int_S \left(u\delta T_x^{(n)} + v\delta T_y^{(n)} + w\delta T_z^{(n)}\right) dS + \int_{Vol} (u\delta B_x + v\delta B_y + w\delta B_z) dV \\ &= \int_{Vol} \begin{bmatrix} \dfrac{\partial U_0^*}{\partial \sigma_{xx}}\delta\sigma_{xx} + \dfrac{\partial U_0^*}{\partial \sigma_{yy}}\delta\sigma_{yy} + \dfrac{\partial U_0^*}{\partial \sigma_{zz}}\delta\sigma_{zz} \\ + \dfrac{\partial U_0^*}{\partial \tau_{xy}}\delta\tau_{xy} + \dfrac{\partial U_0^*}{\partial \tau_{xz}}\delta\tau_{xz} + \dfrac{\partial U_0^*}{\partial \tau_{yz}}\delta\tau_{yz} \end{bmatrix} dV \\ &\equiv \delta \int_{Vol} U_0^* dV = \delta U^* \end{aligned} \tag{6.78}$$

In addition, in a manner similar to the way we defined the potential of the applied loads, we can define a *complementary potential function* as

$$V^* \equiv -\int_{S_u} \left(\bar{u}T_x^{(n)} + \bar{v}T_y^{(n)} + \bar{w}T_z^{(n)}\right) dS - \int_{Vol} (\bar{u}B_x + \bar{v}B_y + \bar{w}B_z)\, dV \tag{6.79}$$

where the $(\bar{u}, \bar{v}, \bar{w})$ are a set of prescribed displacements. Then we define the *total complementary energy* as the sum of Eqs. (6.78) and (6.79):

$$\Pi^* \equiv U^* + V^* \tag{6.80}$$

so that a statement equivalent to Eq. (6.78) is simply that the *first variation of the total complementary energy must be zero*:

$$\begin{aligned} \delta^{(1)}\Pi^* &= \delta^{(1)}(U^* + V^*) = 0 \\ &= \int_{Vol} \begin{bmatrix} \dfrac{\partial U_0^*}{\partial \sigma_{xx}}\delta\sigma_{xx} + \dfrac{\partial U_0^*}{\partial \sigma_{yy}}\delta\sigma_{yy} + \dfrac{\partial U_0^*}{\partial \sigma_{zz}}\delta\sigma_{zz} \\ + \dfrac{\partial U_0^*}{\partial \tau_{xy}}\delta\tau_{xy} + \dfrac{\partial U_0^*}{\partial \tau_{xz}}\delta\tau_{xz} + \dfrac{\partial U_0^*}{\partial \tau_{yz}}\delta\tau_{yz} \end{bmatrix} dV \\ &\quad - \int_S \left(\bar{u}\delta T_x^{(n)} + \bar{v}\delta T_y^{(n)} + \bar{w}\delta T_z^{(n)}\right) dS \\ &\quad - \int_{Vol} (\bar{u}\delta B_x + \bar{v}\delta B_y + \bar{w}\delta B_z)\, dV \end{aligned} \tag{6.81}$$

And while we will not do so here, it is possible to show that the vanishing of the total complementary energy does, in fact, correspond to a *minimum* of the total complementary energy. The proof of this parallels that which we have shown for the principle of minimum total potential energy, only here it depends on the complementary energy being a positive definite function of the stresses.

6.5 Some Comments on the Energy Methods

This chapter has outlined the underlying "theory" material about elastic bodies and structures in fairly general terms. However, it is worth remembering that everything we have done in this chapter is, in some sense, a generalization of the specific energy applications we had developed for discrete models of structures and for axially loaded elements. And we will go on to develop similar models and solve structural problems for beams and frames. However, there are a few points worth making in a more or less conversational way because, while we have seen some and will see more applications of these particular points, their general development requires too much machinery for now. Thus, we will both recap some things already done and look ahead to their future application.

6.5.1 The Castigliano Theorems

We had developed the two Castigliano theorems for axially loaded bars in Sections 4.5, 4.6, and 5.2, and for discrete models of structures in Section 6.2. In fact, the general developments in Sections 6.2.1 and 6.2.2, and the relevant Castigliano's theorems expressed in Eqs. (6.28) and (6.32), are as general and as valid as needed for modeling linear elastic structures. This is because the two theorems are about calculating forces and displacements at discrete points, so that sooner or later, we have to represent the complementary and strain energies for linear structures in the forms of Eqs. (6.16) and (6.25), respectively.

One generalization is to permit the complementary and strain energy densities to represent *nonlinear* material behavior, for example, in one dimension,

$$\sigma = \sigma_0 \varepsilon^m \tag{6.82}$$

where m is an arbitrary exponent. There are problems we can solve (hint!) that could include nonlinear material behavior. How would such behavior affect the application of the Castigliano theorems? Not much. Remember that the specific forms for the complementary and strain energy densities developed in Section 6.2 were for linear elastic materials. The only modifications we would have to make in order to admit nonlinearities would be as follows.

For nonlinear complementary energy, we would only have to say that

$$U^* = U^*(P_i) \tag{6.83}$$

Then the first variation of the complementary energy would appear as

$$\delta^{(1)} U^* = \frac{\partial U^*(P_i)}{\partial P_i} \delta P_i \tag{6.84}$$

and the first Castigliano theorem would show the calculation of the displacement under the specified load to be

$$u_i = \frac{\partial U^*(P_i)}{\partial P_i} \tag{6.85}$$

For nonlinear strain energy, we would only have to say that

$$U = U(u_i) \tag{6.86}$$

Then the first variation of the strain energy would appear as

$$\delta^{(1)}U = \frac{\partial U(u_i)}{\partial u_i}\delta u_i \tag{6.87}$$

and the second Castigliano theorem would show the calculation of the force producing the specified displacement to be

$$P_i = \frac{\partial U(u_i)}{\partial u_i} \tag{6.88}$$

It is clear that the differences between Eqs. (6.28) and (6.85) and Eqs. (6.32) and (6.88) are that the latter forms admit more general functional representations than the quadratic energies that appear for linear structural models. However, as we have said before, even for linear problems the trick is to be able to calculate the flexibility and stiffness coefficients for a given problem. Thus, for structural modeling, whether linear or nonlinear, we still have to know how to *idealize* a structure and how to *discretize* its important aspects so that we can do the calculations we need to do to predict structural behavior.

6.5.2 Direct and Indirect Approaches, and Approximate Solutions

In Section 4.4 we developed exact and approximate solutions to several problems involving axially loaded bars. We outlined an *indirect method*, in which some idealizing assumptions were used to formulate an energy integral that was variationally extremized (usually minimized) to find the governing equation(s) and boundary conditions appropriate to the specific idealization. We also outlined a *direct method* wherein we substituted approximate or trial solutions directly into the energy integral and sought an approximate extremum by minimizing the idealized and discretized energy with respect to arbitrary coefficients in the approximate solution. We also made some assertions (cf. Section 4.4.4) about different kinds of trial functions, and about the kinds of boundary conditions that trial solutions must minimally satisfy.

In that particular discussion we also mentioned that the direct methods we were illustrating represent the foundations of much of the numerical work on structural analysis, such as the algorithms using the finite element method. There is a host of theoretical work that extends the general variational principles just derived into the numerical domain, and an even larger host of applied work solving all manner of structural engineering problems. (There is also extensive, parallel work in computational approaches to other domains, such as geophysical fluid dynamics and electromagnetic field theory – although it appears that the FEM began with the analysis of aircraft structures.) The commentary we have made in this regard (again, see Section 4.4) is valid in general, and we will show further applications and illustrations as we move on.

Perhaps the single most important feature that we can learn from the theory underlying the variational methods is that a *minimum* of the total potential energy results when an admissible displacement field – that is, a field that is continuous, compatible, and consistent with all geometric boundary conditions – is used variationally to extremize that total potential. That the energy extremum is a minimum follows from the fact that the strain energy is a positive definite function of the strains. Further, it is true for both the indirect and the direct (or Rayleigh–Ritz) methods. This also means that there are both techniques and circumstances for which we can put *bounds* on solutions. For example, in buckling and vibration problems (which we do not treat), the critical solutions are *eigenvalues* called, respectively, buckling loads and natural frequencies. These eigenvalues can be bounded from above, because of the minimization property of the variational process, so that we can estimate how well we are doing with some approximate solutions by looking at trends of eigenvalue magnitudes as we refine our approximations.

However, we must be careful not to expect – or promise – too much. What is true for eigenvalues, for example, is *not true* for point-by-point values of stress or deflection in an elastic structure. Numerical refinements based on assessing values of the total potential energy may be possible for certain approximations, but they guarantee very little in terms of local results. Perhaps the main point to keep in mind is that while there is deeper theory and there are very sophisticated applications, *we are still in the modeling business*, so we should always reflect both on what we are trying to model and the meaning of the results we obtain with any particular model.

6.5.3 The Force (Flexibility)–Displacement (Stiffness) Duality

Finally, we comment on the fact that in still another dimension, which we might call the *basic approach paradigm*, we have now seen in general what we had demonstrated in particular. We have developed the principle of *minimum total potential energy*, both for discrete generalizations and for continuous elastic bodies, and we have seen that it is based on displacement variations (theory) or displacement approximations (applications). When we work in the displacement paradigm, the structural and geometrical parameters are normally expressed in terms of stiffness coefficients, which is why the paradigm is often called the *displacement (stiffness) approach*.

We have also developed the principle of *minimum total complementary energy*, both for discrete generalizations and for continuous elastic bodies, and we have seen that it is based on force and stress variations (theory) or force and stress approximations (applications). When we work in the force paradigm, the structural and geometrical parameters are normally expressed in terms of flexibility coefficients, which is why the paradigm is often called the *force (flexibility) approach*.

We summarize the two paradigms in Table 6.1, and we repeat that it is the displacement (stiffness) paradigm that is most often used, both for analytical approximations and in numerical work. However, we can now explore in greater detail why this is the case.

We have already applied the force (flexibility) approach to two indeterminate problems, an elastic bar in Section 4.6 and an internally indeterminate truss in Section 5.2. However, while both of those solutions were based on the force paradigm, they both made good use of our intuitions about displacements. In fact, in both cases we were applying the technique that civil engineers have long called the *principle of consistent deformation*. As applied, this principle basically requires us to enforce any appropriate and evident compatibility

Table 6.1. The summary of the two structural modeling paradigms

ANALYSIS VARIABLE:	Displacement	Force
ENERGY FORM:	Potential: U, Π	Complementary: U^*, Π^*
RESULTING EQUATION:	Equilibrium	Compatibility
CASTIGLIANO THEOREM:	First Theorem: $P_i = \dfrac{\partial U}{\partial u_i}$	Second Theorem: $u_i = \dfrac{\partial U^*}{\partial P_i}$
STRUCTURAL PARAMETER:	Stiffness: k_{ij}	Flexibility: f_{ij}

and continuity conditions. Ironically, this suggests that we may have good intuition about deformation and displacement, but perhaps we don't do as well in estimating force and stress distributions.

In fact, this is the reason that the displacement or stiffness paradigm is used predominantly in the finite element business, as well as with other approximate techniques. Simply put, it is easier to guess at or estimate the shape of a structure under a load than it is to "guesstimate" a force or stress distribution within the interior of that structure. We will see this much more clearly when we do beams and frames than we have when doing bars and trusses.

There is another point that reinforces the use of the displacement paradigm. Recall that when we examining the indeterminate three-segment bar in Section 4.6, we formulated the resulting equations for the displacements at the nodes in matrix form, first in Eqs. (4.73). We were then able to manipulate the matrix equations easily so as to move directly to a solution for the two internal nodes. That is, we were able to partition the problem fairly easily because we could partition and otherwise manipulate the matrix representation of the displacements at the nodes within the structure. This, too, contributes to the predominance in use of the displacement (stiffness) paradigm in structural engineering.

Problems

6.1 Why is the strain energy density of Eq. (6.49) positive definite?

6.2 Apply the definition of stress in Eq. (6.53) to the strain energy density of Eq. (6.52) to obtain the three-dimensional Hooke's law.

6.3 Show first that the definition of normal strain (e.g., Eq. (4.2)) can be used to calculate the change in length of a line element that was, say, originally of length dx:

$$\left(1 + \frac{\partial u}{\partial x}\right) dx = (1 + \varepsilon_{xx})dx$$

Then extend this result for all three dimensions to show that the dilatation of Eq. (6.51) is the relative volume change of the element $dxdydz$.

6.4 Write the strain energy density of Eq. (6.52) explicitly in terms of the stresses and confirm that the result is the complementary energy density of Eqs. (6.77) and (6.78).

6.5 Calculate the total variation of the complementary energy functional of Eqs. (6.80) and (6.81) and show that U^* is positive definite and that the extreme value of Π^* corresponds to a minimum.

6.6 Write the principle of minimum total complementary energy (Eq. (6.81)) in matrix form, after Eqs. (6.67) and (6.68). Use a symmetric matrix $[D^*]$ as the complementary energy counterpart of the matrix $[D]$ used to describe the strain energy.

6.7 For a two-dimensional (x, y) problem, absent any body forces, show that an assumed stress distribution of the form

$$\sigma_{xx} = \frac{\partial^2 \varphi}{\partial y^2}, \qquad \sigma_{yy} = \frac{\partial^2 \varphi}{\partial x^2}, \qquad \tau_{xy} = -\frac{\partial^2 \varphi}{\partial x \partial y}$$

satisfies planar equilibrium. The solution $\varphi(x, y)$ is the two-dimensional specialization of the *Beltrami stress function.*

6.8 Use the solution given in Problem 6.7 to minimize the total complementary energy and show that the resulting compatibility condition is

$$\frac{\partial^2 \varepsilon_{yy}}{\partial x^2} + \frac{\partial^2 \varepsilon_{xx}}{\partial y^2} - 2\frac{\partial^2 \varepsilon_{xy}}{\partial x \partial y} = 0$$

6.9 Use the strain–displacement relations (1.2) and (1.3) to express the result found in Problem 6.8 in terms of displacements. How does this new result bear on the conditions for compatibility in an elastic body?

6.10 Consider a beam pinned at $x = 0$ and $x = L$, on which concentrated loads are applied at interior points, namely, $P_1\delta_D(x - x_1)$ and $P_2\delta_D(x - x_2)$. Use the Maxwell–Betti reciprocal theorem (6.73) to show that the displacement at point $x = x_1$ due to the load P_2 is equal to the displacement at point $x = x_2$ due to the load P_1.

6.11 For a one-dimensional bar having an axial load P at its centerline, with both ends restrained from moving, find the displacement at the centerline of the bar in terms of P, A, σ_0, and m if its stress–strain law is given by Eq. (6.82).

6.12 Find the horizontal and vertical deflections of point C of the truss shown in Fig. 4.18 if the bar behaviors are governed by the stress-strain law of Eq. (6.82). Express the answers in terms of the bar lengths, a common cross sectional area A, P_x, P_y, σ_0, and m.

The Sears Tower in Chicago, Illinois, is the second (1974) of the innovative structural towers designed by Fazlur Kahn of Skidmore, Owings, and Merrill. Kahn wanted this design to be clearly distinct from his Hancock Center (Fig. 1.3c). The tower is thus a bundle of nine tubes, even though each is a separate, structurally complete tower. (Photo by J. Wayman Williams.)

7

Beams: Transversely Loaded Members

We have said before (cf. Section 2.1) that structures can be considered to be divided roughly into two categories. In one category, the applied loads and the resulting stresses and deflections lie in the same plane or surface, the plane or surface of the structure. The axially loaded elements that we treated in Chapter 4 are one-dimensional structures exactly of this type, and their utility arises because we can assemble such elements in configurations that act in much more general (and useful) ways, for example, as trusses.

In the second type of structure, the applied loads are perpendicular to a plane or surface, the deflection will also manifest itself as movement that is normal to the surface, but the stresses that support the applied loads and the resulting displacements appear both in the same direction as the applied loads and within the plane of the structure. The simplest one-dimensional structure of this kind is the elementary or Euler–Bernoulli beam, the modeling of which we turn to now.

The approach we take will be similar to that taken for bars. After briefly reviewing beam theory in Section 7.1, we begin our general derivation of beam theory by formulating a kinematic model of an ideal elastic beam. Then we will use the principle of minimum total potential energy both to derive the governing equation of equilibrium and the corresponding boundary conditions and to develop approximate solutions to various beam problems.

7.1 A Brief Review of Engineering Beam Theory

In order to set our notation and provide some familiar guideposts, we will first review some of the more important results of the engineering theory of beams as they are outlined in the theory of strength of materials. The beam is a long and slender element, typically designed to carry loads that are placed on its top surface, with the beam itself being supported by reactions to earth or ground. For beams whose loading and geometry are symmetric with respect to the y-axis (outwardly normal to the plane of the paper), we describe the stress distribution in the beam in terms of the bending stress σ_b, which is related to the moment resultant $M(x)$ by the classic formula of beam theory

$$\sigma_b = \frac{M(x)z}{I} \tag{7.1}$$

In Eq. (7.1), the coordinate z is positive measured downward and I represents the second moment of the beam's cross-sectional area about the centroidal y-axis. The moment $M(x)$ can be viewed (1) as a stress resultant that represents the integrated effect of the bending stress across the beam's cross-sectional area (as we will demonstrate in Section 7.2) or

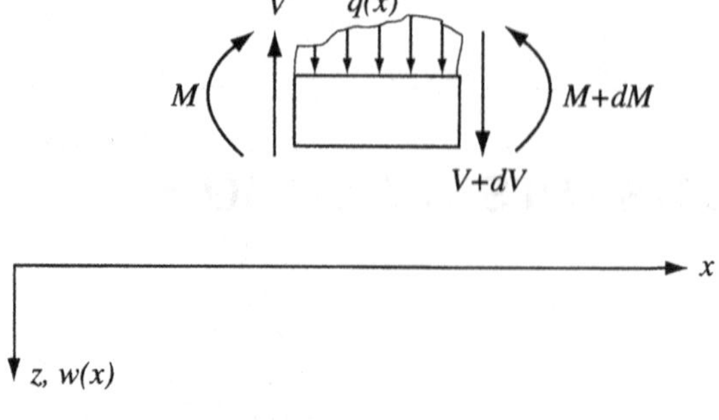

Figure 7.1. Some strength-of-materials nomenclature for elastic beams.

(2) together with the transverse shear force $V(x)$, as one of the resultants required to keep the beam in equilibrium as an arbitrary load per unit length $q(x)$ is applied along the top surface. In fact, a simple look at an element of the beam makes it quite clear that the vertical shear forces alone cannot maintain the element in equilibrium (cf. Fig. 7.1). Now, the (hopefully!) familiar equations for transverse force and moment equilibrium are, repectively,

$$\begin{aligned} \frac{dV(x)}{dx} + q(x) &= 0 \\ \frac{dM(x)}{dx} &= V(x) \end{aligned} \tag{7.2}$$

Equations (7.2) can be *uncoupled*, that is, we can eliminate the shear force between the two equations to find a single equation of equilibrium expressed in terms of the moment:

$$\frac{d^2M(x)}{dx^2} + q(x) = 0 \tag{7.3}$$

Equation (7.3) tells us how the moment varies along the length of a beam, depending of course on how the load varies along the length. Generally, however, we also want to know how the beam deflects or moves under the load. As we will show in Section 7.2, the moment can be related to the second derivative of the deflection $w(x)$ according to

$$M(x) = -EI\frac{d^2w(x)}{dx^2} \tag{7.4}$$

The second derivative of the displacement in Eq. (7.4) is a good approximation of the *curvature* of the deflected shape of the beam, as long as the slope $w'(x)$, the derivative with respect to x of $w(x)$, is small compared to 1. Thus, Eq. (7.4) is a *moment-curvature relation* that tells us how the deflected shape responds or curves with respect to the moment resultant produced in the beam.

Finally, we can eliminate the moment between Eqs. (7.3) and (7.4) to yield the equation of equilibrium expressed in terms of the deflection $w(x)$:

$$EI\frac{d^4w(x)}{dx^4} + q(x) = 0 \tag{7.5}$$

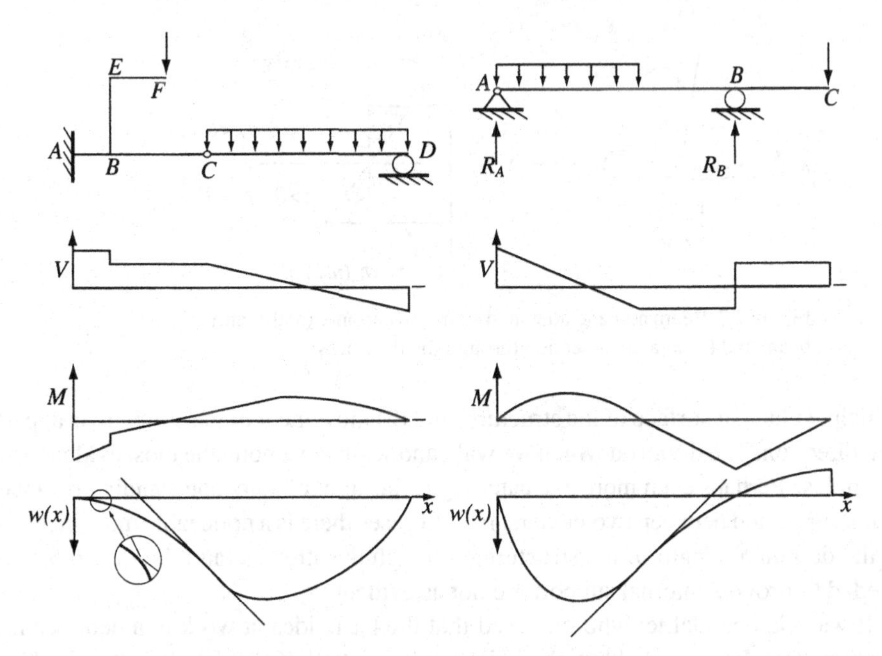

Figure 7.2. Beam shears, bending moments, and deflected shapes.

Thus, by integrating Eq. (7.5) and applying appropriate boundary conditions to ascertain the values of the four arbitrary constants that come with the general solution to a fourth-order differential equation, we can find the deflected shape that corresponds to a given loading on the beam.

In addition to solving differential equations for the moment, Eq. (7.3), and the deflection, Eq. (7.5), we also draw shear and moment diagrams in order to portray visually some aspects of how a beam responds to a given load. Some shear and moment diagrams are shown in Fig. 7.2, along with sketches of the deflected shapes. We bring these two elements together for beams because the visual nature of the deflected shape, taken together with the shear and moment diagrams, offer opportunities for seeing and understanding behaviors that are not so easily portrayed for an element like the axially loaded bar. Further, while we will now derive the beam equations in a rather mathematical way, it is well to remember that our derivation will begin with a model or a mental picture of deflection and stress behavior.

7.2 Deriving Engineering Beam Theory With Potential Energy

We want to apply the principle of minimum total potential energy, in an indirect approach, to derive what is sometimes called the *technical* or *engineering theory of beams*. Thus, we need to formulate an idealized model of what we expect a beam to be and how we expect it to behave. In addition, since we intend to apply the total *potential* energy principle, we expect our idealization to be displacement based, that is, to be founded on an assumption about the type of deflection behavior we anticipate.

We know from our prior experience with beams that we are interested in the kind of structural element where (1) external loads are applied in a direction perpendicular to the

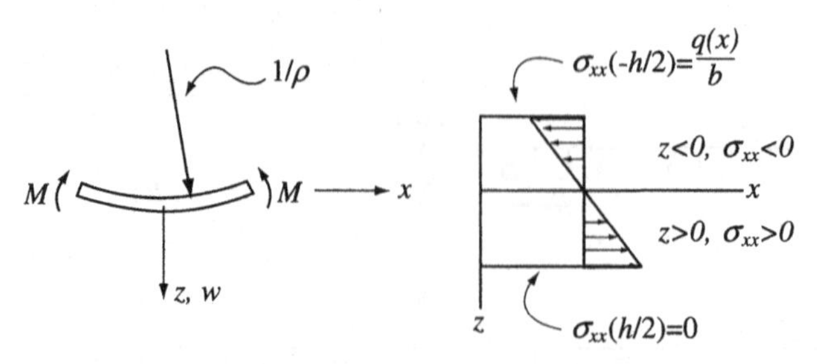

Figure 7.3. Beam nomenclature and sign conventions: (a) the entire beam and (b) at a cross section through the thickness.

principal plane or surface of the structure, and (2) movement of the structure is apparent in that direction. The floors on which we walk and work are among the most evident practical examples. Perhaps even more eye catching is the sight of someone standing on a wooden plank supported between two ladders. In both cases there is a noticeable movement normal to the dominant length of the structure, although the stresses (and their resulting forces) needed to provide internal support are not as evident.

It was Galileo Galilei who observed that the basic idea at work in a bent beam is the Archimedean lever, and although Galileo didn't get all of the details just right, his basic observation was right on. Thus, we are led to construct an idealization that would support some sort of lever action through the actual thickness of the structure, so that it is natural to think that the normal stress in the axial or longitudinal direction should be distributed through the thickness linearly (as shown in Fig. 7.3(b)). This suggests that the longitudinal normal strain should be distributed linearly, which further implies that the longitudinal displacement should also vary linearly through the thickness of the beam.

7.2.1 Kinematic Assumptions of Engineering Beam Theory

We establish a beam coordinate system as shown in Fig. 7.3. We assume that the beam is long and thin, and that it has a rectangular cross-sectional area whose thickness h and width b are small compared to the length L, that is,

$$h, b \ll L \tag{7.6}$$

In addition, we will assume that all loads applied on the top of the beam are uniform across the width, that is, in the y direction (or perpendicular to the plane of the paper). In fact, we will assume that nothing of much interest happens in the y direction, so that the displacement in that direction, $v(x, y, z)$, vanishes, as do all dependencies on, and all derivatives with respect to, the coordinate y.

We further assume that the longitudinal displacement $u(x, z)$ is dependent on the thickness and longitudinal coordinates only and that it varies linearly through the thickness, as follows (see Fig. 7.4):

$$u(x, z) = -zw'(x) \tag{7.7}$$

where $w(x)$ is the transverse displacement (which is assumed to vary only with the coordinate x), that is,

$$w(x, z) = w(x) \tag{7.8}$$

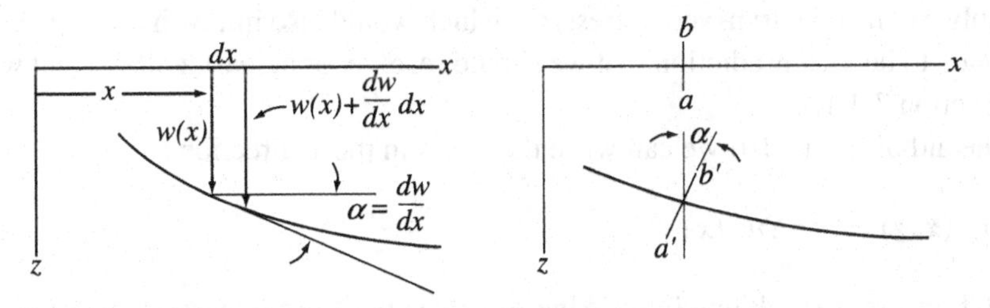

Figure 7.4. Displacement assumptions for an elastic beam.

In a more rigorous analysis these kinematic assumptions are shown to be parts of Taylor series expansions of the displacements in the z-coordinate. However, experience shows that we need keep only the linear term in z in Eq. (7.7) and only the zeroth-order term in Eq. (7.8). Equation (7.8) also signals the idea that the three-dimensional, physical object identified as a beam is being represented or modeled in terms of the displacement of points along a line along the x-axis, and in a direction transverse or normal to that line.

Having established an intuitively plausible displacement pattern, we now use the strain–displacement relations (1.2) and (1.3) to calculate the resulting strains. We then find that there is only one nontrivial strain component:

$$\varepsilon_{xx}(x, z) = -zw''(x) \tag{7.9}$$

We take particular note of the fact that the transverse shear strain γ_{xz} is zero. The vanishing of this shear strain means that a line originally perpendicular to the x-axis will remain so after the beam has deformed because no shear angle is produced between the x- and z-axes.

The assumptions embodied in Eqs. (7.7–7.9) are called the *Euler–Bernoulli hypothesis*, that is, *a line segment that was originally normal to the beam centerline will remain normal to the deformed centerline and will remain unstretched as the beam deforms.* In the context of two-dimensional structural models such as flat *plate* structures and *shells*, which are curved plates, the Euler–Bernouli hypothesis is sometimes called the *hairbrush hypothesis*.

7.2.2 Stresses and Stress Resultants for Engineering Beam Theory

Given our assumption that there is no loading in the y direction (normal to the plane of the paper) and that the beam is relatively thin in this direction, we would expect that the stress component σ_{yy} should be very small or almost zero. Similarly, while we know that over the loaded top surface we would have $\sigma_{zz}(x, -h/2) = -q(x)/b$, we also expect the stress component σ_{zz} in the thickness or z direction to be very small or essentially zero. Thus, we expect to find that the bending of beams is dominated by the axial bending stress σ_{xx}, so that the two remaining transverse stresses should be much smaller than the bending stress, that is, $\sigma_{yy}, \sigma_{zz} \ll \sigma_{xx}$. As a result, we can once again use a one-dimensional stress–strain law:

$$\sigma_{xx} = E\varepsilon_{xx} \tag{7.10}$$

Since all of the shear strains are zero, the stress–strain laws (cf. Problem 6.2) suggest that all of the shear stresses should correspondingly vanish. On the other hand, equilibrium appears to require (cf. Fig. 7.1) that we have a nontrivial transverse shear force, which would imply a nontrivial transverse stress τ_{xz}, which would also imply that $\gamma_{xz} \neq 0$. Thus, there appears to be a contradiction that we should resolve, or at least address, and we will do so in Section 7.4.1.

With the aid of Eq. (7.10) we can write the stress in the x direction as

$$\sigma_{xx}(x,z) = -zEw''(x) \tag{7.11}$$

We can define stress resultants by looking at values of stress integrated over the cross-sectional area of the beam on which they act. First, a straightforward integration of Eq. (7.11) over the cross section yields

$$\int_{-h/2}^{h/2} \sigma_{xx}(x,z)bdz = \int_{-h/2}^{h/2} [-zEw''(x)]bdz \equiv 0 \tag{7.12}$$

Thus, for a symmetric cross section, the axial stress (7.11) produces an axial stress resultant – which here is just the average of the axial stress – of zero.

We can define a *bending stress resultant or moment* by remembering Galileo's lever hypothesis and taking the first moment of Eq. (7.11):

$$M(x) \equiv \int_{-h/2}^{h/2} \sigma_{xx}(x,z)zbdz \tag{7.13}$$

which by virtue of Eq. (7.11) becomes

$$M(x) \equiv \int_{-h/2}^{h/2} [-zEw''(x)]zbdz = -EIw''(x) \tag{7.14}$$

where I is the second moment of the beam's cross-sectional area, also called the moment of inertia of the cross-sectional area. Note that the moment of inertia could depend on x. Further, the moment depends directly on the transverse displacement $w(x)$, which is why $w(x)$ is usually called the *bending deflection*.

It is now interesting to eliminate $w''(x)$ between Eqs. (7.11) and (7.14) to write the stress distribution in terms of the bending stress resultant, that is,

$$\sigma_{xx}(x,z) = \frac{zM(x)}{I} \tag{7.15}$$

This formula reflects our assumption that the bending stress distribution through the thickness is linear in the z-coordinate. Further, we see that the bending term looks just like the familiar result from the strength of materials, Eq. (7.1). We also see in Eq. (7.15) the origin of a sign convention for defining a positive moment acting on a beam's cross section (see Fig. 7.3). That is, a moment taken as positive for a beam that is bent so as to be *concave up* produces a bending stress that is tensile (positive) in the bottom-half of the cross section and compressive (negative) in the upper-half of the beam's cross section.

This is also consistent with the results as expressed in terms of the second derivative of the bending deflection. Recall from the calculus that the term $-w''(x)$ actually represents an approximation of the *curvature* of the deflected shape shown in Fig. 7.3, the more exact result being that

$$\frac{1}{\rho} = \frac{-w''(x)}{[1 + (w'(x))^2]^{3/2}} \cong -w''(x) \tag{7.16}$$

The approximate curvature appears in the right-hand sides of both Eqs. (7.16) and (7.14). It is valid if the slope of the deflected shape is sufficiently small that its square can be neglected compared to 1. This means, of course, that we can identify Eq. (7.14) as the *moment-curvature relation* for a bent elastic beam.

There is another stress resultant to define, namely, the *shear force* $V(x)$. Defining this resultant is a bit of a problem because, as we have already pointed out, our kinematic assumptions lead to the conclusion that there is no transverse shear strain, so that the consequent shear stress and resultant force should also vanish. However, equilibrium (and our experience of beams in the strength-of-materials approach) suggests that both the shear stress and force must be present for equilibrium to be satisfied. Thus, we define the shear force as

$$V(x) \equiv \int_{-h/2}^{h/2} \tau_{xz}(x, z) b dz \tag{7.17}$$

Now, we could do a more detailed analysis of the stresses through the thickness by using the two-dimensional equations of elasticity. If we did that for a beam with a square cross section, we would find that the transverse shear stress is related to the moment by

$$\tau_{xz} = \frac{M'(x)}{2I}\left(\left(\frac{h}{2}\right)^2 - z^2\right) \tag{7.18}$$

Equation (7.18) shows that the shear stress is distributed symmetrically and parabolically through the thickness, and that it takes on its maximum value at the center of the beam's cross section. Further, we can now calculate the shear force by substituting Eq. (7.18) into Eq. (7.17) and performing the indicated integration through the thickness, after which we obtain another result familiar from the strength of materials,

$$V(x) \equiv \frac{dM(x)}{dx} \tag{7.19}$$

Now, having defined the stress resultants and satisfied planar equilibrium (at the very least in some average sense) in the x direction, have we not completed the whole business? Do we need to use the total potential energy principle?

Yes, we do, for two reasons. The first is to develop the equilibrium equation and its matching boundary conditions in terms of the bending deflection. The second is to justify our modeling assumptions about the relative magnitudes of the various stress and strain components, which we will do in Section 7.4.

7.2.3 Strain and Total Potential Energies for Engineering Beam Theory

The starting point for the total potential energy formulation is the strain energy density for our model of an elastic beam. For the case where we have only one nontrivial strain,

$$U_0 = \frac{1}{2}\sigma_{xx}\varepsilon_{xx} = \frac{1}{2}E\varepsilon_{xx}^2 \tag{7.20}$$

In view of the strain–displacement result (7.9), the strain energy (7.20) becomes

$$
\begin{aligned}
U &\equiv \frac{1}{2}\int_0^L \int_{-h/2}^{h/2} E\varepsilon_{xx}^2(x,z)bdzdx \\
&= \frac{1}{2}\int_0^L \int_{-h/2}^{h/2} E(-zw''(x))^2 bdzdx \\
&= \frac{1}{2}\int_0^L EI(w''(x))^2 dx
\end{aligned}
\tag{7.21}
$$

To complete the total potential energy we need the potential of the applied loads. For our beam model we limit ourselves to normally applied loads acting on the top surface of the beam. Thus, in the most general case we would have

$$
V = -\int_0^L (q(x)/b)w(x,-h/2)bdx = -\int_0^L q(x)w(x)dx \tag{7.22}
$$

In formulating and completing Eq. (7.22) we have noted that the applied force is expressed as a line load or a load per unit length, so that the first integral represents the potential due to the pressure or stress it creates, integrated over the top surface. We have also explicitly recognized that the bending deflection is constant through the thickness by virtue of our kinematic assumptions.

Thus, the total potential energy for our elementary beam model is

$$
\Pi = U + V = \frac{1}{2}\int_0^L EI(w''(x))^2 dx - \int_0^L q(x)w(x)dx \tag{7.23}
$$

This result is very consistent in appearance and structure with all of the total potential energy formulations we have seen before, including discrete structural models (Eqs. (3.19) and (3.34)) and axially loaded bars (Eq. (4.13)).

7.2.4 Minimum Total Potential Energy for Engineering Beam Theory

Since we are now used to minimizing the total potential energy, we will apply here the operator technique to obtain the first variation of Eq. (7.23) and leave as an exercise the analysis that confirms the nature of the extremum defined by that first variation. So, for small variations of the bending deflection, $\delta w(x)$,

$$
\delta^{(1)}\Pi = \int_0^L EIw''(x)\delta w''(x)dx - \int_0^L q(x)\delta w(x)dx \tag{7.24}
$$

This looks much like the other first variations of the total potential energy that we have seen, save for the higher-order derivatives on the bending deflections. This means that we need to perform an additional round of integrating by parts, after assuming as usual that the delta and derivative operators commute. Thus, after the integrations we obtain

$$
\begin{aligned}
\delta^{(1)}\Pi &= [EIw''(x)\delta w'(x)]_0^L - \left[\frac{d(EIw''(x))}{dx}\delta w(x)\right]_0^L \\
&\quad + \int_0^L \left[\frac{d^2(EIw''(x))}{dx^2} - q(x)\right]\delta w(x)dx
\end{aligned}
\tag{7.25}
$$

The results of setting this first variation to zero are easier to interpret if they are written in terms of the stress resultants defined in Section 7.2.2. We defined the moment-curvature relation in Eq. (7.14), and we can write the shear force in terms of the bending deflection by combining it with Eq. (7.19), that is,

$$V(x) = \frac{dM(x)}{dx} = -\frac{d(EIw''(x))}{dx} \tag{7.26}$$

Thus, in terms of the shear force and the bending moment, Eq. (7.25) is

$$\begin{aligned}\delta^{(1)}\Pi = &-[M(x)\delta w'(x)]_0^L + [V(x)\delta w(x)]_0^L \\ &- \int_0^L [M''(x) + q(x)]\delta w(x)dx\end{aligned} \tag{7.27}$$

Setting the first variation (7.27) to zero yields the equation of equilibrium in bending that results from the vanishing of its integral:

$$\frac{d^2M(x)}{dx^2} + q(x) = 0 \tag{7.28}$$

which is exactly the same as our strength-of-materials result, Eq. (7.3). We can write Eq. (7.28) in terms of $w(x)$ by substituting from Eq. (7.14) to find

$$\frac{d^2(EIw''(x))}{dx^2} - q(x) = 0 \tag{7.29}$$

which for a uniform beam made of a homogeneous material reduces to

$$EIw^{iv}(x) - q(x) = 0 \tag{7.30}$$

From the bracketed terms in Eq. (7.27) we find a pair of boundary condition choices at both ends, $x = 0$ and $x = L$, providing the total of four boundary conditions needed for the fourth-order differential equations (7.29) and (7.30):

$$\text{Either} \quad M(x) = 0 \quad \text{or} \quad \delta w'(x) = 0 \tag{7.31}$$

and

$$\text{Either} \quad V(x) = 0 \quad \text{or} \quad \delta w(x) = 0 \tag{7.32}$$

Equations (7.31) and (7.32) represent a pair of *generalized force–displacement dualities*, that is, at a given boundary point we prescribe (1) either a moment or a slope *and* (2) either a shear force or a displacement. In other words, at each point we have a pair of choices, of choosing between *kinematic* or *geometric* (e.g., slope, displacement) and *force* (e.g., moment, shear) conditions. We will present some examples of how these dualities are developed in the next section.

In the meantime, though, we should observe a central point of the above energy-based derivation, namely, that we have derived a model of bending that is totally uncoupled from any axial problem. Thus, to continue our examination of beam bending, we can focus entirely on the bending deflection and its consequent forces and moments.

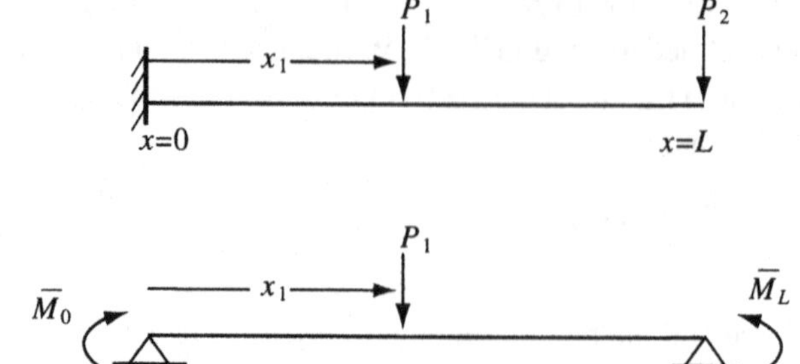

Figure 7.5. Two elastic beams with boundary loads.

7.2.5 Some Typical Boundary Conditions for Engineering Beams

We want to finish our energy-based derivation of beam theory with some illustrations of how boundary conditions are incorporated and interpreted. We begin with the first of the beams shown in Fig. 7.5, the cantilever beam with concentrated loads at an intermediate point, $x = x_1$, and at the tip. The total potential energy for the bending of this beam is similar to Eq. (7.23), except that we will express the load at $x = x_1$ with a Dirac delta function and include the tip load as a boundary term as follows:

$$\Pi_{cant} = \frac{1}{2}\int_0^L EI(w''(x))^2 dx - \int_0^L P_1 \delta_D(x - x_1) w(x) dx - P_2 w(L) \tag{7.33}$$

If we apply the variational delta operator to the total potential (7.33), we obtain the following first variation:

$$\begin{aligned} \delta^{(1)}\Pi_{cant} &= \int_0^L [EI w^{\text{iv}}(x) - P_1 \delta_D(x - x_1)]\delta w(x) dx \\ &\quad - [M(x)\delta w'(x)]_0^L + [V(x)\delta w(x)]_0^L - P_2 \delta w(L) \end{aligned} \tag{7.34}$$

Thus, the appropriate equilibrium equation we need solve is

$$EI w^{iv}(x) - P_1 \delta_D(x - x_1) = 0 \tag{7.35}$$

The corresponding boundary conditions arise from the pair choices inherent in Eq. (7.34) as follows:

$$\begin{aligned} w(0) = 0 \quad &\Rightarrow \quad \delta w(0) = 0 \\ w'(0) = 0 \quad &\Rightarrow \quad \delta w'(0) = 0 \\ M(L) = 0 \quad &\Rightarrow \quad -EI w''(L) = 0 \\ V(L) - P_2 = 0 \quad &\Rightarrow \quad -EI w'''(L) - P_2 = 0 \end{aligned} \tag{7.36}$$

The first two conditions are evident kinematic or geometric constraints due to the built-in or clamped support at the left end of the beam, $x = 0$. The third condition follows from the absence of a moment at the free end of the cantilever, while the last boundary condition indicates how the applied shear force is properly accounted for at the boundary point where it is placed. Note that the first two of conditions (7.36) are geometric boundary conditions, which means that the variations are zero there because the deflection and slope are both prescribed, while the last two requirements are force boundary conditions because we are assigning values to the moment and shear at the tip of the beam.

The second of the beams pictured in Fig. 7.5 would be handled in a similar way. We need only take care that we account for the signs for the potentials of the moments applied at the left and right pinned supports. Thus, the total potential energy here takes the form

$$\begin{aligned}\Pi_{pin} &= \frac{1}{2}\int_0^L EI(w''(x))^2 dx - \int_0^L P_1\delta_D(x - x_1)w(x)dx \\ &\quad - \bar{M}_0 w'(0) + \bar{M}_L w'(L)\end{aligned} \tag{7.37}$$

Applying the delta operator to the total potential (7.37) yields the first variation:

$$\begin{aligned}\delta^{(1)}\Pi_{pin} &= \int_0^L [EIw^{iv}(x) - P_1\delta_D(x - x_1)]\delta w(x)dx + [V(x)\delta w(x)]_0^L \\ &\quad - [M(x)\delta w'(x)]_0^L - \bar{M}_0\delta w'(0) + \bar{M}_L\delta w'(L)\end{aligned} \tag{7.38}$$

The appropriate equilibrium equation we must solve is now familiar:

$$EIw^{iv}(x) - P_1\delta_D(x - x_1) = 0 \tag{7.39}$$

The corresponding boundary conditions arise from the pair choices inherent in Eq. (7.38) and are

$$\begin{aligned} w(0) = 0 \quad &\Rightarrow \quad \delta w(0) = 0 \\ M(0) - \bar{M}_0 = 0 \quad &\Rightarrow \quad -EIw''(0) - \bar{M}_0 = 0 \\ w(L) = 0 \quad &\Rightarrow \quad \delta w(L) = 0 \\ -M(L) + \bar{M}_L = 0 \quad &\Rightarrow \quad +EIw''(L) + \bar{M}_L = 0 \end{aligned} \tag{7.40}$$

Here we see two sets of parallel conditions at each end of the beam, one confirming the constraint against vertical motion at the support, the other reflecting the fact that the internal moment of the beam at the support must equal the moment applied externally at that point. If the potential of the applied moments was carefully constructed, the signs automatically work out properly.

Finally, for both sets of boundary conditions reviewed and displayed in Eqs. (7.36) and (7.40), note that we have restated the force boundary conditions in terms of appropriate derivatives of the bending deflection because, after all, it is in terms of $w(x)$ that we have expressed the equations of equilibrium.

7.3 Using the Total Potential Energy Principle for Beam Analysis

Having derived a theory for the bending of beams by demonstrating one facet of the energy approach, we now apply that theory in order to show that the results it yields are both internally consistent and practically useful. It is also useful to show the other facet of the energy approach, the direct method, to see what we can achieve by developing approximate solutions to various problems.

7.3.1 The Indirect Approach and Exact Solutions

Now we develop some exact solutions to the equation of equilibrium of a beam (and its associated boundary conditions) that we derived in Section 7.2.

Example 7.1. Find the bending deflection and the bending stress resultant for the classic simply supported beam with a concentrated load placed at the center.

From Fig. 7.5, with $x_1 = L/2$, $P_1 = P$, and $\bar{M}_0 = \bar{M}_L = 0$, we have the following differential equation for $w(x)$:

$$EIw^{iv}(x) - P\delta_D(x - L/2) = 0 \tag{7.41}$$

while boundary conditions for this simply supported beam are

$$\begin{aligned} w(0) &= 0, \qquad & w(L) &= 0 \\ EIw''(0) &= 0, \qquad & EIw''(L) &= 0 \end{aligned} \tag{7.42}$$

Equation (7.41) is a linear differential equation with constant coefficients. Subject to the boundary conditions (7.42), its solution can be found to be

$$w(x) = \frac{PL^3}{48EI}\left[3\left(\frac{x}{L}\right) - 4\left(\frac{x}{L}\right)^3 + 8\left(\frac{x}{L} - \frac{1}{2}\right)^3 H\left(x - \frac{L}{2}\right)\right] \tag{7.43}$$

where $H(x - L/2)$ is, again, the *Heaviside* or *unit step function* that we introduced in Section 4.4.1 when we discussed bars loaded by axial point loads. In view of the definition of the unit step function, the solution (7.43) can also be written in the form

$$w(x) = \begin{cases} \dfrac{PL^3}{48EI}\left[3\left(\dfrac{x}{L}\right) - 4\left(\dfrac{x}{L}\right)^3\right] & x \leq L/2 \\[2ex] \dfrac{PL^3}{48EI}\left[4\left(\dfrac{x}{L}\right)^3 - 12\left(\dfrac{x}{L}\right)^2 + 9\left(\dfrac{x}{L}\right) - 1\right] & x \geq L/2 \end{cases} \tag{7.44}$$

Both expressions of the bending deflection (Eqs. (7.43) and (7.44)) show that it is symmetric about the center of the beam. Thus, we could also simply represent the solution as the first of Eqs. (7.44), together with the caveat that the solution is symmetric about $x = L/2$.

To analyze the stresses for this problem, we must also calculate the moment in the beam. We can do this by differentiating the above results twice and using the moment-curvature relation, Eq. (7.14), from which we get an expected result:

$$M(x) = \begin{cases} \dfrac{Px}{2} & x \leq L/2 \\[2ex] \dfrac{P(L - x)}{2} & x \geq L/2 \end{cases} \tag{7.45}$$

■

Now we can also calculate the moment for Example 7.1 by applying the principles of statics directly to the appropriate version of the beam pictured in Fig. 7.5, from which we would get exactly the same results as in Eqs. (7.45). This raises an interesting question. If we can get the moment distribution directly from statics, can we not just invert

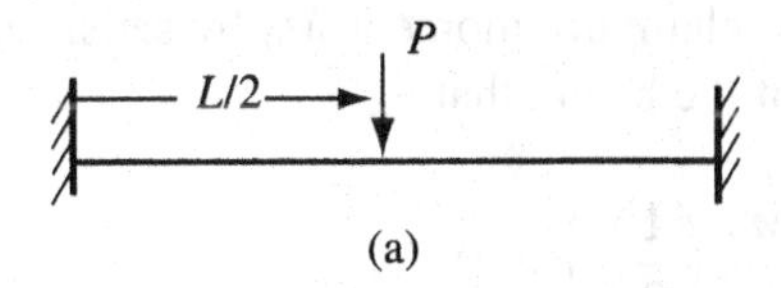

(a)

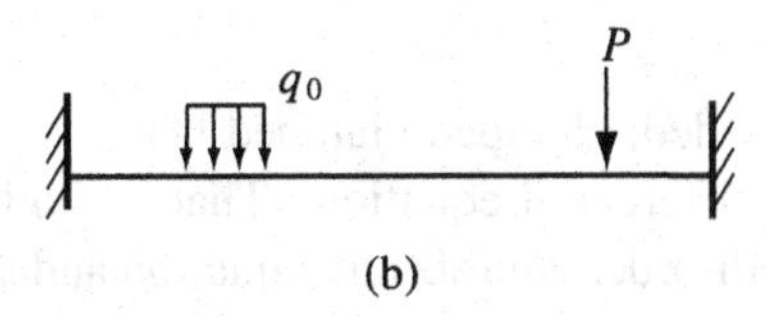

(b)

Figure 7.6. Fixed-ended beams with (a) symmetric and (b) asymmetric loading.

the process we've just gone through and calculate the bending deflection by using the moment-curvature relationship together with results obtained from statics? That is, if we had obtained Eqs. (7.45) by applying the principles of statics, perhaps we could find the bending deflection by integrating the second-order equation

$$w''(x) = -\frac{M(x)}{EI} = -\frac{Px}{2EI} \quad x \leq L/2 \tag{7.46}$$

This is a simpler differential equation to solve because it is only second order. It automatically satisfies the moment condition at $x = 0$, so we have only the deflection boundary condition to satisfy, that is, $w(0) = 0$. Our second condition is a symmetry condition, namely, that $w'(L/2) = 0$. The solution to the differential equation (7.46) that satisfies these two conditions is exactly the one displayed as the first of Eqs. (7.44), so we have found the correct solution by solving a simpler version of the problem. Can't we do this in general?

7.3.2 Boundary Conditions and Indeterminate Beams

In general we cannot take the approach just outlined because the simpler, second-order approach to beam problems works only for statically determinate beams. Occasionally, we can use the second-order equation for indeterminate beams where some symmetry or other special circumstance is evident.

Example 7.2. Find the support or clamping moments for the *fixed-ended* or *clamped-clamped* beam pictured in Fig. 7.6(a).

In this case, since the beam and its loading are symmetric, the clamping moment M_0 at the left end has an obvious and symmetric counterpart at the right end. The moment along the beam is easily calculated, so that the second-order equation for the deflection is

$$w''(x) = -\frac{M(x)}{EI} = -\frac{1}{EI}\left(\frac{Px}{2} - M_0\right) \quad x \leq L/2 \tag{7.47}$$

The solution to Eq. (7.47) that satisfies kinematic conditions at $x = 0$ is

$$w(x) = -\frac{PL^3}{12EI}\left(\left(\frac{x}{L}\right)^3 - \frac{6M_0}{PL}\left(\frac{x}{L}\right)^2\right) \quad x \leq L/2 \tag{7.48}$$

Finally, for this problem, we can determine the clamping moment M_0 by satisfying the symmetry condition on the slope at the center of the beam, that is,

$$w'(L/2) = 0 \quad \Rightarrow \quad \left(3\left(\frac{1}{2}\right)^2 - \frac{12M_0}{PL}\left(\frac{1}{2}\right)\right) = 0 \tag{7.49}$$

■

On the other hand, for the asymmetrically loaded, clamped-clamped beam shown in Fig. 7.6(b), we cannot apply the second-order differential equation. That is, both here and in general, we have to integrate the full fourth-order equilibrium equation and satisfy all four appropriate boundary conditions in order to find the bending deflection of an indeterminate beam.

7.3.3 Discretization and the Direct Displacement Approach

Recall that in Sections 4.4.2 and 4.4.3 we obtained approximate solutions for axially loaded bars. Similarly, we now want to apply the principle of minimum potential energy in the Rayleigh–Ritz or direct approach to obtain approximate representations of the expected deflected shape of bent beams. That is, we want to explore how we construct approximate solutions of the form

$$w(x) = \sum_{n=0}^{N} W_n \phi_n(x) \tag{7.50}$$

Note that in using Eq. (7.50) to represent the deflection we are returning again to the *discretization* phase of modeling because it is the values of the discrete constants W_n that we determine by applying the potential energy principle.

As we have observed in our discussion of axially loaded members, the single most pressing requirement is that the individual trial functions, the $\phi_n(x)$, satisfy *at least* the geometric boundary conditions. True, it is highly desirable that the $\phi_n x$ satisfy as many of the natural or force conditions as possible, as we will see in the examples that follow, but it is an absolute requirement that the trial functions satisfy all kinematic conditions. And, again, it is also very helpful, although not required, if the trial functions form an *orthogonal set*, that is, if

$$\int_0^L \phi_n(x)\phi_m(x)dx = \begin{cases} 0 & \text{for } n \neq m \\ a_n & \text{for } n = m \end{cases} \tag{7.51}$$

This property is very attractive because the resulting linear equations for the coefficients W_n are *uncoupled*, so that the equations governing the W_n are diagonal in matrix notation. Further, as we will see in Example 7.3, it is also helpful if the trial functions are such that they can be *normalized* with respect to the same constant for all values of n, that is, if the a_n are independent of n.

While the principles applied are identical, obtaining approximate solutions for beams is more complicated than for axially loaded bars because the governing equilibrium equation is of higher order – fourth, rather than second – and we have two conditions at each boundary point, rather than one at each for bars. The kinematic conditions for beams are the "varied" ones in Eqs. (7.27), that is, the conditions wherein we choose to prescribe

(1) the displacement $w(x)$ rather than the shear force and (2) the slope $w'(x)$ rather than the moment.

There is another feature of what we're about to do that is very, very important. In the discussion we've just started, we have said not a word about whether the beam we are modeling is determinate or indeterminate. This is a natural consequence of the fact that we are using the displacement or stiffness approach in a direct manner, and so we begin with trial functions that are required to satisfy all of the geometric or kinematic boundary conditions of the problem at hand. Thus, from the very beginning we are making sure that we are accounting for the geometric constraints that make a structure indeterminate. One of the results we then obtain is an approximation of equilibrium that includes the resultant forces needed to maintain these geometric constraints.

7.3.3.1 A simply supported beam with a concentrated load at the midpoint

As a first illustration, let us find approximate solutions for the (determinate) centrally loaded, simply supported beam. The geometric conditions that we must satisfy are that the displacements vanish at both ends of the beam. The remaining two boundary conditions are force conditions that require the moment to vanish at both supports. This is a problem whose deflected shape and resulting moment distribution should be symmetric about the centerline. We can use symmetry both to guide us in choosing trial functions and to check on the reasonableness of our answers.

Guided by what we know of standard solutions to beam problems, we will start by using polynomials in x as trial functions. The drawback of polynomials is that they are not typically orthogonal, except for a very restricted class of polynomial-based functions. On the other hand, polynomials do appear to be intuitively useful.

Example 7.3. Find an approximate solution for the deflection of a centrally loaded, simply supported beam, based on the polynomial approximation:

$$w_1(x) = A' + B'x + C'x^2 + D'x^3 \tag{7.52}$$

We start by rewriting this polynomial in an entirely equivalent form:

$$w_1(x) = \frac{PL^3}{EI}\left(A + B\frac{x}{L} + C\left(\frac{x}{L}\right)^2 + D\left(\frac{x}{L}\right)^3\right) \tag{7.53}$$

Equations (7.52) and (7.53) are equivalent because they each contain four arbitrary constants. They are different because we are using some prior knowledge about the forms of our solutions to make the arbitrary constants (A, B, C, D) in Eq. (7.53) dimensionless, while the constants (A', B', C', D') are not. This makes it easier to use physical dimensions to guide and check solutions, which is useful for calculations, whether done by hand or an a computer.

All of the geometric conditions must be satisfied, namely, $w(0) = w(L) = 0$. After doing the algebra needed to make the solution (7.53) satisfy these we find

$$w_1(x) = \frac{PL^3}{EI}\left(C\left(\frac{x}{L}\right)^2 + D\left(\frac{x}{L}\right)^3 - (C+D)\left(\frac{x}{L}\right)\right) \tag{7.54}$$

Now we have only two unknown constants to determine. We could find values for both C and D by minimizing the total potential energy with respect to both, or we could take a slightly different tack and apply a symmetry condition first. In the former instance, keeping both constants, we substitute the trial function (7.54) into the total potential energy for this beam, Eq. (7.23), that is,

$$\Pi_{12} = \frac{1}{2}\int_0^L EI(w''(x))^2 dx - \int_0^L P\delta_D(x - L/2)w_1(x)dx \tag{7.55}$$

After the approximation (7.54) is substituted, the total potential energy becomes

$$\Pi_{12} = \left(\frac{P^2L^3}{2EI}\right)\int_0^1 [2C + 6D\xi]^2 d\xi + \left(\frac{P^2L^3}{2EI}\right)\left[\frac{C}{2} + \frac{3D}{4}\right] \tag{7.56}$$

wherein we have again used a dimensionless axial coordinate ξ in the integral. After integration Eq. (7.56) becomes a quadratic algebraic form in C and D:

$$\Pi_{12} = \left(\frac{P^2L^3}{2EI}\right)\left[4C^2 + 12CD + 12D^2 + \frac{C}{2} + \frac{3D}{4}\right] \tag{7.57}$$

We find the minimum of this total potential energy by setting to zero its partial derivatives with respect to C and D, which yields two linear algebraic equations:

$$\begin{bmatrix} 8 & 12 \\ 12 & 24 \end{bmatrix}\begin{Bmatrix} C \\ D \end{Bmatrix} = -\begin{Bmatrix} 1/2 \\ 3/4 \end{Bmatrix} \tag{7.58}$$

Note that the coefficient and right-hand column matrices of Eq. (7.58) consist simply of numbers. This is a consequence of the nondimensionalization that we introduced at the beginning of this example, and it makes the "grunt" work easier – even for simple problems, never mind more complex cases. We can then find the displacement coefficients to be

$$\begin{Bmatrix} C \\ D \end{Bmatrix} = \begin{Bmatrix} -1/16 \\ 0 \end{Bmatrix}$$

so that the displacement approximation we have found is

$$w_1(x) = \frac{PL^3}{48EI}\left(3\left(\frac{x}{L}\right) - 3\left(\frac{x}{L}\right)^2\right) \tag{7.59}$$

■

Is the solution (7.59) of Example 7.3 a good approximation to the exact solution (which we have already found as Eqs. (7.43) or (7.44))? Could it even be another version of the exact solution? And how would we know?

First we can check the boundary conditions. Equation (7.59) clearly satisfies the geometric conditions, but the force conditions follow from the moment,

$$M_1(x) = -EIw_1''(x) = \frac{PL}{8} \tag{7.60}$$

This result differs from the exact moment (cf. Eq. (7.45)). Indeed, it is quite clear that Eq. (7.60) doesn't vanish at either end of the beam, that is, it doesn't satisfy either force

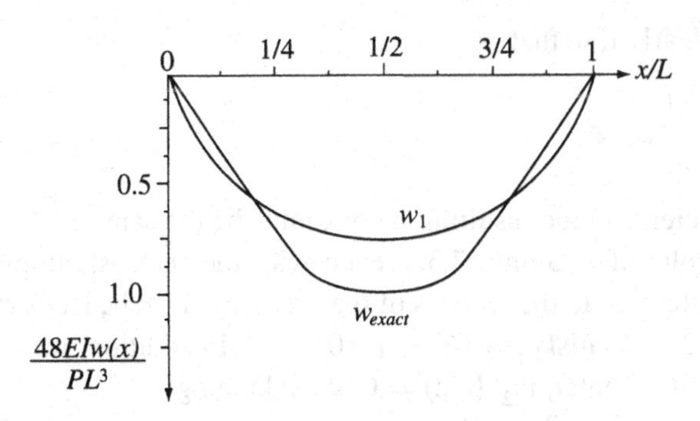

Figure 7.7. Plots (highly exaggerated) of the exact (Eq. (7.44)) and an approximate deflection (from Eq. (7.59) of Example 7.3) for the centrally loaded, pinned-pinned beam.

boundary condition. In addition, although the moment represented by Eq. (7.60) is symmetric about the center of the beam because it is constant along the entire length of the beam, recall that the exact moment varies linearly with the beam coordinate.

We can also compare the deflected shape given by Eq. (7.59) with the exact result. We can calculate from Eq. (7.59) a deflection at the center of the beam that is three-fourths of the exact solution, and its slope vanishes there because of symmetry, so this approximate shape does mimic the exact solution, as we show in Fig. 7.7. However, this approximate deflection has a constant curvature, and thus a constant radius of curvature, which we can also see from Eq. (7.60). This doesn't make sense because we wouldn't expect this beam deflection to be a circular segment. Thus, on balance we would not regard the approximation $w_1(x)$ of Eq. (7.54) as being really good because it neither replicates the moment distribution very well nor satisfies the force conditions, even though we are not required to satisfy these conditions.

A further comment on the results of Example 7.3. The square matrix in Eq. (7.58) is symmetric, even for this approximate solution. Why is this so? Well, here it follows from a mathematical property of the integral in Eq. (7.56), which is actually the stored strain energy for our displacement approximation, that is,

$$U_{12} = \left(\frac{2P^2L^3}{EI}\right)\int_0^1 [2C + 6D\xi]^2 d\xi$$

$$= \left(\frac{2P^2L^3}{EI}\right)\int_0^1 [2C + 6D\xi]\begin{Bmatrix} 2C \\ 6D\xi \end{Bmatrix} d\xi \tag{7.61}$$

While the integral of a squared series of terms such as appear in the first of Eqs. (7.61) is not generally expressible as a perfect square of some integral of the individual terms, it clearly must be true that this commutative property holds for the cross-product terms, as their term-by-term integrations must perforce be the same. That is, while

$$\int_0^1 [A + B\xi]^2 d\xi \neq [A + \alpha B]^2 \tag{7.62}$$

for any real constant α, it is clearly true that

$$\int_0^1 2AB\xi d\xi = AB\int_0^1 2\xi d\xi \tag{7.63}$$

so that the off-diagonal coefficients in the resulting matrix must be the same.

We could improve the results of Example 7.3 – even with the same starting point, Eq. (7.53) – by paying more attention to the physics of the problem. That is, let us restrict its domain to be $0 \leq x \leq L/2$ and satisfy $w_2(0) = w_2''(0) = 0$. In addition, we apply a symmetry condition at the beam's center, $w_2'(L/2) = 0$, thus defining

$$w_2(x) = C\frac{PL^3}{EI}\left(3\left(\frac{x}{L}\right) - 4\left(\frac{x}{L}\right)^3\right) \tag{7.64}$$

Minimizing the total potential energy with respect to C for the trial solution (7.64) produces, in fact, the exact solution. Here, of course, we did everything right in terms of paying attention to the boundary conditions and to symmetry.

Still another approach is to choose a trial function that is part of an orthogonal set, such as the sinusoids we tried for the bars in Chapter 4.

Example 7.4. Find a one-term sinusoidal approximate solution for the deflection of a centrally loaded, simply supported beam.

For the simple support boundary conditions, we start with the approximation

$$w_3(x) = W\frac{PL^3}{EI}\sin\frac{\pi x}{L} \tag{7.65}$$

which has a corresponding moment along the beam of

$$M_3(x) = W(\pi^2 PL)\sin\frac{\pi x}{L} \tag{7.66}$$

Clearly, the solution $w_3(x)$ is symmetric about $x = L/2$, and it, together with its resulting moment, satisfies all of the boundary conditions at each beam end. As we will shortly show in a more general form, minimizing the total potential energy for the approximation (7.65) produces a displacement coefficient

$$W = \frac{2}{\pi^4}$$

which yields the following deflection and moment results:

$$\begin{aligned} w_3(x) &= \frac{2}{\pi^4}\left(\frac{PL^3}{EI}\right)\sin\frac{\pi x}{L} \\ M_3(x) &= \frac{2}{\pi^2}(PL)\sin\frac{\pi x}{L} \end{aligned} \tag{7.67}$$

We have shown a comparison of these results with the exact solution in Table 7.1, and we see that the deflection result is quite accurate, but the moment result – while better than what we found with $w_1(x)$ – is still not as good as we might like. That is, while the approximate deflections are within some two percent of the exact deflections, the moments

Table 7.1. Exact (Eqs. (7.44) and (7.45)) and approximate (Eqs. (7.67)) deflections and moments for a centrally loaded, pinned beam.

	$x=0$	$x=L/4$	$x=L/2$	Variation in x
$\frac{48EIw_3(x)}{PL^3}$	0	0.697	0.986	sinusoidal
$\frac{48EIw_{exact}(x)}{PL^3}$	0	0.688	1.000	cubic
$\frac{4M_3(x)}{PL}$	0	0.573	0.811	sinusoidal
$\frac{4M_{exact}(x)}{PL}$	0	0.500	1.000	linear/triangular

(and thus the stresses) are off by almost 20%. The reason for the discrepancy in the moment result is the same as that which we pointed out for approximate solutions of axially loaded members, namely, the moments require double differentiation of the deflection, thus giving us reason to expect a much more ragged result than with the deflection itself. ■

We can get better moments and stresses than those found in Example 7.4 by sharpening or refining our approximation. The sinusoids form a set of orthonormal functions, so it makes sense for us to combine the generic form (7.50) with the sinusoids as the trial functions, that is,

$$w_4(x) = \left(\frac{PL^3}{EI}\right)\sum_{n=1}^{\infty} W_n \sin\frac{n\pi x}{L} \tag{7.68}$$

The total potential energy for this assumed solution is:

$$\begin{aligned}\Pi_{14} &= \left(\frac{P^2L^3}{2EI}\right)\int_0^1\left[\sum_{n=1}^{\infty} W_n(n\pi)^2\sin n\pi\xi\right]^2 d\xi - \left(\frac{P^2L^3}{EI}\right)\sum_{n=1}^{\infty} W_n\sin\left(\frac{n\pi}{2}\right)\\ &= \left(\frac{\pi^4P^2L^3}{4EI}\right)\sum_{n=1}^{\infty} W_n^2n^4 - \left(\frac{P^2L^3}{EI}\right)\sum_{n\text{ odd}}^{\infty} W_n\sin\left(\frac{n\pi}{2}\right)\end{aligned} \tag{7.69}$$

where we have used the orthogonality and the normalization of the sinusoids in integrating the strain energy, that is,

$$\begin{aligned}\int_0^1\left[\sum_{n=1}^{\infty} W_n n^2\sin n\pi\xi\right]^2 d\xi &= \int_0^1\left[\sum_{n=1}^{\infty} W_n^2 n^4\sin^2 n\pi\xi\right]d\xi\\ &\quad + \int_0^1\left[\sum_{m\neq n}^{\infty}\sum_{n\neq m}^{\infty} W_mW_nm^2n^2\sin m\pi\xi\,\sin n\pi\xi\right]d\xi\\ &= \frac{1}{2}\sum_{n=1}^{\infty} W_n^2n^4 + 0\end{aligned} \tag{7.70}$$

We also see that the potential of the load P produces only odd terms in the series in the total potential (7.69), which means that all of the even terms vanish. Therefore, the sums

Table 7.2. Exact (Eq. (7.45)) and approximate (Eq. (7.73)) moments for a centrally loaded, pinned beam

	$x = 0$	$x = L/4$	$x = L/2$
$\frac{4M_{exact}(x)}{PL}$	0	0.500	1.000
$\frac{4M_4(x)}{PL}(n = 1)$	0	0.573	0.811
$\frac{4M_4(x)}{PL}(n = 1, 3)$	0	0.510	0.901
$\frac{4M_4(x)}{PL}(n = 1, 3, 5)$	0	0.486	0.933
$\frac{4M_4(x)}{PL}(n = 1, 3, 5, 7)$	0	0.498	0.950

can be combined so that the total potential becomes

$$\Pi_{14} = \left(\frac{P^2L^3}{EI}\right)\sum_{nodd}^{\infty}\left[\frac{(n\pi)^4}{4}W_n^2 - W_n \sin\frac{n\pi}{2}\right] \tag{7.71}$$

Thus, we can minimize the total potential energy (7.71) with respect to each of the W_n as an independent parameter, so that this minimization produces the following coefficients:

$$W_n = \frac{2}{(n\pi)^4}\sin\frac{n\pi}{2} \tag{7.72}$$

which yields the following deflection and moment results:

$$\begin{aligned} w_4(x) &= \left(\frac{PL^3}{EI}\right)\sum_{nodd}^{\infty}\frac{2}{(n\pi)^4}\sin\frac{n\pi}{2}\sin\frac{n\pi x}{L} \\ M_4(x) &= (PL)\sum_{nodd}^{\infty}\frac{2}{(n\pi)^2}\sin\frac{n\pi}{2}\sin\frac{n\pi x}{L} \end{aligned} \tag{7.73}$$

Since we have already shown that the deflection approximation converges rather rapidly – recall Table 7.1 – we will focus on the moment now, for which we display results in Table 7.2. We see that the convergence is fairly rapid, although we should also note the slight oscillation in how the moment at the quarter-span approaches the exact result, that is, from above for smaller values of n, and then from below as n increases. This is clearly a result of the sinusoidal aspect of the solution, so it should not be a surprise. But the magnitude of the difference between the exact moment and a given nth-order approximation decreases rapidly with increasing n, as we should expect from the form of the solutions (7.73). That is, we can see that the magnitude of the deflection coefficients are proportional to $1/n^4$, while the terms in the moment expansion are proportional to $1/n^2$. Further, the series itself includes only odd terms. If we include only four terms in our approximation ($n = 1, 3, 5, 7$), the ratio of the fourth term to the first is only $1/49$. Thus, we see that it doesn't take very many terms to get to a very satisfactory approximation.

Admittedly, the problem just solved is about as simple a beam problem as one can do, but the principles involved in doing *idealization* and *discretization* are very important and broadly applicable. It is interesting that we have chosen trial functions that span the entire domain of interest, that is, the entire length of a bent beam. On the other hand, computational discretization also allows the analyst or modeler to break up a beam or other structure into smaller pieces, into "finite elements," and to appropriately match solutions for each element (i.e., transmit equilibrating forces and enforce continuity of deformation) at every inter-element boundary. Remember that our analysis of trusses is in this vein. Indeed, since trusses often serve (in their entirety) as beams, their makeup as assemblies of bars amounts to assembling smaller finite elements (i.e., the bars) to make up the entire structure. In fact, the use of matrix mathematics to formulate truss problems was one of the main paths toward the finite element method, and the variational approaches we describe supplied theoretical support for developing approximations with finite elements.

However, even with the ever-expanding availability of computational power, developing the kinds of solutions we are doing here is an important and useful skill. This is partly because it is good to understand both what the computer can do for us and how its results may be limited, but also because the kinds of approximations we develop here can be used to get back-of-the-envelope solutions to do real-time estimates and to provide a means for checking and assessing computer-based results.

7.3.3.2 A fixed-ended beam with a concentrated load at the midpoint

Let us do two more examples, here with the idea of showing off the variational approach to discretization in structural analysis in the context of an indeterminate structure. Suppose the centrally loaded beam were now *clamped* or *fixed* at both ends (see Fig. 7.6(a)). We have already noted (in Section 7.3.2) that for indeterminate beams we generally must solve the full fourth-order equilibrium equation to obtain the beam deflection. It is reasonable to inquire as to whether a similar requirement of increased work is needed in a variational approach to indeterminate structures. The answer is, as they say, "not exactly."

Example 7.5. Find a one-term polynomial approximation for the deflection of the indeterminate, centrally loaded, fixed-ended beam.

Satisfying all the geometric boundary conditions for this problem requires that $w(0) = w'(0) = w(L) = w'(L) = 0$. If we had a trial function that satisfied these conditions, our solution would proceed exactly as in Section 7.3.3.1 for the simply supported beam. So, we will start our direct approach with the familiar trial function used in Example 7.4, that is,

$$w_{c1}(x) = \frac{PL^3}{EI}\left(A + B\left(\frac{x}{L}\right) + C\left(\frac{x}{L}\right)^2 + D\left(\frac{x}{L}\right)^3\right) \tag{7.74}$$

We will satisfy both geometric conditions at the left support and the symmetry condition on the slope at the center, after which we will have only one unknown constant to determine. After satisfying these conditions we find that

$$w_{c1}(x) = C\frac{PL^3}{EI}\left(\left(\frac{x}{L}\right)^2 - \frac{4}{3}\left(\frac{x}{L}\right)^3\right) \tag{7.75}$$

The total potential energy for this beam is

$$\Pi_{c1} = \int_0^{L/2} EI(w''_{c1}(x))^2 dx - Pw_{c1}(L/2) \tag{7.76}$$

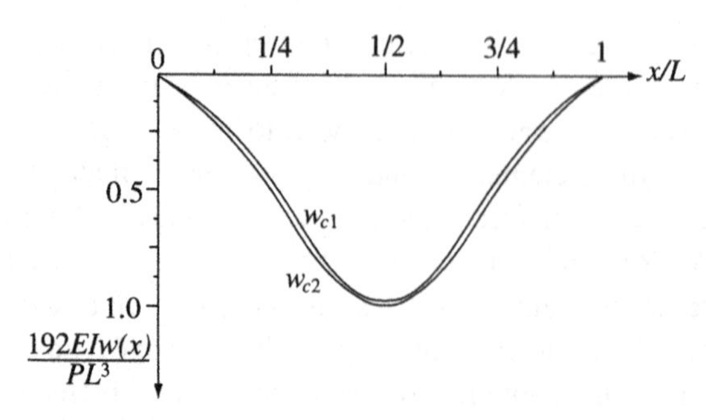

Figure 7.8. Plots (highly exaggerated) of the exact (Eq. (7.79) and a one-term cosinusoidal approximate deflection (Eq. (7.85) of Example 7.6) for the centrally loaded, fixed-ended beam.

which becomes, after we substitute the approximation (7.75),

$$\Pi_{c1} = \left(\frac{C^2P^2L^3}{EI}\right)\int_0^{1/2}[2 - 8\xi]^2 d\xi - \left(\frac{CP^2L^3}{12EI}\right) \tag{7.77}$$

After performing the integration we find

$$\Pi_{c1} = \left(\frac{P^2L^3}{EI}\right)\left(\frac{2C^2}{3} - \frac{C}{12}\right) \tag{7.78}$$

We minimize this total potential expression with respect to the coefficient C, which produces the result that $C = 1/16$, which means in turn that for this fixed-ended beam the approximate deflection is

$$w_{c1}(x) = \frac{PL^3}{16EI}\left(\left(\frac{x}{L}\right)^2 - \frac{4}{3}\left(\frac{x}{L}\right)^3\right), \quad 0 \le x \le L/2 \tag{7.79}$$

Equation (7.79) is, not surprisingly, the exact solution that we would obtain with the indirect method, that is, by solving the fourth-order beam deflection equation and satisfying the boundary conditions. We show a sketch of this deflected shape in Fig. 7.8. Note the vanishing deflection and slope at both supports require that there exist two *inflection points*, symmetrically located within each half of the beam. These inflection points reflect *extreme values of the slope*, which means that they represent points at which the curvature – and thus also the moment – must vanish, that is,

$$\frac{d}{dx}w'_{c1}(x) = 0 \quad \Rightarrow \quad w''_{c1}(x) = -\frac{M(x)}{EI} = 0 \tag{7.80}$$

The inflection points occur at the quarter points, that is, at $x = L/4$ and at $x = 3L/4$. We often estimate the locations of inflection points of beams (and frames) because they serve as useful points about which to sum moments.

The corresponding exact moment is found by substituting Eq. (7.79) into the moment-curvature relation, which yields

$$M_{c1}(x) = M_{exact}(x) = -\frac{PL}{8}\left(1 - 4\left(\frac{x}{L}\right)\right), \quad 0 \le x \le L/2 \tag{7.81}$$

For this case, the moment at the supports, the *fixed-end moment*, is

$$M_{c1}(0) = -EIw_{c1}''(0) = -\frac{PL}{8} \tag{7.82}$$

■

Example 7.6. Find a one-term trigonometric approximation for the deflection of the indeterminate, centrally loaded, fixed-ended beam.

In constructing a trigonometric trial function, we should note that neither the sine nor the cosine is zero at any point where its own derivative vanishes. Thus, elementary trigonometric functions cannot be used on their own because they cannot satisfy all of the geometric boundary conditions for a fixed-ended beam. However, by mapping the argument of a complete cycle of a cosine curve (from 0 to 2π) onto the beam length ($0 \le x \le L$) and comparing its shape to our exact solution – and to our intuition about what we expect here – we can see that the following trial function is a reasonable approximation:

$$w_{c2}(x) = W\frac{PL^3}{EI}\left(1 - \cos\frac{2\pi x}{L}\right) \tag{7.83}$$

This trial shape satisfies all of the geometric conditions (and thus all of the boundary conditions), is symmetric about $x = L/2$, and strongly resembles the exact shape found in Example 7.5. The total potential energy for this solution is

$$\begin{aligned}\Pi_{c2} &= \left(\frac{P^2L^7}{2EI}\right)\int_0^1 \left[W\left(\frac{2\pi}{L}\right)^2 \cos 2\pi\xi\right]^2 d\xi - \left(\frac{P^2L^3}{EI}\right)W(1 - \cos\pi) \\ &= \left(\frac{4\pi^4P^2L^3}{EI}\right)W^2 - \left(\frac{2P^2L^3}{EI}\right)W \end{aligned} \tag{7.84}$$

Minimization of Eq. (7.84) with respect to W then produces the result that $W = 1/4\pi^4$, from which we can find the following deflection and moment:

$$\begin{aligned} w_{c2}(x) &= \frac{PL^3}{4\pi^4EI}\left(1 - \cos\frac{2\pi x}{L}\right) \\ M_{c2}(x) &= -\frac{PL}{\pi^2}\cos\frac{2\pi x}{L} \end{aligned} \tag{7.85}$$

The deflection given by the approximation (7.85) is also shown in Fig. 7.8, and we see that it really replicates the exact solution very well – within two percent – because it satisfies all of the boundary conditions.

■

If we compare the approximate (Eq. (7.85)) and the exact (Eq. (7.81)) moments for the centrally loaded, fixed-ended beam, we see that the inflection points are still located at the quarter points in both the exact and approximate solutions. However, the magnitudes of the (maximum) moments differ both at the supports and the center, that is, $PL/8$ for the exact solution against PL/π^2 for the approximate, which is a difference of almost 20%. Thus, even though we have a very good match for the deflection in Example 7.6, the differentiation devil bites us again by making the approximate moment distribution rougher than its exact counterpart. Thus, even for this simple problem, finding an approximate

Table 7.3. Maximum deflections and moments for centrally loaded, simply supported and fixed-ended beams

	Simple supports	Clamped supports
$\frac{48EI\,w_{max}(x)}{PL^3}$	1.000 at $L/2$	0.250 at $L/2$
$\frac{4M_{max}(x)}{PL}$	1.000 at $L/2$	0.500 at 0, $L/2$, L

moment or stress distribution in which we can have confidence requires greater effort than finding a good approximation for a deflected shape.

7.3.4 Further Observations on Determinate and Indeterminate Beams

The two problems we have just solved in Sections 7.3.3.1 and 7.3.3.2 allow us to reinforce some observations made in Section 2.3 about the reasons we design indeterminate structures. In Table 7.3 we show normalized maximum values of the midspan deflection and the moment (at three points) for both simply supported and fixed-ended centrally loaded beams. The maximum deflection is decreased by 75% and the maximum moment (and so the bending stress) is decreased by 50% for the indeterminate, fixed-ended beam in comparison with its determinate, simply supported counterpart. Thus, we again see clearly that the redundant supports provide significant added stiffness.

Further, while perhaps less evident, this example also illustrates the safety or redundancy introduced by the redundant supports. Consider what happens if we were to lose one of the vertical reactions of each beam. For the simply supported beam, failure is unavoidable because the one remaining support cannot maintain equilibrium. The once fixed-ended beam, however, remains in equilibrium, although it remains indeterminate and its reactions are rather different than those of their symmetric counterpart. We can readily imagine how this effect multiplies in the frame of a large high-rise building, thus providing a variety of alternate loadpaths should one or more connections fail.

7.4 Validating and Estimating in Elementary Beam Theory

We have asserted that the elementary model of engineering beam theory works quite well for long, slender beams. The widespread use of elementary beam models in structural engineering certainly supports this assertion. However, we can also demonstrate how well the model works in terms of how consistently it obeys the relevant modeling assumptions. We will now do this by example.

7.4.1 Validating the Assumptions of Elementary Beam Theory

We start by using the solution for the centrally loaded, pinned-pinned beam to calculate the magnitudes of some terms we have neglected. We calculate the strain energy stored in the beam due to the bending displacement (or stress):

$$U_{bend} = EI\int_0^{L/2}(w''(x))^2dx = \frac{1}{EI}\int_0^{L/2}(M(x))^2dx \tag{7.86}$$

where we have noted the symmetry of the loading and the solution. For the bending moment for this problem, we find

$$U_{bend} = \frac{1}{EI}\int_0^{L/2}(Px/2)^2 dx = \frac{P^2L^3}{96EI} \tag{7.87}$$

We note in passing that because the strain and complementary energies are equal for this linear elastic beam, we can use Eq. (7.87) and Castigliano's second theorem to calculate the deflection under the centrally applied load, and in so doing we confirm the correctness of our calculations so far.

In deriving beam theory we ignored the strain energy stored in shear because our kinematic assumptions dictated that the shear strain would vanish. We know, however, that equilibrium requires a nontrivial transverse shear stress, as provided by Eq. (7.18). Thus, for the centrally loaded, pinned-pinned beam,

$$\tau_{xz} = \frac{P}{4I}((h/2)^2 - z^2) \tag{7.88}$$

The strain energy developed by this shear stress is (again noting symmetry)

$$U_{shear} = \frac{1}{G}\int_0^{L/2}\int_{-h/2}^{h/2}\tau_{xz}^2(x,z)bdzdx \tag{7.89}$$

We can substitute from Eq. (7.88) and integrate, assuming that the beam's cross-section is rectangular, thus finding

$$U_{shear} = \frac{9P^2L^3}{160EI}\left[(1+v)\left(\frac{h}{L}\right)^2\right] \tag{7.90}$$

The ratio of the two strain energy results, that is, of Eq. (7.90) to Eq. (7.87), is

$$\frac{U_{shear}}{U_{bend}} = \frac{27}{5}\left[(1+v)\left(\frac{h}{L}\right)^2\right] \tag{7.91}$$

So we see that for long, slender beams, for which $(h/L)^2 \ll 1$, the strain energy due to the shear stress is indeed much smaller than that due to bending and thus we are consistent in neglecting it as we formulate beam bending problems.

We can also verify that we are consistent in neglecting the normal stress through the beam's thickness. Since the strain energy is proportional to the square of the strains, or stresses, we can expect that the ratio of the transverse shear stress to the bending stress should be of order of magnitude

$$\frac{\tau_{xz}}{\sigma_{xx}} \propto \sqrt{\frac{U_{shear}}{U_{bend}}} \propto \left(\frac{h}{L}\right) \ll 1 \tag{7.92}$$

In fact, this reasoning can be carried a step further by examining the two-dimensional equations of equilibrium in the axial and thickness directions in a form where the axial and thickness coordinates are rendered dimensionless by normalizing them with respect to the lengths of each of their domains. If we did such an analysis, we would confirm that

Eq. (7.92) is correct and that

$$\frac{\sigma_{zz}}{\sigma_{xx}} \approx \left(\frac{h}{L}\right)^2 \ll 1 \tag{7.93}$$

Thus, the normal stress through the thickness is smaller than the bending stress by two orders of magnitude, that is, the effects of this stress component are still less important than the effects of the (small) transverse shear stress, which, recall, we do need to maintain equilibrium (cf. Section 7.2.2).

It is also worth noting that there are other confirmations of our modeling assumptions, and they appear both as *higher-order* approximations and as "exact" solutions to elasticity problems that are formulated with fewer, less restrictive idealization assumptions. For example, the idealizations of beam kinematics that we presented in Section 7.2.1 can be relaxed by including shear deformation and thus allowing greater flexibility in the ability of the beam to deflect. One way to do this is to distinguish the rotation of a line element from the slope created by the bending deflection, that is, to relax the Euler–Bernoulli hypothesis and allow a line originally normal to the centerline of the beam to be other than perpendicular to the deformed centerline. Thus, in place of the deflection assumptions embodied in Eqs. (7.7) and (7.8), we could assume that

$$\begin{aligned} u(x,z) &= -z\psi(x) \equiv -z[w'(x) - \beta(x)] \\ w(x,z) &= w(x) \end{aligned} \tag{7.94}$$

which produces the following field or set of nonzero strains:

$$\begin{aligned} \varepsilon_{xx}(x,z) &= -z\psi'(x) = -z[w''(x) - \beta(x)] \\ \gamma_{xz}(x,z) &= \beta(x) \end{aligned} \tag{7.95}$$

We have introduced a shear angle $\beta(x)$ in Eqs. (7.94) and (7.95) – which measures the change from 90° of a line originally normal to the centerline – as well as the net rotation $\psi(x) \equiv [w'(x) - \beta(x)]$. While we are not going to pursue this line here, we mention it to point out that more elaborate theories have been developed, and in this case we would have two dependent variables instead of the single variable of the engineering theory of beams, that is, either the pair $w(x)$ and $\beta(x)$ or, equivalently, $w(x)$ and $\psi(x)$. In either event there will be a coupled pair of differential equations of equilibrium, and the shear strain will not be zero. This theory produces results that are quite useful for short, stubby beams – beams that are frequently called *shear beams*. Equations (7.94) and (7.95) are the foundations of the *Timoshenko theory of beams*, which is named after their developer, the great mechanician, Stephen P. Timoshenko.

Another way to validate the assumptions of elementary beam theory is to solve the relevant equations of elasticity theory for the stresses and deformation of a plane beam, that is, solve the axial and transverse equations of equilibrium for the three stress components, after which we can integrate further to find the displacement field. Such an elasticity solution would show that the magnitudes of the stresses obtained in both Timoshenko beam analyses and elasticity problems behave exactly as indicated in Eqs. (7.92) and (7.93). Further, in both kinds of formulations, it also turns out that the equations describing the deflections have a very similar appearance to prior results (e.g., Eqs. (7.44)) and show very similar behavior. For example, for a centrally loaded, pinned-pinned beam, we could apply the shear beam theory to find the midspan deflection as follows:

$$w_{shear}(L/2) = \frac{PL^3}{48EI}\left[1 + \frac{(12+11\nu)}{5}\left(\frac{h}{L}\right)^2\right]$$
$$= w_{elem}(L/2)\left[1 + \frac{(12+11\nu)}{5}\left(\frac{h}{L}\right)^2\right] \tag{7.96}$$

It is clear from Eq. (7.96) that the maximum deflection of such a beam differs from that found with elementary beam theory only by a very small amount – except when the beam truly is short and stubby.

7.4.2 On the Magnitudes of Beam Deflections and Stresses

Just how large is the bending deflection of a beam, and in comparison to what? This answer would be useful to us as a guide to judging the validity of beam deflections that we may be calculating. However, it turns out that the precise answer to the question about the size of bending deflections depends on considerations derived from nonlinear elasticity theory, a subject which is well beyond the scope of our discussions. The only direct approximation we have made so far has been that we assumed the slope of the deflected shape to be small when we identified the curvature approximation (Eq. (7.16)) in the moment-curvature relation (7.14). In this context we could use the dimensionless axial coordinate ξ and write the bending deflection in the form

$$w(x) = w_0 f\left(\frac{x}{L}\right) = w_0 f(\xi) \tag{7.97}$$

where w_0 is the maximum displacement of the bent beam under consideration, so that all of the remaining terms in $w(x)$ embodied in $f(x/L) = f(\xi)$ are of order unity. By the chain rule of the calculus, the slope is then

$$\frac{dw(x)}{dx} = \frac{1}{L}\frac{dw(\xi)}{d\xi} = \frac{w_0}{L}f'(\xi) \tag{7.98}$$

so that the small slope restriction can be written as:

$$\frac{dw(x)}{dx} \ll 1 \quad \Rightarrow \quad \frac{w_0}{L} \ll 1 \tag{7.99}$$

which suggests that a beam's bending deflection is small compared to its length.

However, there is a tighter restriction that derives from the fact that a completely general kinematic description of motion in an elastic body depends on some geometrically nonlinear terms (which form the basis for nonlinear elasticity theory, including structural stability theory). It can be shown that a necessary consequence of these nonlinearities is that they can be ignored only when a beam's bending deflection is small compared to its thickness, that is,

$$\frac{w(x)}{h} \ll 1 \quad \text{or} \quad \frac{w_0}{h} \ll 1 \tag{7.100}$$

Of course, we have implicitly applied this assumption in using the Taylor series expansions (7.7) and (7.8) to develop the kinematic field that we used to begin our energy-based derivation of beam theory. However, in order to be both conservative and safe, we now formally declare that the *bending deflection of an elementary beam should be taken as smaller than the thickness in virtually all circumstances.* Exceptions to this rule are possible, but they require much more elaborate analysis than we need (or want) here.

Finally, on magnitudes of the physical parameters, we can use results similar to the above estimates of the magnitudes of the deflections to estimate the bending stresses that are generated in a bent beam. We know that the bending stress is related to the moment by Eq. (7.15), and that the moment is related to the second derivative of the displacement through the moment-curvature relation (7.14). Thus, eliminating the moment between the two yields

$$\sigma_b = -Ezw''(x) \tag{7.101}$$

Now we can substitute from Eq. (7.97) to find

$$\left(\frac{\sigma_b}{E}\right) = -\left(\frac{z}{h}\right)\left(\frac{h}{L}\right)^2\left(\frac{w_0}{h}\right) f''\left(\frac{x}{L}\right) \tag{7.102}$$

In terms, then, of orders of magnitude of behavior, we can eliminate the functions that are of magnitude of unity to find

$$\left(\frac{\sigma_b}{E}\right) \approx -\left(\frac{h}{L}\right)^2\left(\frac{w_0}{h}\right) \tag{7.103}$$

This result suggests that the *bending stress we calculate for a bent beam should be very much smaller than the Young's modulus* of the material of which the beam is made – perhaps by three (or more) orders of magnitude.

Problems

7.1 Find the exact solution for a simply supported beam under a uniform load q_0 and from it calculate both the maximum deflection and the maximum moment in the beam.

7.2 Find the exact solution for a fixed-ended beam under a uniform load q_0 and from it calculate both the maximum deflection and the maximum moment in the beam. Where are the inflection points on this beam? How do these results compare with those obtained in Problem 7.1.

7.3 Calculate the exact deflection solution for a simply supported beam with a single load P_1 applied at $x = a = L - b$ on the beam.

7.4 For the beam shown in Fig. 7.P4, draw the shear and moment diagrams and sketch the deflected shape.

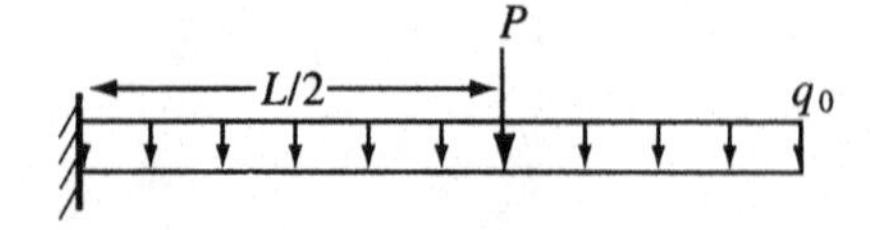

Figure 7.P4. Figure for Problem 7.4.

7.5 Find the exact deflection for the beam of Problem 7.4 and compare the two deflected shapes.

7.6 For the beam shown in Fig. 7.P6, draw the shear and moment diagrams and sketch the deflected shape.

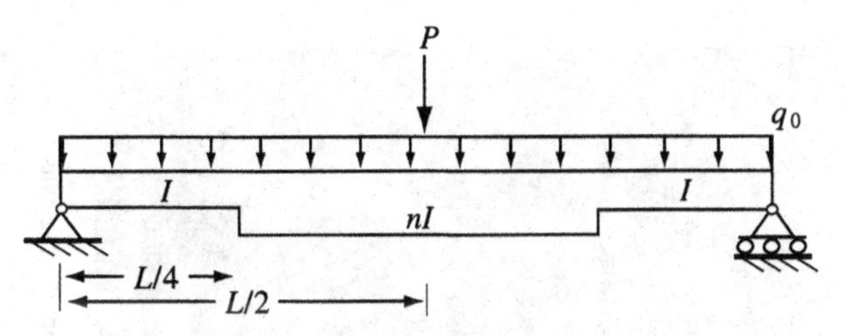

Figure 7.P6. Figure for Problem 7.6.

7.7 Find the exact deflection for the beam of Problem 7.6 and compare the two deflected shapes.

7.8 For the beam shown in Fig. 7.P8, draw the shear and moment diagrams and sketch the deflected shape.

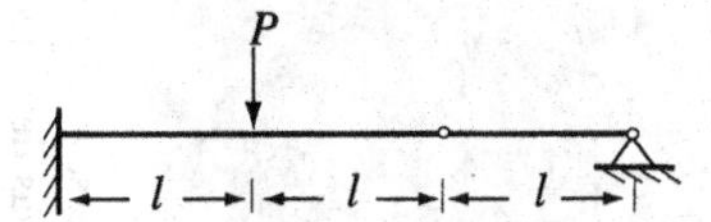

Figure 7.P8. Figure for Problem 7.8.

7.9 Find the exact deflection for the beam of Problem 7.9 and compare the two deflected shapes.

7.10 For the beam and loading of Problem 7.8, find an approximate one-term solution for the maximum deflection using a sinusoid as a trial function. Compare your result with that found in Problem 7.9.

7.11 To extend the solution to Problem 7.10 by adding another term to the trial function, how would you choose the next term? Calculate the maximum deflection with this solution and compare it with the answers to Problems 7.9 and 7.10.

Spanning the Hudson River between Manhattan and the New Jersey palisades, the George Washington Bridge was built in 1931 with a span twice that of any prior suspension bridge. It was designed by Othmar Ammann to be slender, with only a single deck. The second deck and its stiffening truss structure was added in the 1950s. During that added construction the author was able, as a youngster, to watch the change in the bridge's deflected shape due to the weight of the second deck as it crept slowly from each shore toward the center of the bridge. (Photo by J. Wayman Williams.)

8

Calculating Beam Deflections

In this chapter we will focus on ways to calculate beam deflections and force resultants with energy methods. We will apply both the force and displacement methods, solve indeterminate problems with the principle of least work, describe the unit load method, and lay the foundations for Chapter 9 in which we will model structural frames as assemblies of beamlike members.

8.1 Castigliano's Theorems for Beams

We can apply two Castigliano theorems in the context of beam analysis, that is, we can use these theorems to calculate forces and deflections for bent beams. As we discussed in Section 6.5, among other places, the force method makes use of Castigliano's second theorem:

$$u_i = \frac{\partial U^*}{\partial P_i} \tag{8.1}$$

while the displacement method makes use of Castigliano's first theorem:

$$P_i = \frac{\partial U}{\partial u_i} \tag{8.2}$$

We will continue to use the nomenclature of U for strain energy and U^* for complementary energy. Even though these two quantities must be equal in magnitude for linearly elastic materials, we distinguish between them according to whether they are written in terms of forces (U^*) or displacements (U).

8.1.1 First Theorem versus Second Theorem

As a practical matter, in the remaining discussions on beams and frames we will use Castigliano's second theorem. Why will we do that? The first theorem works quite well when it makes sense to write the strain energy in terms of the unknown displacements at discrete points, which is part of finite element modeling, for example. Remember the three-segment axially loaded bar we modeled in Section 4.6? Here we cast the strain energy) of the bar in terms of the displacements of the (discrete) *movable* endpoints of each bar segment (compare this matrix formulation with Eq. (4.77)):

$$U = \frac{1}{2}\begin{Bmatrix} u_2 \\ u_3 \end{Bmatrix}^T \begin{bmatrix} k_1 + k_2 & -k_2 \\ -k_2 & k_2 + k_3 \end{bmatrix} \begin{Bmatrix} u_2 \\ u_3 \end{Bmatrix} \tag{8.3}$$

where the first vector on the right-hand side of Eq. (8.3) is the *transpose* of the column vector of the displacements (see the Appendix).

We will go into more detail about these matrix calculations in Chapter 10, but for now suffice it to say that we can differentiate Eq. (8.3) with respect to the nodal displacements u_2 and u_3 and then apply Castigliano's first theorem (Eq. (8.2)), from which we would find the matrix result

$$\begin{Bmatrix} \dfrac{\partial U}{\partial u_2} \\ \dfrac{\partial U}{\partial u_3} \end{Bmatrix} = \begin{bmatrix} k_1 + k_2 & -k_2 \\ -k_2 & k_2 + k_3 \end{bmatrix} \begin{Bmatrix} u_2 \\ u_3 \end{Bmatrix} \equiv \begin{Bmatrix} P_2 \\ P_3 \end{Bmatrix} \tag{8.4}$$

or, in terms of the *stiffness matrix* $[k]$ for the bar (compare to Eq. (4.77)),

$$\begin{Bmatrix} \dfrac{\partial U}{\partial u_2} \\ \dfrac{\partial U}{\partial u_3} \end{Bmatrix} = [k] \begin{Bmatrix} u_2 \\ u_3 \end{Bmatrix} \equiv \begin{Bmatrix} P_2 \\ P_3 \end{Bmatrix} \tag{8.5}$$

Why have we reprised this "old material" from Chapter 4 to show why (and where) Castigliano's first theorem is useful? For the axially loaded bars we could easily write the displacement field and its corresponding strain energy in terms of discrete nodal values because the displacement between two points in such a bar can be written as a linear function of the coordinates, with the coefficients expressed in terms of the nodal displacement values. This is something that we cannot do as easily or as meaningfully for beams and other structures, as you can see by looking back at the beam analyses we have already done. It may suggest that we approximate beams and other structures as assemblies of very small, finite elements for which we can write the displacements as we do for bars, but as we've already said, such finite elements are the stuff of a different book.

With Castigliano's second theorem we can write the stored energy in terms of discrete forces, and then we are able to calculate *displacements at discrete points* by differentiating with respect to particular forces. This idea has several very useful implementations, including calculating: deflections at points under specified loads, deflections at points at which no loads are given or prescribed, and values of redundant forces for indeterminate structures. We have already done this for axially loaded bars (in Section 4.5), and now we will do it for beams. This is why we are choosing Castigliano's second theorem over his first.

8.1.2 Castigliano's Second Theorem for Beam Deflections

In this section we will use second theorem to find the deflections of bent beams at points where they are loaded and at other points on the beams.

Example 8.1. Find the deflection under the load P_1 and the slope at the tip of the cantilever beam loaded as shown in Fig. 8.1.

Since we will be applying Castigliano's second theorem, we must formulate the complementary energy for this bent beam. For this linear elastic system, we can derive the complementary energy by solving the moment-curvature relation (7.14) for the curvature

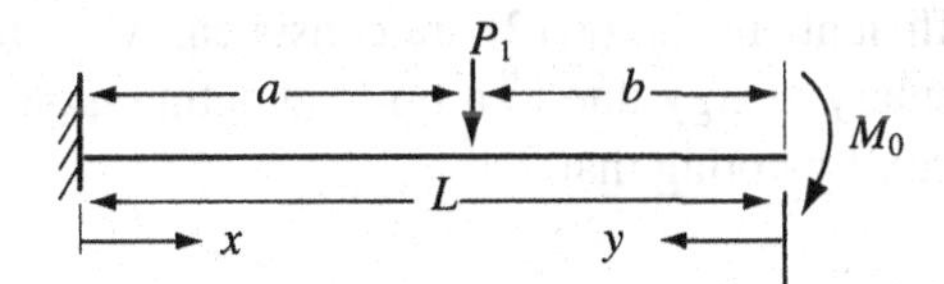

Figure 8.1. A cantilever beam with concentrated loads.

and substituting that into the strain energy:

$$U^* = \frac{EI}{2}\int_0^L \left(\frac{-M(x)}{EI}\right)^2 dx = \frac{1}{2EI}\int_0^L (M(x))^2 dx \tag{8.6}$$

Thus, for the beam shown in Fig. 8.1 we need to find the moment $M(x)$ so that we can evaluate the integral in Eq. (8.6). This is a statically determinate beam for which both the support reactions and the distribution of moment are easily found. We can choose a coordinate sytem to start at either end of the beam, so that the moment can be calculated as if we started from the tip as

$$\begin{aligned} M(x) &= -M_0 & 0 \le x \le b \\ &= -M_0 - P_1(x-b) & b \le x \le L \end{aligned} \tag{8.7}$$

or, equivalently, in a coordinate starting at the support, the moment looks like

$$\begin{aligned} M(y) &= -M_0 + P_1(y-a) & 0 \le y \le a \\ &= -M_0 & a \le y \le L \end{aligned} \tag{8.8}$$

We can evaluate the complementary energy integral (8.6) in terms of y, say,

$$U^* = \frac{1}{2EI}\int_0^a (M_0 - P_1(y-a))^2 dy + \frac{1}{2EI}\int_a^L (M_0)^2 dy \tag{8.9}$$

and so in a straightforward fashion we find that

$$U^* = \frac{a^3}{6EI}P_1^2 + \frac{a^2}{2EI}P_1 M_0 + \frac{L}{2EI}M_0^2 \tag{8.10}$$

In this form, the complementary energy is very clearly a discrete quadratic form, the kind we discussed in Section 6.2. We can write it in a similar form by introducing the following influence coefficients:

$$f_{11} = \frac{a^3}{3EI}, \qquad f_{12} = \frac{a^2}{2EI} = f_{21}, \qquad f_{22} = \frac{L}{EI} \tag{8.11}$$

so that the complementary energy can be written as

$$U^* = \frac{1}{2}f_{11}P_1^2 + f_{12}P_1M_0 + \frac{1}{2}f_{22}M_0^2 \tag{8.12}$$

Both forms of the complementary energy, Eq. (8.10) and Eq. (8.12), are dimensionally correct, as are the corresponding influence coefficients. Also, the influence or flexibility coefficients are symmetric, as will be the stiffness coefficients we find below. We have identified this property as the Maxwell reciprocal principle (cf. Section 6.2).

We can verify that the influence coefficients in Eq. (8.12) are consistent with simpler results. The first term is the complementary energy due to a tip load acting alone on a cantilever of length a, which we can verify by noting that

$$\left.\frac{\partial U^*}{\partial P_1}\right|_{M_0=0} = \delta_a|_{M_0=0} = \frac{P_1 a^3}{3EI} \equiv w(x = b \to 0;\, y = L \to a) \tag{8.13}$$

Similarly, the last term of Eq. (8.12) corresponds to the complementary energy due to a moment acting alone on a cantilever L. That this is true can be verified by noting that

$$\left.\frac{\partial U^*}{\partial M_0}\right|_{P_1=0} = \theta_L|_{P_1=0} = \frac{M_0 L}{EI} \equiv w'(x = 0;\, y = L) \tag{8.14}$$

In the presence of both the concentrated load and the moment at the tip, the total vertical deflection under the load P_1 follows from applying Castigliano's second theorem to the entire complementary energy:

$$\delta_a = \left.\frac{\partial U^*}{\partial P_1}\right|_{y=a} = \frac{a^3}{3EI}P_1 + \frac{a^2}{2EI}M_0 = f_{11}P_1 + f_{12}M_0 \tag{8.15}$$

while the slope at the tip produced by both loads also can be found by the second theorem, that is,

$$\theta_L = \left.\frac{\partial U^*}{\partial M_0}\right|_{y=L} = \frac{a^2}{2EI}P_1 + \frac{L}{EI}M_0 = f_{21}P_1 + f_{22}M_0 \tag{8.16}$$

■

Now the solutions in Eqs. (8.15) and (8.16) can be written in matrix form as

$$\begin{Bmatrix} \delta_a \\ \theta_L \end{Bmatrix} = \begin{bmatrix} f_{11} & f_{12} \\ f_{21} & f_{22} \end{bmatrix} \begin{Bmatrix} P_1 \\ M_0 \end{Bmatrix} \tag{8.17}$$

The square matrix in Eq. (8.17) is the symmetric *flexibility matrix*, which is the inverse of the corresponding stiffness matrix:

$$\begin{bmatrix} f_{11} & f_{12} \\ f_{21} & f_{22} \end{bmatrix} = \begin{bmatrix} k_{11} & k_{12} \\ k_{21} & k_{22} \end{bmatrix}^{-1} \tag{8.18}$$

With the stiffness matrix, which we can calculate either by using determinants to solve the linear equations (8.17) or by inverting their square matrix, we can write the complementary energy as its counterpart strain energy:

$$U = \frac{1}{2}k_{11}\delta_a^2 + k_{12}\delta_a\theta_L + \frac{1}{2}k_{22}\theta_L^2 \tag{8.19}$$

And, as we have just said, it is this form that we cannot get with the first theorem of Castigliano because we have no way of calculating Eq. (8.19) in terms of arbitrary and unknown displacements.

Example 8.2. Find the deflection at the tip of the cantilever shown in Fig. 8.1 when only the load $P = P_1$ is acting.

We can't get that result from the answer we have just worked out in Example 8.1, although we can certainly find the slope at the tip for this case by simply setting M_0 to zero in Eq. (8.16). If we want to use Castigliano's second theorem to find the tip *deflection*, we would put a load at the tip, after which we could find the limiting value of letting that load go to zero to find the effect of the other two loads on the deflection at the end. In other words, we would add a tip load P_2 to the beam shown in Fig. 8.1, so that we have loads acting at both $y = a$ and $y = L$. The moment for this case would then be

$$\begin{aligned} M(y) &= -M_0 + P_1(y-a) + P_2(y-L) \quad & 0 \le y \le a \\ &= -M_0 + P_2(y-L) \quad & a \le y \le L \end{aligned} \tag{8.20}$$

The complementary energy then follows as

$$\begin{aligned} U^* &= \frac{1}{2EI}\int_0^a (M_0 - P_1(y-a) - P_2(y-L))^2 dy \\ &+ \frac{1}{2EI}\int_a^L (M_0 - P_2(y-L))^2 dy \end{aligned} \tag{8.21}$$

The integrals in Eq. (8.21) can be evaluated to find the influence coefficients that allow us to write the quadratic form for the complementary energy:

$$U^* = \frac{1}{2}f_{11}P_1^2 + f_{12}P_1P_2 + f_{13}P_1M_0 + \frac{1}{2}f_{22}P_2^2 + f_{23}P_2M_0 + \frac{1}{2}f_{33}M_0^2 \tag{8.22}$$

or, in matrix form,

$$U^* = \frac{1}{2}\begin{Bmatrix} P_1 \\ P_2 \\ M_0 \end{Bmatrix}^T \begin{bmatrix} f_{11} & f_{12} & f_{13} \\ f_{21} & f_{22} & f_{23} \\ f_{31} & f_{32} & f_{33} \end{bmatrix} \begin{Bmatrix} P_1 \\ P_2 \\ M_0 \end{Bmatrix} \tag{8.23}$$

Now we can apply Castigliano's second theorem to calculate the displacement components at those points where we have discrete loads, that is,

$$\begin{Bmatrix} \delta_a \\ \delta_L \\ \theta_L \end{Bmatrix} = \begin{Bmatrix} \dfrac{\partial U^*}{\partial P_1} \\ \dfrac{\partial U^*}{\partial P_2} \\ \dfrac{\partial U^*}{\partial M_0} \end{Bmatrix} = \begin{bmatrix} f_{11} & f_{12} & f_{13} \\ f_{21} & f_{22} & f_{23} \\ f_{31} & f_{32} & f_{33} \end{bmatrix} \begin{Bmatrix} P_1 \\ P_2 \\ M_0 \end{Bmatrix} \tag{8.24}$$

Now we can find the deflection at the tip of the beam, with or without a load acting there. Thus, without a concentrated load at the tip,

$$\delta_L|_{P_2=0} = \left.\frac{\partial U^*}{\partial P_2}\right|_{P_2=0} = f_{21}P_1 + f_{23}M_0 \tag{8.25}$$

■

Note that we write the complementary energy in terms of *generalized forces*, so we often find that the influence coefficients for a given problem have different physical dimensions.

Thus, for example, the physical dimensions of two of the influence coefficients (8.11) are

$$[f_{11}] = \left[\frac{a^3}{3EI}\right] = \frac{\text{length}}{\text{force}}, \qquad [f_{12}] = \left[\frac{a^2}{2EI}\right] = \frac{1}{\text{force}} \tag{8.26}$$

This is also consistent with respect to the *generalized deflections* we calculate with the second theorem. Thus, for the beam pictured in Fig. 8.1,

$$\delta_a\big|_{P_1=0} = \left.\frac{\partial U^*}{\partial P_1}\right|_{\substack{y=a \\ P_1=0}} = \frac{a^2}{2EI} M_0 = f_{12} M_0 \tag{8.27}$$

and

$$\theta_L\big|_{M_0=0} = \left.\frac{\partial U^*}{\partial M_0}\right|_{\substack{y=L \\ M_0=0}} = \frac{a^2}{2EI} P_1 = f_{21} P_1 \tag{8.28}$$

It is evident that the dimensions of the flexibility coefficient as defined in Eq. (8.26) are appropriate to calculating the deflection with Eq. (8.27) and the slope with Eq. (8.28). Further, we can use these results to characterize in a different way the Maxwell reciprocity principle expressed in Eq. (6.21), that is,

$$f_{12} = \frac{\delta_a\big|_{P_1=0}}{M_0} = f_{21} = \frac{\theta_L\big|_{M_0=0}}{P_1} \tag{8.29}$$

Thus, given the load P_1 at the point $y = a$ and the moment M_0 at the tip of the cantilever, $y = L$, the deflection-per-unit-moment at $y = a$ is equal to the slope-per-unit-force at the tip, In other words, the off-diagonal influence coefficients allow us to characterize the behaviors of pairs of points of a loaded structure.

8.1.3 Consistent Deformations and Least Work for Indeterminate Beams

In the direct displacement method we demonstrated in Chapter 7, we always represented the beam deflection with trial functions that, at a minimum, satisfied all of the geometric conditions. This meant that we didn't have to worry about the determinacy of a given beam. However, having said this, it turns out that we often want to find the redundant forces supporting indeterminate beams (as we did for indeterminate trusses in Section 5.2), and we can apply Castigliano's second theorem to calculate redundant reactions in two ways.

The first method consists of decomposing an indeterminate beam into a set of determinate beams by removing the redundant(s). The beam that remains after we remove the redundant(s) becomes the determinate *primary structure*, and we calculate its deflection at each point where there had been a redundant support. We then remove all of the external loads from the primary structure and separately apply unknown force(s) at the location of each removed redundant. We apply the second theorem and superposition to sum to zero the net deflection for the entire set of decomposed beams at each of those points where redundants were removed. As we now demonstrate, this ensures *consistent deformation* of the fully loaded beam with all supports in place.

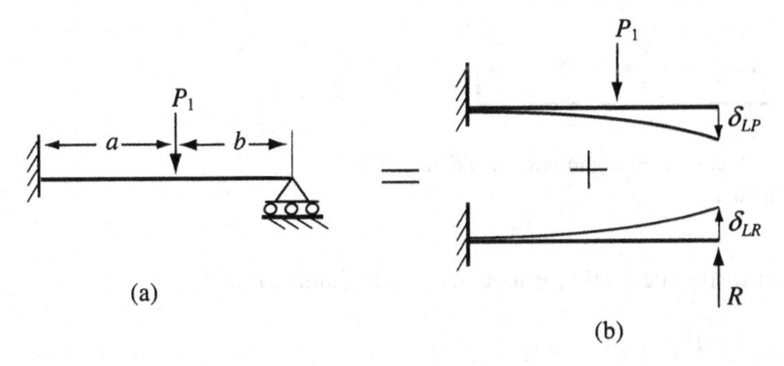

Figure 8.2. (a) An indeterminate beam and (b) a decomposed representation.

Example 8.3. Find the reaction R at the pinned end of the indeterminate beam shown in Fig. 8.2(a).

First we decompose the given beam by removing the redundant reaction to leave the primary structure as the first of the beams in Fig. 8.2(b). With a little imagination, we can use the flexibility coefficients derived in the previous section to calculate the tip deflection of the primary beam by recognizing that the tip deflection is equal to that under the load plus that produced by the projection of the constant slope on the right-half of the beam. Thus, here,

$$\delta_{LP} = \frac{Pa^3}{3EI} + (L-a)\frac{Pa^2}{2EI} \quad \downarrow \tag{8.30}$$

And the deflection due to the upward-directed tip load R is

$$\delta_{LR} = \frac{RL^3}{3EI} \quad \uparrow \tag{8.31}$$

For consistent deformations, the net deflection at the right support must be zero, so, with due regard for the directions shown in Eqs. (8.30) and (8.31), we set

$$\delta_{LP} - \delta_{LR} = 0 \tag{8.32}$$

which thus produces an equation for the redundant reaction, that is,

$$R = \left(\frac{a}{L}\right)^2 \left(\frac{3L-a}{2L}\right) P \tag{8.33}$$

This result is certainly dimensionally correct, and we can also find support for its validity by examining the limiting case of $b \to 0$ or, equivalently, $a \to L$. ■

Example 8.4. Find the redundant reaction R of Example 8.3 by calculating the deflection at the tip of the cantilever beam shown in Fig. 8.3.

This problem is clearly a subset of the problem solved in Example 8.2. Here, however, the moment at the tip, M_0, is zero. We can find the deflection at the tip from the matrix of displacements given in Eq. (8.24), so that now we have

$$\delta_L|_{M_0=0} = \left.\frac{\partial U^*}{\partial P_2}\right|_{M_0=0} = f_{21}P_1 + f_{22}P_2 \tag{8.34}$$

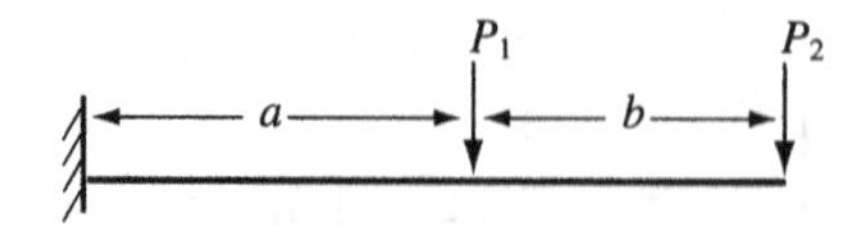

Figure 8.3. A cantilever beam with two concentrated forces as loads.

where the relevant influence coefficients can be calculated to be

$$f_{21} = \frac{L^3}{3EI}\left(\frac{a}{L}\right)^2\left(\frac{3L-a}{2L}\right)$$
$$f_{22} = \frac{L^3}{3EI} \tag{8.35}$$

If we require that the tip deflection vanishes, Eqs. (8.34) and (8.35) tell us that

$$\begin{aligned} P_2 &= -\frac{f_{21}}{f_{22}}P_1 \\ &= -\left(\frac{a}{L}\right)^2\left(\frac{3L-a}{2L}\right)P_1 \end{aligned} \tag{8.36}$$

Thus, a tip load directed upward and of the magnitude calculated in Eq. (8.36) produces the same magnitude (cf. Eq. (8.33)) and has the same effect as a support reaction at the tip, that is, it makes the beam deflection zero at that point. Thus, the deflected shape of our beam is once again shown to be consistent. ■

We now want to show a second method of calculating redundant forces with the aid of Castigliano's second theorem, here with a slighly different twist on calculating the deflection at points that can't move because of the action of the redundant supports. We note that we when we have used the second theorem to calculate the deflections of points on beams, we were following the second paradigm or approach to problems in solid and structural mechanics wherein we use stress or force variables as unknowns (cf. Section 6.5.3). These force variables are chosen to represent an equilibrium state and are then used to minimize the complementary energy (more generally, the total complementary energy), from which process emerges a set of force variables that satisfy compatibility. That is, the minimization of the complementary energy produces an equilibrium set of forces that do not violate any geometrical constraints.

It is in this vein that we now demonstrate the *principle of least work* in which we choose redundant forces or reactions at support points as the basis for minimizing the complementary energy. We know what values we want the deflections to have at such support points. In fact, support movement is typically zero, although it could be otherwise, say, for an elastic support, and we will illustrate how we handle that kind of situation in Example 8.6.

Example 8.5. Find the redundant reaction R at the pinned end of the beam shown in Fig. 8.2(a).

This beam is indeterminate to the first degree, and the right-hand reaction has already been chosen as the redundant. With a coordinate system measured from the tip, the

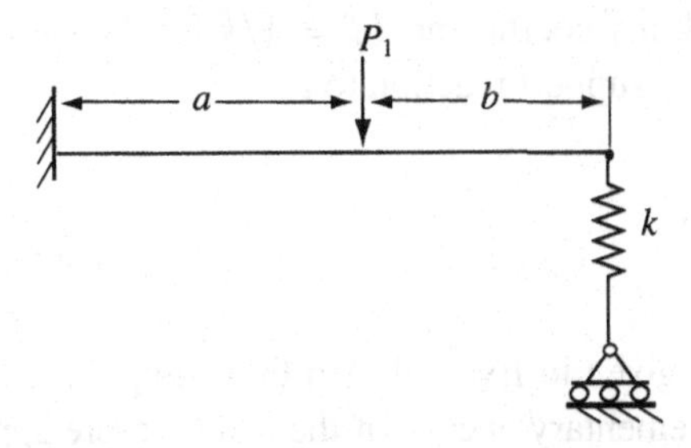

Figure 8.4. A propped cantilever with an elastic support.

moment in the beam is

$$\begin{aligned} M(x) &= Rx & 0 \le x \le b \\ &= Rx - P_1(x-b) & b \le x \le L \end{aligned} \tag{8.37}$$

The corresponding complementary energy integral expressed in terms of x is

$$U^* = \frac{1}{2EI}\int_0^b (Rx)^2 dx + \frac{1}{2EI}\int_b^L (Rx - P_1(x-b))^2 dx \tag{8.38}$$

After performing the integrations in Eq. (8.38) and recalling that $b = L - a$, we find

$$U^* = \frac{1}{2}\left(\frac{a^3}{3EI}\right)P_1^2 - \left(\frac{a^2(3L-a)}{6EI}\right)P_1 R + \frac{1}{2}\left(\frac{L^3}{3EI}\right)R^2 \tag{8.39}$$

Now, we can use Castigliano's second theroem and Eq. (8.39) to calculate the deflection at the right end, and we know that we want that deflection to be zero if the support is truly a rigid support. Thus, we say that

$$\delta_L = \frac{\partial U^*}{\partial R} = -\left(\frac{a^2(3L-a)}{6EI}\right)P_1 + \left(\frac{L^3}{3EI}\right)R = 0 \tag{8.40}$$

Equation (8.40) represents still another equation for the redundant reaction, now determined directly by applying the principle of least work. It takes only a minor bit of algebra to show that Eq. (8.40) produces the same value of the reaction R as we have found previously in Eqs. (8.33) and (8.36). ■

Example 8.6. Find the redundant reaction R at the elastically supported end of the beam shown in Fig. 8.4.

Instead of the rigid support of Fig. 8.2(a), the redundant support of this propped cantilever is an elastic spring of stiffness k. Such structural supports are common because structural members are often supported on other elastic members, which can themselves be modeled as elastic springs. In the style of applying the principle of least work, we would say in this instance that the deflection at the propped end is not zero but, instead, is equal to $-R/k$, where the sign follows from the fact that we are taking the reaction force as positive upward, so that a positive spring deflection would be above the beam's unloaded position. Thus, in place of Eq. (8.40) we would have a modified statement of least work:

$$\delta_L = \frac{\partial U^*}{\partial R} = -\left(\frac{a^2(3L-a)}{6EI}\right)P_1 + \left(\frac{L^3}{3EI}\right)R = -\frac{R}{k} \tag{8.41}$$

Here, and in terms of the appropriate flexibility coefficients ($f = 1/k$ for the spring and Eqs. (8.35) for the beam), we would find the support reaction to be

$$R = \frac{f_{21}}{f_{22} + (1/k)} P_1 = \frac{f_{21}}{f_{22} + f} P_1 \tag{8.42}$$

■

We could also have obtained the result given in Example 8.6 by adding the complementary energy of the spring to the complementary energy of the beam before applying Castigliano's second theorem. In this case we would consider that we are calculating the *net* deflection of the *combined* elastic system, so that we would write the total complementary energy as

$$U^*_{total} = \frac{1}{2} f_{11} P_1^2 - f_{12} P_1 R + \frac{1}{2} f_{22} R^2 + \frac{1}{2} f R^2 \tag{8.43}$$

Since we want the *net* deflection to be zero so that the elastic spring is properly and compatibly connected to the beam, we say that

$$\delta_L^{net} = \frac{\partial U^*_{total}}{\partial R} = -f_{21} P_1 + f_{22} R + f R = 0 \tag{8.44}$$

Equation (8.44) clearly produces the same result as the one given in Eq. (8.42).

We might add that the results for the spring support seem intuitively pleasing, as well as correct. They are, of course, dimensionally correct and consistent. In addition, their limiting behaviors in the case of no spring ($f \to \infty$) and of an infinite spring ($f \to 0$) conform exactly to the limit models we would expect and have found in our prior results in this section.

We can also apply the principle of least work to problems with more than one unknown. The process here is a straightforward extension of what we have done, as we will soon see both in Section 8.2 and when we discuss some indeterminate frames in Chapter 9 (and, of course, there are some problems at the end of the chapter).

Finally, it might seem curious that we have a minus sign in the representations of the complementary energy for the propped-cantilever problem, that is, in Eqs. (8.39) and (8.43). These signs reflect something that is evident from the corresponding deflection equations, that is, Eqs. (8.40) and (8.44), namely, that the loads on the beam, P_1 and R, are in different directions, and so their contributions to the deflections calculated should have different sign.

The appearance of the minus signs may also make you wonder whether the complementary energy is truly positive definite (as we claimed in Chapter 6) if such sign differences appear. Well, it turns out that there is a simple mathematical test that the coefficients of quadratic forms must pass in order to ensure that they are positive definite. Consider the following quadratic form (which is a slightly rewritten version of Eq. (8.43)):

$$U^*_{total} = \frac{1}{2} f_{11} P_1^2 - f_{12} P_1 R + \frac{1}{2} (f_{22} + f) R^2 \tag{8.45}$$

A necessary and sufficient condition for the form (8.45) to be positive definite is that *its principal minors are each positive*. The *principal minors* are determinants developed by forming subdeterminants along the diagonal of the matrix of the coefficients of the quadratic form. In this case that means

$$
\begin{aligned}
&|f_{11}| > 0 \\
&\begin{vmatrix} f_{11} & f_{12} \\ f_{12} & f_{22} + f \end{vmatrix} > 0
\end{aligned}
\tag{8.46}
$$

That is, each of the principal minor determinants shown in Eq. (8.46) must itself be positive. You can easily verify that these determinants of our flexibility coefficients do pass the tests of Eqs. (8.46), so that the complementary energy is indeed positive definite, as advertised.

8.2 Calculating and Using Beam Deflections

In this section we will do some beam problems that are intended to show ways of using the methods we have described to calculate deflections and moments (or stresses) and ways of using these results to model some additional kinds of physical behavior.

8.2.1 Deflections and Flexibility and Stiffness Coefficients

Consider the simply supported beam shown in Fig. 8.5. It is clearly a variant of the classic problem we solved in Section 7.3.1. In fact, we can derive one exact solution to the current problem by slightly perturbing the exact solution (7.43) to the centrally loaded beam. It is easy to show that if there were only a single load P_1 applied to the beam at $x = a = L - b$, the deflection could be written as

$$
\begin{aligned}
w(x;\, P_1@a) = \frac{P_1L^3}{6EI}\Bigg[&\left(\frac{b}{L}\right)\left(1-\left(\frac{b}{L}\right)^2\right)\left(\frac{x}{L}\right)-\left(\frac{b}{L}\right)\left(\frac{x}{L}\right)^3 \\
&+\left(\frac{x}{L}-\frac{a}{L}\right)^3 H(x-a)\Bigg]
\end{aligned}
\tag{8.47}
$$

Clearly, for a beam with the same physical and geometric properties, but on which there is but a single load P_2 applied to the beam at $x = c = L - d$, the corresponding solution would be

$$
\begin{aligned}
w(x;\, P_2@c) = \frac{P_2L^3}{6EI}\Bigg[&\left(\frac{d}{L}\right)\left(1-\left(\frac{d}{L}\right)^2\right)\left(\frac{x}{L}\right)-\left(\frac{d}{L}\right)\left(\frac{x}{L}\right)^3 \\
&+\left(\frac{x}{L}-\frac{c}{L}\right)^3 H(x-c)\Bigg]
\end{aligned}
\tag{8.48}
$$

Equally clearly, for the case of two loads shown in Fig. 8.5, the total solution is found by superposing the two solutions to the individual problems, that is,

$$
w(x;\, P_1@a;\, P_2@c) = w(x;\, P_1@a) + w(x;\, P_2@c) \tag{8.49}
$$

Now we can calculate the deflections at both $x = a$ and $x = c$, that is, under the two

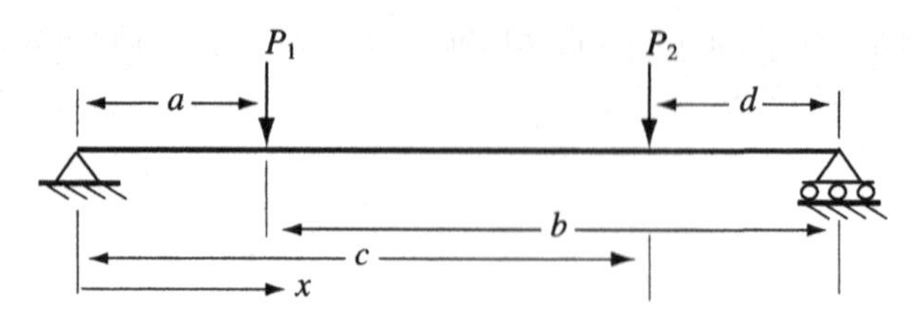

Figure 8.5. A simply supported beam with two loads.

loads, by simply evaluating Eqs. (8.49) in conjunction with Eqs. (8.47) and (8.48):

$$\begin{aligned}\delta_1 &= w(a;\, P_1@a;\, P_2@c) = f_{11}P_1 + f_{12}P_2\\ \delta_2 &= w(c;\, P_1@a;\, P_2@c) = f_{21}P_1 + f_{22}P_2\end{aligned} \tag{8.50}$$

where the flexibility or influence coefficients can be shown to be

$$\begin{aligned}f_{11} &= \frac{a^2(L-a)^2}{3EIL} = \frac{a^2b^2}{3EIL}\\ f_{12} &= \frac{a(L-c)(2Lc-a^2-c^2)}{6EIL} = \frac{ad(L^2-a^2-d^2)}{6EIL} = f_{21}\\ f_{22} &= \frac{c^2(L-c)^2}{3EIL} = \frac{c^2d^2}{3EIL}\end{aligned} \tag{8.51}$$

It is clear that there are several ways of expressing the flexibility coefficients in Eqs. (8.51). For example, we can always write $L-b$ for a and $L-d$ for c. This provides still another reason that we should habitually check both the physical dimensions of each term and any accessible limiting cases. The dimensions of all of the flexibility coefficients of Eqs. (8.51) are all consistent (see Eqs. (8.26) for a reminder), and the coefficients themselves turn out to be just what we should expect for a centrally-loaded beam.

We sometimes have a need for approximate or ad hoc flexibility or stiffness coefficients to do interim calculations. In experimental studies of the vibration response of already built structures or structural models, for example, having such flexibility coefficients allows us to assess the effects on the dynamic response of adding mass or placing loads. We recall that the direct displacement method worked pretty well for calculating displacements, although not as well for calculating stresses, so it would be interesting to see if we can use the direct displacement method to generate approximate flexibility coefficients.

Example 8.7. Find approximate flexibility coefficients for the beam shown in Fig. 8.5 by using a sinusoidal trial function to represent the beam deflection.

For that simply supported beam we take a familiar trial function that clearly satisfies both geometric and force conditions:

$$\tilde{w}(x) = \tilde{W}\sin\frac{\pi x}{L} \tag{8.52}$$

The total potential energy corresponding to this problem is

$$\begin{aligned}\tilde{\Pi} = \frac{1}{2}\int_0^L EI(\tilde{w}''(x))^2dx &- \int_0^L P_1\delta_D(x-a)\tilde{w}(x)dx\\ &- \int_0^L P_2\delta_D(x-c)\tilde{w}(x)dx\end{aligned} \tag{8.53}$$

After substituting the trial function (8.52) and integrating, we obtain

$$\tilde{\Pi} = \frac{\pi^4 EI\tilde{W}^2}{4L^3} - P_1\tilde{W}\sin\frac{\pi a}{L} - P_2\tilde{W}\sin\frac{\pi c}{L} \tag{8.54}$$

We find the trial function's amplitude by minimizing Eq. (8.54), which yields

$$\tilde{W} = \frac{2P_1L^3}{\pi^4 EI}\sin\frac{\pi a}{L} + \frac{2P_2L^3}{\pi^4 EI}\sin\frac{\pi c}{L} \tag{8.55}$$

so that we find our assumed, approximate deflected shape to be

$$\tilde{w}(x) = \left(\frac{2P_1L^3}{\pi^4 EI}\sin\frac{\pi a}{L} + \frac{2P_2L^3}{\pi^4 EI}\sin\frac{\pi c}{L}\right)\sin\frac{\pi x}{L} \tag{8.56}$$

Then, by evaluating Eq. (8.56) at the appropriate coordinates, we can identify approximations of the influence coefficients, that is,

$$\begin{aligned}
\tilde{\delta}_a = \tilde{w}(a) &= \left(\frac{2L^3}{\pi^4 EI}\sin^2\frac{\pi a}{L}\right)P_1 \\
&\quad + \left(\frac{2L^3}{\pi^4 EI}\sin\frac{\pi a}{L}\sin\frac{\pi c}{L}\right)P_2 = \tilde{f}_{11}P_1 + \tilde{f}_{12}P_2 \\
\tilde{\delta}_c = \tilde{w}(c) &= \left(\frac{2L^3}{\pi^4 EI}\sin\frac{\pi c}{L}\sin\frac{\pi a}{L}\right)P_1 \\
&\quad + \left(\frac{2L^3}{\pi^4 EI}\sin^2\frac{\pi c}{L}\right)P_2 = \tilde{f}_{21}P_1 + \tilde{f}_{22}P_2
\end{aligned} \tag{8.57}$$

so that the approximate flexibility coefficients are

$$\begin{aligned}
\tilde{f}_{11} &= \frac{2L^3}{\pi^4 EI}\sin^2\frac{\pi a}{L} \\
\tilde{f}_{12} &= \frac{2L^3}{\pi^4 EI}\sin\frac{\pi a}{L}\sin\frac{\pi c}{L} = \tilde{f}_{21} \\
\tilde{f}_{22} &= \frac{2L^3}{\pi^4 EI}\sin^2\frac{\pi c}{L}
\end{aligned} \tag{8.58}$$

■

We can now compare the approximate flexibility coefficients found in Example 8.7 to their exact counterparts (Eqs. (8.51)). For example, if two equal loads are applied at symmetric quarter points, that is, at $x = L/4$ and at $x = 3L/4$, the ratios of the influence coefficents are

$$\begin{aligned}
\frac{\tilde{f}_{11}}{f_{11}} &= \frac{8}{9}\left(\frac{96}{\pi^4}\right) = 0.876 = \frac{\tilde{f}_{22}}{f_{22}} \\
\frac{\tilde{f}_{12}}{f_{12}} &= \frac{8}{7}\left(\frac{96}{\pi^4}\right) = 1.126 = \frac{\tilde{f}_{21}}{f_{21}}
\end{aligned} \tag{8.59}$$

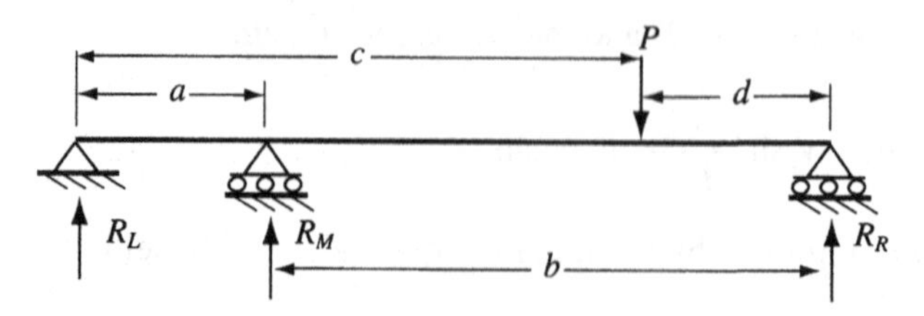

Figure 8.6. An indeterminate beam with three supports and one load.

Thus, we have found pretty good approximations of these coefficients, although we typically don't need approximate flexibility coefficients when we can calculate them exactly. However, it suggests that we might be able to approximate the stiffness coefficients by inverting the approximate flexibility coefficients, that is,

$$\begin{bmatrix} \tilde{k}_{11} & \tilde{k}_{12} \\ \tilde{k}_{21} & \tilde{k}_{22} \end{bmatrix} = \begin{bmatrix} \tilde{f}_{11} & \tilde{f}_{12} \\ \tilde{f}_{21} & \tilde{f}_{22} \end{bmatrix}^{-1} \tag{8.60}$$

Remember that we cannot calculate exact or approximate stiffness coefficients as easily because we cannot cast our stored energy results in terms of discrete, *known* values of displacements at particular points. However, these results suggest a way of calculating approximate stiffness coefficients.

8.2.2 Using Flexibility Coefficients for Indeterminate Beams

We now solve an indeterminate problem that resembles the determinate problem we have solved exactly and approximately in Section 8.2.1. The point is to use influence coefficients to calculate the redundants of indeterminate structures.

Example 8.8. Determine the reactions of the beam shown in Fig. 8.6.

This indeterminate beam looks something like the beam shown in Fig. 8.5. In particular, we could regard the left-hand load $P_1 = -R_M$ as the unknown middle reaction (located at $x = a$) whose value will be determined by requiring the deformation to be consistent. If $P_2 = P$ is the only applied load, the middle reaction of the beam is determined by setting the first of Eqs. (8.50) to zero:

$$\delta_1 = 0 \quad \Rightarrow \quad R_M = \frac{f_{12}}{f_{11}} P_2 = \frac{f_{12}}{f_{11}} P \quad \uparrow \tag{8.61}$$

We are using the arrow in Eq. (8.61) as a reminder that the reaction will act upward. We can find the two remaining reactions by summing vertical forces and moments about either end reaction (and what do we find if we add the following two equations?):

$$\begin{aligned} R_L &= \left(1 - \frac{c}{L}\right) P - \left(1 - \frac{a}{L}\right) R_M \\ R_R &= \frac{c}{L} P - \frac{a}{L} R_M \end{aligned} \tag{8.62}$$

For the case of the middle reaction occurring at the center of the original beam (i.e., $a = L/2$) and the remaining external load being applied at the center of its span (i.e.,

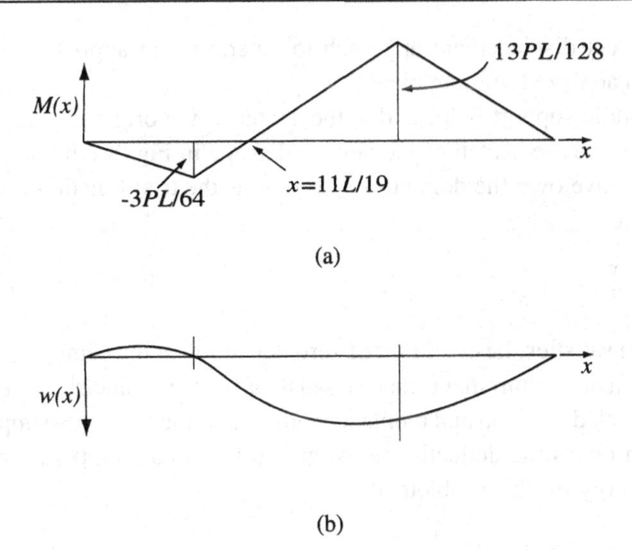

Figure 8.7. (a) The moment diagram and (b) the sketched deflected shape of the beam in Fig. 8.6.

$c = 3L/4$), the three reactions are

$$R_L = -\frac{3}{32}P, \qquad R_M = \frac{11}{16}P, \qquad R_R = \frac{13}{32}P \tag{8.63}$$

We can quickly calculate that these reactions add up to support the load P, but we also note that the leftmost reaction is negative, which means that it is pulling the left end of the beam downward. Is this reasonable? Yes, it is, because the load on the right-hand span pushes that span down, which means there is a negative moment just to the right of the middle support that rotates the beam on that side of the support down, so that on the other (left-hand) side of the support the beam is forced to rotate up to maintain continuity or compatibility. We have sketched the moment diagram and the deflected shape in Fig. 8.7. ■

We can also calculate the deflection δ_2 under the load P for the beam analyzed in Example 8.8 by applying the results given in the second of Eqs. (8.50), with the load $P_1 = -R_M$ and the value of that reaction being given by Eq. (8.61) or Eq. (8.63). Hence we find for both the general and specific cases we have been considering that

$$\begin{aligned}\delta_2 &= -f_{21}R_M + f_{22}P = \left(-f_{21}\frac{f_{12}}{f_{11}} + f_{22}\right)P \\ &= \left(\frac{f_{11}f_{22} - f_{12}^2}{f_{11}}\right)P = \frac{23}{256}f_{11}P = \frac{23}{32}\left(\frac{P(L/2)^3}{48EI}\right)\end{aligned} \tag{8.64}$$

Note that the term in parentheses in the last form of Eq. (8.64) is the deflection of a simply supported beam of length $L/2$ with a concentrated load applied at its center. Thus, the left-hand span provides stiffening to the right-hand span and forces its deflection under the load to be less than what it would be were the left-hand span to disappear. So, after starting with the flexibility coefficients of a simply supported, determinate beam (i.e., Eqs. (8.50)), we have finished by calculating the supporting reactions of an indeterminate beam as well as the deflection of that beam under its applied load.

Example 8.9. Use the direct displacement approach to determine an approximate deflected shape for the beam analyzed in Example 8.8.

We assume that the middle support is located at the center of the original beam (i.e., $x = L/2$). This, together with the sketch of the deflected shape in Fig. 8.7(b), suggests choosing a complete sine wave over the domain $0 \leq x \leq L$ as the trial function for that deflected shape in the form

$$\hat{w}(x) = \hat{W} \sin \frac{2\pi x}{L} \tag{8.65}$$

Note that this trial function satisfies the condition of zero deflection at the center support, which is also consistent with our requirements that we satisfy all of the geometric boundary conditions, in addition to both deflection and moment conditions at the two outer supports. Because our trial function obeys the deflection constraint at the center support, we can write the total potential energy for this problem as

$$\hat{\Pi} = \frac{1}{2} \int_0^L EI(\hat{w}''(x))^2 dx - \int_0^L P\delta_D(x - 3L/4)\hat{w}(x)dx \tag{8.66}$$

If we didn't choose to limit our choice of trial functions as stated, we would have to append to the total potential energy formulation (8.66) a condition (called a *Lagrange multiplier*) to enforce the kinematic condition on our trial function. However, in the light of our choice of Eq. (8.65), we needn't do that. We can simply carry out the usual direct displacement approach and find the amplitude of our trial function to be

$$\hat{W} = -\frac{PL^3}{8\pi^4 EI} \tag{8.67}$$

after which we can evaluate Eq. (8.65) at $x = 3L/4$ to get the following approximate value for the deflection under the load:

$$\hat{\delta}_2 = \frac{1}{2}\left(\frac{P(L/2)^3}{(\pi^4/2)EI}\right) = \frac{(32)(48)}{23\pi^4}\delta_2 \cong 0.686\delta_2 \tag{8.68}$$

This is an approximation that would be acceptable for order-of-magnitude, back-of-the-envelope calculations. We shouldn't be surprised, however, that we are not any closer with only one term because the chosen shape produces peak deflections at $x = L/4$ and $x = 3L/4$ that are equal in magnitude (but not in sign), which we would not expect where we are loading only one of two spans. ■

It is also interesting to compare the beam's true moment diagram (cf. Fig. 8.7(a)) with the approximate moment corresponding to our trial function (8.65):

$$\hat{M}(x) \propto \sin \frac{2\pi x}{L} \tag{8.69}$$

This approximate moment vanishes at the central support, rather than at the true inflection point that is located at $x = (11/19)L \cong 0.579L$. Thus, while we can improve our approximation by taking more terms in a series-type trial function, we should pay attention to the asymmetry of the loading as well as to the symmetry of the geometry. It may be that using trial functions of the form (8.65) may not be the most advantageous for this or similar problems. In fact, it may be that the deflected shape of this

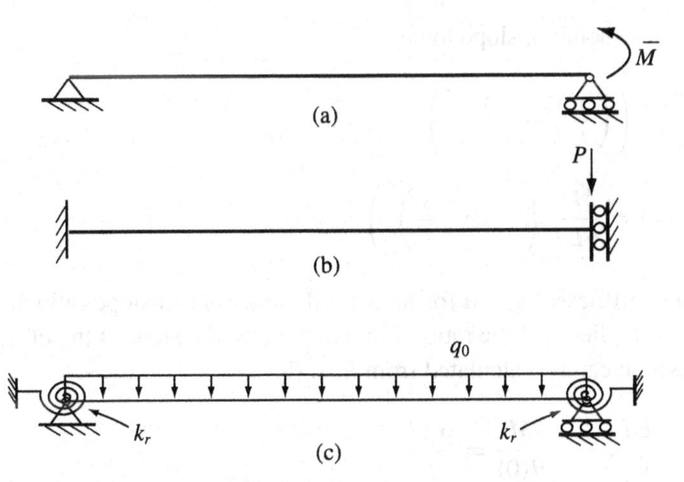

Figure 8.8. Beams and springs: (a) pinned with end moment, (b) clamped with displaced support, and (c) with two rotational springs.

problem might be better approximated with a polynomial or something like a finite element approximation.

8.2.3 Stiffness Coefficients for Beams as Springs and Beams on Springs

We have observed before that elastic structures can be characterized as springs whose stiffnesses depend on the particular circumstances of the element type, its geometry, and its loading. We often use beams as springs, and even more often we characterize their behavior with various stiffnesses. Perhaps the best-known example is the relationship between a force applied at the tip of a cantilever to the deflection under that load, that is,

$$\frac{P}{w_{cant}(L)} = k_{cant} = \frac{3EI}{L^3} \tag{8.70}$$

So, we will solve some beam problems that provide the often used examples of this genre pictured in Fig. 8.8 (and, as usual, we leave still others for you). The results we will obtain in these examples are very useful in modeling and estimating the bending response of connected elastic beams, such as those we find in structural frames.

Example 8.10. Find the applied moment-to-slope ratios at both ends of the beam shown in Fig. 8.8(a).

In this example we are simply applying a given moment $\bar{M}$ at one end of a simply supported beam. The hardest part of solving this problem is ensuring that the correct sign is used in the boundary condition, which here means simply that $M(L) = \bar{M}$. Further, since this is clearly a determinate beam, we need solve only the second-order equation for the deflection because the moment and the reactions are easily determined. Hence, the deflected shape is the solution of

$$M(x) = -EIw''(x) = \frac{\bar{M}}{L}x \tag{8.71}$$

subject to the deflection vanishing at both ends. We can integrate Eq. (8.71) to find the deflection and the corresponding slope to be

$$w(x) = \frac{\bar{M}L^2}{6EI}\left(\left(\frac{x}{L}\right) - \left(\frac{x}{L}\right)^3\right)$$
$$\theta(x) \equiv w'(x) = \frac{\bar{M}L}{6EI}\left(1 - 3\left(\frac{x}{L}\right)^2\right) \tag{8.72}$$

And, finally, the two stiffnesses asked for here are the moment-to-slope ratio at the end where the moment is applied and the ratio of the moment to the slope at the other end of the beam, both of which can be calculated from Eqs. (8.72):

$$\frac{\bar{M}}{\theta(L)} = -\frac{3EI}{L}, \qquad \frac{\bar{M}}{\theta(0)} = \frac{6EI}{L} \tag{8.73}$$

Note that both stiffnesses have the appropriate dimensions of force $\times$ length. Further, we see that a given moment produces half as large an angle at the far end of the beam (away from the point of application) than at its point of application. That is, the far end is twice as stiff or one-half as flexible as the beam at the point of application. ■

Example 8.11. Find the applied moment-to-slope ratios at both ends of a beam like that of Fig. 8.8(a) but for which the left pin is now fixed or clamped.

This beam is indeterminate to the first degree, so that finding a corresponding solution requires solving the homogeneous fourth-order equation subject to the set of boundary conditions

$$w(0) = 0, \qquad w(L) = 0$$
$$w'(0) = 0, \qquad M(L) = -EIw''(L) = \bar{M} \tag{8.74}$$

and for which a solution is easily found. It, in turn, yields the following stiffness:

$$\frac{\bar{M}}{\theta(L)} = -\frac{4EI}{L} \tag{8.75}$$

and at the far end of the beam ($x = 0$) it produces a moment of

$$M(0) = -\frac{\bar{M}}{2} = \frac{2EI}{L}\theta(L) \tag{8.76}$$

Thus, the moment induced at the fixed end of the beam is one-half of that applied over the pinned support. We would also note that there is no particular significance to the minus signs that appear in Eqs. (8.73) and (8.75), other than a slavish adherence to sign conventions for both moment and slope. ■

Example 8.12. Find the effective stiffness that resists (or is seen by) the load at the movable end of the modified fixed-ended beam shown in Fig. 8.8(b).

In this problem there is still a restriction against vertical motion at the right end, although it is the slope of the beam that is required to remain zero there. As with the problem we dealt with in Example 8.11, it is governed by the homogeneous fourth-order equation subject to the following boundary conditions:

$$\begin{aligned} w(0) &= 0, \qquad w'(0) = 0 \\ w'(L) &= 0, \qquad EIw'''(L) + P = 0 \end{aligned} \tag{8.77}$$

Again, we can easily find the solution and calculate the following stiffness:

$$\frac{P}{w(L)} = \frac{12EI}{L^3} \tag{8.78}$$

The deflected shape that produces this interesting result also has an inflection point at the center of the beam – and we will recall both features when we discuss frames in Chapter 9. ■

Example 8.13. Find the exact deflected shape of the shown in Fig. 8.8(c). How do the rotational springs affect the deflection at the center of the beam?

We assume that the two rotational springs shown have stiffness k_r. Here we find the exact solution by integrating the fourth-order beam equation with a constant forcing function q_0 and satisfying four boundary conditions. These four conditions – including two that are the equivalent of prescribing moments that restrict rotation in direct proportion to whatever rotation actually occurs, thus reflecting the rotational restraint introduced by the two springs – are

$$\begin{aligned} w(0) &= 0, \qquad EIw''(0) - k_r w'(0) = 0 \\ w(L) &= 0, \qquad EIw''(L) + k_r w'(L) = 0 \end{aligned} \tag{8.79}$$

After integrating and satisfying the boundary conditions, we find the solution

$$w(x) = \frac{q_0 L^4}{24EI}\left[\left(\frac{x}{L}\right)^4 - 2\left(\frac{x}{L}\right)^3 + \frac{\alpha}{\alpha+2}\left(\frac{x}{L}\right)^2 + \frac{2}{\alpha+2}\left(\frac{x}{L}\right)\right] \tag{8.80}$$

where we have introduced a dimensionless spring ratio α defined as

$$\alpha \equiv \frac{k_r L}{EI} \tag{8.81}$$

Finally, the deflection at the center of the beam is

$$w(L/2) = \frac{q_0 L^4}{384EI}\left[\frac{\alpha+10}{\alpha+2}\right] \tag{8.82}$$

This result clearly portrays the effect of the rotational springs because the term in the brackets includes both the simply supported beam ($\alpha \to 0$) and the fixed-ended beam ($\alpha \to \infty$), and the midspan deflection in each case is exactly what we obtain from the standard analyses of each. ■

8.2.4 Deflections via Unit Loads and Virtual Work

In Section 5.2 we showed that Castigliano's second theorem for calculating truss deflections could be extended to the concept of virtual work and the unit load method. We now apply this idea to calculating the deflections of bent beams. Imagine, for example, that we are designing a wooden diving board. Thus, the right first-order idealization would be the elementary cantilever beam, and we would probably want to limit the tip's deflection and slope by requiring

$$\begin{aligned} w(L) &\leq w_0 \text{ m} \\ \theta(L) &\leq \theta_0 \text{ rad} \end{aligned} \tag{8.83}$$

Thus, we need to know the deflection and slope at the tip in order to design the board, that is, in order to choose the right material and geometric properties.

We start by temporarily placing a load $\bar{P}$ and a moment $\bar{M}$ at the board's tip, and then formulate the complementary energy so that we can take appropriate limits. In formulating the complementary energy, we recognize that the beam's moment distribution is a function of the actual tip load W and our temporary loads:

$$\begin{aligned} M(x) &= -Wx - \bar{P}x - \bar{M} \\ &\equiv M(x; W, \bar{P}, \bar{M}) \end{aligned} \tag{8.84}$$

where we measure the x-coordinate from the tip. The complementary energy is

$$U^* = \int_0^L \frac{M^2(x; W, \bar{P}, \bar{M})}{2EI} dx \tag{8.85}$$

The deflection and slope of the tip of the diving board are then calculated by evaluating the following partially differentiated integrals:

$$\begin{aligned} w(L) &= \left.\frac{\partial U^*}{\partial \bar{P}}\right|_{\bar{P}=\bar{M}=0} = \frac{1}{EI}\int_0^L M(x)\frac{\partial M(x)}{\partial \bar{P}} dx \\ \theta(L) &= \left.\frac{\partial U^*}{\partial \bar{M}}\right|_{\bar{P}=\bar{M}=0} = \frac{1}{EI}\int_0^L M(x)\frac{\partial M(x)}{\partial \bar{M}} dx \end{aligned} \tag{8.86}$$

Before we substitute Eq. (8.84) into Eq. (8.86) to evaluate these integrals, we note that the partial derivative terms in the integrals occur – and are legitimate – because we can interchange the differentiation with respect to the loads $\bar{P}$ and $\bar{M}$ with the integration because the latter is over the spatial coordinate and not over the load. Also, having taken the limits indicated in Eqs. (8.86), we understand $M(x)$ to be the actual moment produced just by the externally applied load(s). Further, we can also see that

$$\begin{aligned} \frac{\partial M(x)}{\partial \bar{P}} &= -x \equiv m_p(x) \\ \frac{\partial M(x)}{\partial \bar{M}} &= -1 \equiv m_m(x) \end{aligned} \tag{8.87}$$

That is, the derivatives of the moment with respect to the loads $\bar{P}$ and $\bar{M}$ are exactly the moment distributions that would have been obtained by simply placing a unit force and a unit moment, respectively, on the otherwise unloaded beam, in the same place and direction of the generalized forces $\bar{P}$ and $\bar{M}$. We see then that the moments produced by the generalized unit loads m_p and m_m are much like the bar force "gradients" we saw in

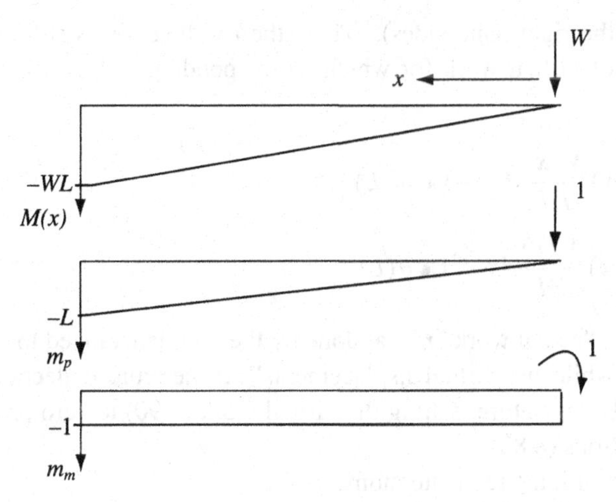

Figure 8.9. Actual and unit moments on a tip-loaded cantilever beam.

Eqs. (5.10) and play the same role as the bar forces produced by unit loads in Eqs. (5.11). Thus, we can rewrite Eqs. (8.86) in the form

$$w(L) = \frac{1}{EI}\int_0^L M(x)m_p(x)dx$$
$$\theta(L) = \frac{1}{EI}\int_0^L M(x)m_m(x)dx \tag{8.88}$$

And so we have the *unit load method* or *dummy load method* for beams.

We show the actual and unit moments of this beam in Fig. 8.9. Note that we use the same sign conventions for calculating unit moments as for moments due to applied loads. As a consequence, the signs on the resulting deflections will conform to the same sign conventions we have been using throughout. Further, there clearly are unit shears produced by the generalized unit loads. However, since we continue to assume that beams are sufficiently slender that shear deformation can be neglected, we do not calculate or use them.

We can also see in Eqs. (8.88) an expression of the concept of virtual work applied to beams, just as we have seen in their truss counterparts, Eqs. (5.14) and (5.15), as well as in Section 3.5. Thus, multiplying Eqs. (8.88) by their unit generalized loads and rearranging the integrals somewhat produces the following results, which we recognize as embodying statements of work:

$$1 \bullet w(L) = \int_0^L m_p(x)\frac{M(x)}{EI}dx$$
$$1 \bullet \theta(L) = \int_0^L m_m(x)\frac{M(x)}{EI}dx \tag{8.89}$$

The work done on both sides of the equations involves unit generalized forces (on the left-hand sides) or *their* consequent moments (e.g., the $m(x)$ on the right-hand sides) multiplying the actual generalized deflections produced at the points where they are sought (on the left-hand sides) or the beam curvatures (from deflections) produced by the applied loads (the $(M(x)/EI)$ on the right-hand sides). Thus, the result (8.89) is still another statement of the principle of virtual work for which, corresponding to Eqs. (3.39) and (3.40), we can write

$$\begin{aligned} \delta W_{virt/p} &= \int_0^L m_p(x)\frac{M(x)}{EI}dx - 1 \bullet w(L) \\ \delta W_{virt/m} &= \int_0^L m_m(x)\frac{M(x)}{EI}dx - 1 \bullet \theta(L) \end{aligned} \tag{8.90}$$

As we have seen before, the "actual work" is that done by the unit generalized loads and their consequent moments, while the "virtual displacements" are the actual deflections and curvatures in the beam. Also as before, setting the virtual work (8.90) to zero produces the desired deflection equations (8.88).

To return to our diving board, the requisite moments are

$$\begin{aligned} M(x) &= -Wx \\ m_p(x) &= -x \\ m_m(x) &= -1 \end{aligned} \tag{8.91}$$

so that the deflection and slope at the tip of the board are the familiar looking

$$\begin{aligned} w(L) &= \frac{WL^3}{3EI} \\ \theta(L) &= \frac{WL^2}{2EI} \end{aligned} \tag{8.92}$$

Thus, in view of the design specification (8.83), we need a board such that

$$\begin{aligned} I &\geq \frac{WL^3}{3Ew_0} \\ I &\geq \frac{WL^2}{2E\theta_0} \end{aligned} \tag{8.93}$$

We are (implicitly) assuming in Eqs. (8.93) that the length of the board and the material of which it is made are fixed, and in fact they are typically prescribed in equipment standards that are set by competitive swimming associations. Thus, there is actually only one "design variable," the second moment of the cross section, I. Can we design a product to meet two requirements – on deflection and slope – if we have only one design variable?

This question is part of another subject, that of design objectives and design optimization. For our present purposes, for a fixed board length, we would simply take the more restrictive design objective, that is, we would take the larger of the pair $(w_0, 2L\theta_0/3)$ and compute a value of I accordingly.

Finally, as we might guess from our prior experience with trusses and the simple example we have just done, the unit load method outlined works for any determinate structure, even if the loading and supports are more complicated. For example, for the beam and loading shown in Fig. 8.10, we can calculate the deflection at the hinge by applying a unit load there. However, in order to calculate a slope at the hinge we must

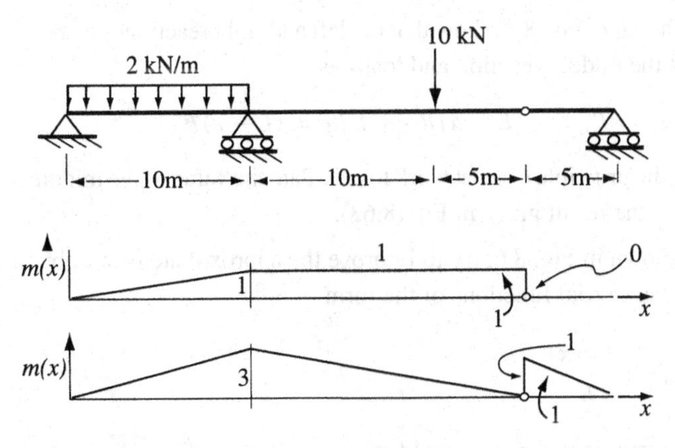

Figure 8.10. Actual and unit moments on a beam with a hinge.

apply two unit moments – because there are actually two different slopes, one on each side of the hinge. The hinge is a point of discontinuity in the slope of the beam, so while the moment is clearly zero there, it is *not* a point of maximum slope (or an inflection point), such as when the moment vanishes at an otherwise "regular" point in a beam.

Problems

8.1 For the beam shown in Fig. 8.4 but having an elastic support at the right end, verify that the flexibility coefficients do satisfy Eqs. (8.42).

8.2 Determine the central reaction for the beam shown in Fig. 8.P2.

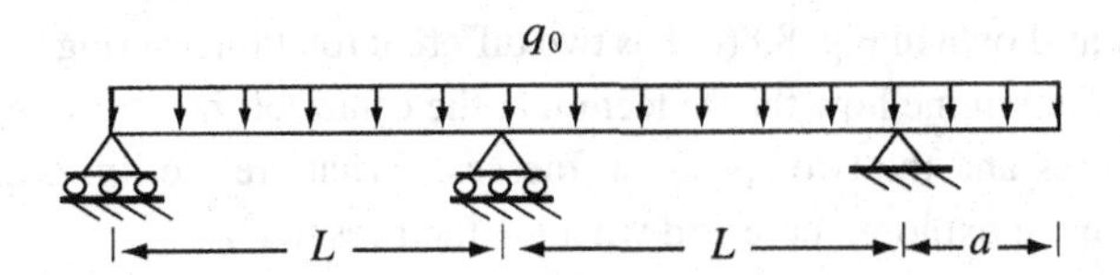

Figure 8.P2. Figure for Problem 8.2.

8.3 Apply the principle of consistent deformations to the indeterminate beam shown in Fig. 8.P3 to demonstrate that the moment at the center support is

$$M_C = -\frac{3\left(P_1 L_1^2 + P_2 L_2^2\right)}{16(L_1 + L_2)}$$

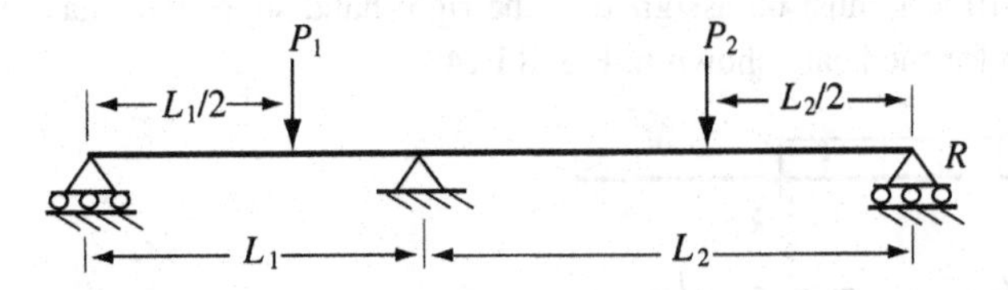

Figure 8.P3. Figure for Problem 8.3.

8.4 For the beam shown in Fig. 8.6, show that the left and right reactions can be expressed as functions of the middle reaction and loads as

$$aR_{\mathrm{M}} + LR_{\mathrm{R}} = cP, \qquad (L - a)R_{\mathrm{M}} + LR_{\mathrm{L}} = (L - c)P$$

and then apply the principle of least work to calculate the value of the middle reaction. Compare it with the result given in Eq. (8.63).

8.5 For the beam shown in Fig. 8.6, try to improve the approximate result obtained with assuming successive trial functions of the form

$$\bar{w} = \sum_{n=1}^{N} \bar{W}_n \sin \frac{2n\pi x}{L}$$

How does this result look (or improve) for large values of n?

8.6 For the beam of Fig. 8.8(b), change the translating support so that it also allows a finite rotation governed by a rotational spring of stiffness k_r. Find the deflected shape for this problem and show that at the tip

$$w(L) = \frac{PL^3}{3EI}\left[1 + \frac{3EI}{k_r L}\right]$$

8.7 Find the solution to the beam bending problem defined by the boundary conditions (8.74) and use it to verify the stiffness given in Eq. (8.75).

8.8 Find the solution to the beam bending problem defined by the boundary conditions (8.77). Use it to confirm the existence and location of an inflection point and to verify the stiffness given in Eq. (8.78).

8.9 Formulate the total potential energy for the problem of Fig. 8.8(c) and minimize it to confirm the differential equation and the boundary conditions (8.79).

8.10 Assume that the beam shown in Fig. 8.8(c) has two different rotational springs at each end, say, k_l and k_r. Determine how the deflection at the center of the beam depends on the beam properties and the two spring stiffnesses. What are the limiting case beam models that can be explored here and what are their results?

8.11 Find the slopes on either side of the hinge of the beam shown in Fig. 8.10. Sketch the beam's deflected shape.

8.12 Find the tip deflection for the beam of Problem 8.4 assuming it is made of steel and that $q_0 = 20$ kN/m, $I = 60{,}000$ cm^4, $L = 10$ m, and $a = 5$ m.

8.13 How would the answer to Problem 8.12 change if the beam were made of aluminum? What cross-sectional properties would be need to keep the deflection the same? Could the beam be made of wood?

8.14 What value of stiffness must be assigned to the right-hand support to make the tip deflection vanish for the beam shown in Fig. 8.P14?

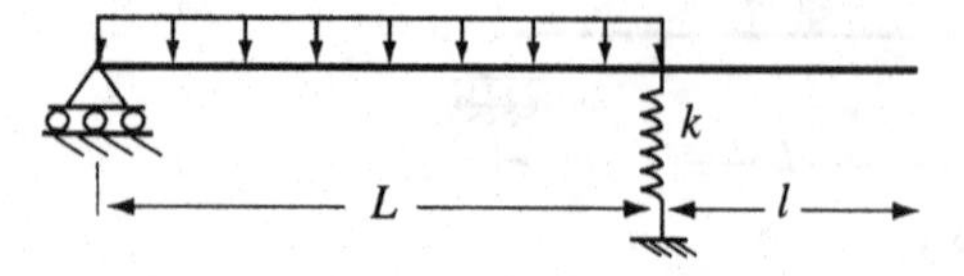

Figure 8.P14. Figure for Problem 8.14.

8.15 Find the flexibility or influence coefficients for the beam shown in Fig. 8.P15.

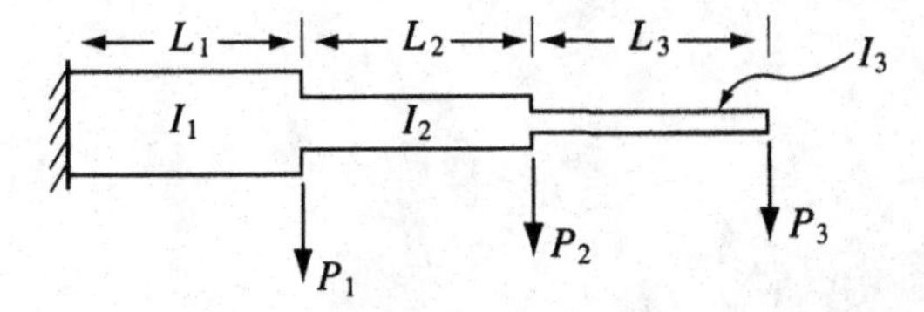

Figure 8.P15. Figure for Problem 8.15.

The towers of the Golden Gate Bridge across San Francisco Bay, built in 1937. The towers are, in fact, multistory portal frames writ very large, and rather dramatically. They are designed to carry the lateral bridge forces that result from wind blowing against the entire suspension structure and roadway. (Photo by Clive L. Dym.)

9

Frames: Assemblages of Beams

We now look at frames as examples of how beams can be applied in more general configurations. In so doing we will have to account for axial loads to maintain equilibrium, even while we are configuring frames with beams, for which the axial loading and deformation uncouple from the bending problem (cf. Section 9.2). Our work here will focus heavily on applying Castigliano's second theorem, so we will introduce matrix notation more frequently, again with the view that it is useful for arithmetic manipulation and as a backdrop for modern structural computation techniques, such as finite element methods.

9.1 Modeling Stored Energy in Frames Assembled of Beamlike Elements

As the first topic in this chapter we introduce assemblages of beams called frames. Frames are very important in structural engineering because they form the skeletons on which most buildings are hung. The most notable examples are the frames of high-rise buildings, which are assemblies of beams that support lateral or transverse loads and columns that carry axial loads down through the structural frame into the foundation and, thus, to ground.

We intend to introduce the principal ideas of how we assemble beamlike elements to form frames. As a starting point, consider the right-angled frame shown in Fig. 9.1(a). This frame is assembled by joining two beams together at the corner, and we assume that *the connection there is rigid*, by which we mean that the angle between the two beam elements, AB and BC, remains at 90° when the frame bends. There clearly must be three reactions at the one support in order for the frame to remain upright, and we can calculate their values (as shown in the figure) with the same principles of statics we use for beams.

However, life gets a little more complicated when it comes to finding the internal forces required to keep each of the beam elements in equilibrium. We see this immediately upon breaking up the frame into its constituent beamlike elements (Figs. 9.1(b,c)). For the horizontal beam BC, for example, there is a horizontal force that we normally do not have for beams, and yet the force Q being applied at point C must be transmitted through the horizontal beam BC to the vertical beam or column AB to get to the support. Further, we must transmit moments and forces in the two directions across the rigid connection at joint B.

The calculations required to start this analysis are not hard, but we must be careful to be consistent throughout. Each element, whether horizontal or vertical, is regarded as a beam that transmits a shear $V(x)$, a moment $M(x)$, and an axial force resultant $N(x)$. Axial force resultants will here be taken as positive in tension, while sign conventions for the shear and moment will be taken exactly as we have done for elementary beams. For

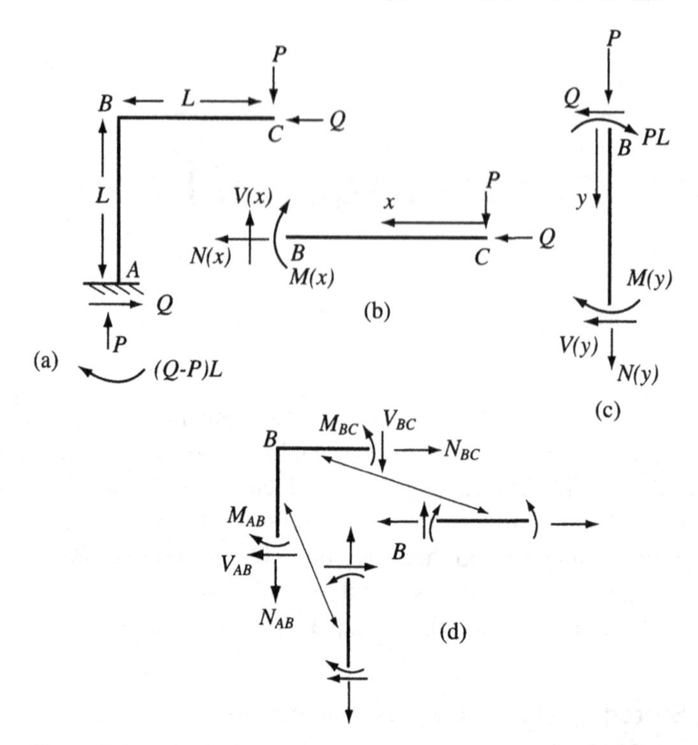

Figure 9.1. A simple frame: (a) overall equilibrium; (b) its horizontal beam, BC; (c) its vertical "beam" or leg, AB; and (d) the rigid joint B.

vertical "beams" or columns, however, the sign issue is a bit ambiguous, so we will take the view that the outside of a column of a frame should be regarded as the top of its beam representation (as in Fig. 9.1(c)).

Now for a sidebar on the nomenclature, because we will label vertical beams as *columns*. We consider vertical legs to be columns in a rather restrictive way because we preclude here the stability analysis of columns through which their *buckling* or *loss of stability* is explored. This is a rather complex subject in its own right, and certainly an important one for structures such as high-rise buildings in which columns at the lower levels must carry the entire weight of the many stories of structure, equipment, and people above them. Here we will consider columns as vertical beams that carry the axial loads as described above, but for which stability analyses are assumed to be investigated separately.

Returning to the frame, we need to adopt a running coordinate for each of its legs, so, as shown in Figs. 9.1(b,c), we will put an x-coordinate along member BC, starting at the tip, and we will run a y-coordinate down the vertical leg from joint B to support A. Further, at the joint B, we must be consistent (and careful) in transmitting forces because shears on a column become axial forces on its neighboring (horizontal) beam, and moments must be transmitted consistently with regard to their direction. We show in Fig. 9.1(d) the beam conventions on the two members at the corner B, together with the corner "exploded."

We now calculate the moment and the shear and axial forces for each of the frame members (cf. Figs. 9.1(b,c)). For the horizontal beam BC we find

$$\begin{aligned} M(x) &= -Px \\ V(x) &= P \\ N(x) &= -Q \end{aligned} \tag{9.1}$$

while for the column AB we find

$$\begin{aligned} M(y) &= Qy - PL \\ V(y) &= -Q \\ N(y) &= -P \end{aligned} \tag{9.2}$$

Notice how the axial force on the horizontal beam becomes a shear force on the column, and the shear force in leg BC is transmitted to the frame support as an axial force in the column AB. This observation should reinforce a most important point: We must account for *both* the shear and axial force resultants, as well as the moments, to ensure equilibrium of each of the elements that make up the frame and of the frame in its entirety.

To model the stored energy properly and consistently, we begin by writing down all of the stored energy terms that appear to be relevant, including the complementary energy due to bending (cf. Eq. (8.6)), axial motion (after Eq. (4.56)), and shear deformation (from Eq. (7.89)). The complementary energy for each leg takes the form

$$U^* = \int_0^L \frac{(M(x))^2 dx}{2EI} + \int_0^L \frac{(N(x))^2 dx}{2EA} + \int_0^L \frac{(V(x))^2 dx}{2\kappa GA} \tag{9.3}$$

We have introduced here a *shape factor*, κ, to account for the fact that shear approximations, such as those discussed in Section 7.4.1, have limitations, one of which is that shape factors are needed to account more precisely for different beam cross sections. There are several reasons why we don't need to explore this in detail here beyond saying that κ is of order unity and often depends on the Poisson's ratio of the material of which the beam or frame is made. However, we will now show that the entire integral is negligible for frames made up of long, slender beams, so we don't need to worry about the details of this term.

We now examine the relative sizes of the integrals in the complementary energy (9.3). For convenience we take the cross sections and the moduli to be uniform over the legs of the frame, in which case we can write that

$$\frac{2EIU^*}{L} = \int_0^1 (M(\zeta))^2 d\zeta + \frac{I}{A}\left[\int_0^1 (N(\zeta))^2 d\zeta + \frac{E}{\kappa G}\int_0^1 (V(\zeta))^2 d\zeta\right] \tag{9.4}$$

where we have also introduced once again a dimensionless axial coordinate, $\zeta = x/L$, along the length of the frame leg. Note that the dimensions of all of the terms are equal to those of energy squared.

From the moment and force distributions in Eqs. (9.1) and (9.2), we see that the forces in the columns are proportional to the applied forces P and Q, while the moments are of order PL and QL. Recall also from elementary calculus that

$$I \equiv Ar^2 \tag{9.5}$$

where r is the radius of gyration of the beam's cross section and of order of magnitude of the thickness h of the beam. Thus, if we divide Eq. (9.4) by $(PL)^2$, we would find that

$$\begin{aligned} \frac{2EIU^*}{P^2L^3} = \int_0^1 \left(\frac{M(\zeta)}{PL}\right)^2 d\zeta + \left(\frac{I}{AL^2}\right) \\ \times \left[\int_0^1 \left(\frac{N(\zeta)}{P}\right)^2 d\zeta + \frac{E}{\kappa G}\int_0^1 \left(\frac{V(\zeta)}{P}\right)^2 d\zeta\right] \end{aligned} \tag{9.6}$$

The complementary energy (9.6) is now dimensionless. All of the integrals in Eq. (9.6) are approximately the same size, and each of them is of order unity. The one distinctive feature is that the axial and shear force integrals are multiplied by the common factor:

$$\frac{I}{AL^2} = \left(\frac{r}{L}\right)^2 \approx \left(\frac{h}{L}\right)^2 \ll 1 \tag{9.7}$$

Thus, if the legs of a frame are long, slender beams, we can ignore the axial and shear energy terms and simply take the complementary energy as

$$\frac{2EIU^*}{P^2L^3} \cong \int_0^1 \left(\frac{M(\zeta)}{PL}\right)^2 d\zeta \tag{9.8}$$

or, in dimensional terms

$$U^* \cong \int_0^L \frac{(M(x))^2 dx}{2EI} \tag{9.9}$$

Remember, however, that it is absolutely essential that we account for the axial and shear resultants when we perform the equilibrium analysis that leads to the moments that go into Eq. (9.9).

Another way of stating this approximation is to note that the deformations produced by the axial and shear deformations are so small, relatively speaking, that the net work done by these forces is also very small. As a consequence, the net energy stored as a result of this work is negligible. We are *not* saying that the axial and shear forces are negligible, only that the work they do is negligibly small. We will verify this in Example 9.1 (below) when we actually calculate the deflection of the tip of a loaded frame.

9.2 Calculating the Deflections of Frames

The movement of one or more points on a frame can be determined by using Castigliano's second theorem, just as we have done for trusses and beams.

Example 9.1. Find the vertical and horizontal deflections of point C for the frame shown in Fig. 9.1.

We can calculate the desired deflections by constructing the complementary energy and differentiating it with respect to the loads P and Q, respectively. Since we can use this simple model to confirm the order-of-magnitude analysis we've just completed, we will also account for the axial and shear energies here.

To calculate the complete complementary energy for the frame we simply substitute the resultant distributions given in Eqs. (9.1) and (9.2) for the two frame members into the complementary energy as given in either the dimensional (Eq. (9.4)) or dimensionless (Eq. (9.6)) forms. We find the former to be easier, so that we have here

$$\begin{aligned}\frac{2EIU^*}{L} &= \int_0^1 (-P\zeta)^2 d\zeta + \frac{I}{A}\left[\int_0^1 (-Q)^2 d\zeta + \frac{E}{\kappa G}\int_0^1 (P)^2 d\zeta\right] \\ &+ \int_0^1 (Q\zeta - P)^2 d\zeta + \frac{I}{A}\left[\int_0^1 (-P)^2 d\zeta + \frac{E}{\kappa G}\int_0^1 (-P)^2 d\zeta\right]\end{aligned} \tag{9.10}$$

so that after the integrations and some algebra we find a simple looking form that makes it quite easy to pick out the origin of individual terms:

$$\frac{2EIU^*}{L} = \left(\frac{4}{3}P^2 - PQ + \frac{1}{3}Q^2\right)L^2 + \frac{I}{A}\left[(P^2 + Q^2) + \frac{E}{\kappa G}(P^2 + Q^2)\right] \tag{9.11}$$

Inspection of Eq. (9.11) reveals that the first set of terms are due to the bending of the legs, the second to the axial deformation, and the last to the shear deformation.

Now we can straightforwardly apply Castigliano's second theorem to find the horizontal and vertical components of the deflection of point C as

$$\begin{Bmatrix} \delta_{Cv} \\ \delta_{Ch} \end{Bmatrix} = \begin{Bmatrix} \dfrac{\partial U^*}{\partial P} \\ \dfrac{\partial U^*}{\partial Q} \end{Bmatrix}$$

$$= \frac{L^3}{EI}\begin{bmatrix} \dfrac{4}{3} + \dfrac{1}{AL^2}\left(1 + \dfrac{E}{\kappa G}\right) & -\dfrac{1}{2} \\ -\dfrac{1}{2} & \dfrac{1}{3} + \dfrac{1}{AL^2}\left(1 + \dfrac{E}{\kappa G}\right) \end{bmatrix}\begin{Bmatrix} P \\ Q \end{Bmatrix} \tag{9.12}$$

The flexibility matrix in Eq. (9.12) is properly symmetric, and the dimensions on the right-hand side are clearly those of length. We can easily identify the contributions due to axial and shear deformation in the individual flexibilities (cf. the diagonal terms), and, in accord with our slenderness assumption, these contributions are negligible. In scalar form, the vertical displacement is

$$\delta_{Cv} = \left[\frac{4}{3} + \frac{1}{AL^2}\left(1 + \frac{E}{\kappa G}\right)\right]\left(\frac{PL^3}{EI}\right) - \frac{1}{2}\left(\frac{QL^3}{EI}\right) \tag{9.13}$$

which for a pair of slender members we can approximate as

$$\delta_{Cv} \cong \frac{4}{3}\left(\frac{PL^3}{EI}\right) - \frac{1}{2}\left(\frac{QL^3}{EI}\right) \tag{9.14}$$

This confirms our analysis of what effects are likely to be important for frames made up of slender members, and it shows that the motion of the tip C is clearly a function of the bending in both members due to the two forces applied there. ■

Now, in the real structural world, there are two broad classes of forces that act on structural frames. One class consists of the vertical loads that frames are expected to carry down to their foundations. In this instance, important as it is, the beams act very much like the beams we analyzed earlier in this chapter, although they are invariably supported in an indeterminate fashion, and the columns must both carry the axial forces and be checked for their ability to avoid buckling under the combined effects of the large axial forces and the bending and shear transmitted from the beams to which they are connected.

The second class of forces are lateral or "sideways" forces that arise due to wind pressure, ground motion, and earthquakes. An important aspect of this is to ensure a minimum *sidesway* or lateral deflection, especially for a structure such as the *portal frame* shown in Fig. 9.2(a). Such portal frames – the name derives from the Latin word root

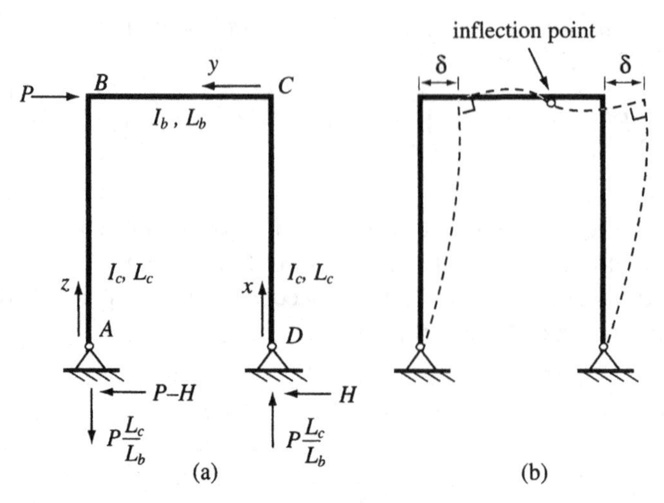

Figure 9.2. An indeterminate portal frame: (a) overall equilibrium and (b) a sketch of the deflected shape (greatly exaggerated).

"port," meaning gate or passageway – are ubiquitous in structures, including the door frames in our homes, the "building blocks" of high-rise buildings, and the cross-bridge structures that make up the towers of elegant suspension bridges. There they transmit the formidable wind loads on the suspension cables and the towers down to the bridge abutments.

Our simple portal frame is also evocative of another kind of structure, namely, facilities that need large spans that are uninterrupted by columns or other structural features. As we will see later, structures so designed tend to need top or "roof" beams that are both very stiff and very efficient in terms of their weight. As with the frame in Fig. 9.2(a), these structures must be designed for their ability to support a lateral load P as shown, and to transmit that load down to the two reaction pins. This frame is indeterminate to the first degree because there are two reaction forces at each of the support points.

We anticipate that the deflected shape of the frame will be as sketched in Fig. 9.2(b). Remember that the joints at points B and C must remain at 90°, as a result of which we should also anticipate that there will be an inflection point – that is, a point where the moment vanishes as it changes sign – in the top (horizontal) beam. We could also expect inflection points in each in of the columns if the base supports were clamped, as we will show in the next section. In the next two examples we will determine the reactions of this frame as well as its *sidesway*, that is, the lateral movement of the top of the frame.

Example 9.2. Find the reactions of the frame shown in Fig. 9.2(a).

The two vertical reactions for the given frame are easily determined by summing vertical forces and moments. However, one of the two horizontal reactions will be our redundant, and we will determine its value by applying the least work principle. In order to formulate the complementary energy for the frame we need apply static equilibrium, and before doing that we need a set of member coordinates. Thus, for member CD we choose the x-coordinate starting at D, for member BC the y-coordinate starting at C, and for member AB the z-coordinate starting at A. Then we can calculate the resultants needed to maintain static equilibrium, and here we report the only the moments:

$$\begin{aligned} M_{DC}(x) &= -Hx & 0 \le x \le L_c \\ M_{CB}(y) &= P\frac{L_c}{L_b}y - HL_c & 0 \le y \le L_b \\ M_{AB}(z) &= (P-H)z & 0 \le z \le L_c \end{aligned} \tag{9.15}$$

where P is the applied load and H is the horizontal reaction at the support D, which we take as the redundant. We are also adopting an informal convention of identifying the member for which a moment is valid with a subscript pair, the first of which is the point from which the relevant coordinate is measured. Then we can substitute these moments into the complementary energy (9.9) to find

$$\begin{aligned} U^* &= \int_0^{L_c} \frac{(-Hx)^2 dx}{2E_c I_c} + \int_0^{L_b} \left(\frac{L_c}{L_b}\right)^2 \frac{(Py - HL_b)^2 dy}{2E_b I_b} \\ &+ \int_0^{L_c} \frac{((P-H)z)^2 dz}{2E_c I_c} \end{aligned} \tag{9.16}$$

Note that in Eq. (9.16) we have maintained separate properties for the top beam, as distinct from the two symmetric columns.

After integration the complementary energy becomes

$$U^* = \frac{L_c^3}{6E_c I_c}H^2 + \frac{L_b^3}{6E_b I_b}\left(\frac{L_c}{L_b}\right)^2 (P^2 - 3PH + 3H^2) + \frac{L_c^3}{6E_c I_c}(P-H)^2 \tag{9.17}$$

We can now apply the least work theorem to Eq. (9.17), which means that

$$\frac{\partial U^*}{\partial H} = 0 = \frac{L_c^3}{3E_c I_c}\left[H + \frac{E_c I_c}{L_c}\frac{L_b}{E_b I_b}\left(-\frac{3}{2}P + 3H\right) - (P-H)\right] \tag{9.18}$$

from which we find a result that seems surprising at first glance, that is,

$$H = \frac{P}{2} \tag{9.19}$$

Thus, the redundant is independent of the relative stiffness. As a consequence, the horizontal reactions respond symmetrically to the lateral load P. ■

Now, we can rewrite the complementary energy (9.17) in the following, more interesting form (which has been foreshadowed in Eq. (9.18)):

$$U^* = \frac{1}{2}\left(\frac{L_c^3}{3E_c I_c}\right)\left[H^2 + \frac{E_c I_c}{L_c}\frac{L_b}{E_b I_b}(P^2 - 3PH + 3H^2) + (P-H)^2\right] \tag{9.20}$$

The coefficient in parentheses in the front of this result clearly reflects the bending stiffness of a cantilever beam with a concentrated load at its tip. This suggests that a dominant element in this frame is how each leg – but particularly the columns – responds to transverse loads at its respective end. And if we look at the first and third moment expressions in Eqs. (9.15) we should not be surprised if the columns act like cantilevers with the indicated tip loads. Thus, too, the first and third terms within the brackets in Eq. (9.20) reflect the rest of the stored energy expressions for these legs.

The middle term in the brackets of Eqs. (9.18) and (9.20) is also quite interesting because it clearly indicates how the stiffnesses of the beam and the columns relate to each other, here through the dimensionless relative stiffness ratio that we will call k_r,

$$k_r \equiv \frac{E_c I_c}{L_c} \frac{L_b}{E_b I_b} = \frac{E_c I_c / L_c}{E_b I_b / L_b} \tag{9.21}$$

Why does this ratio have this form, that is, why doesn't length appear raised to a higher power, say, to the third? The answer can be seen by looking at some of the beam results we developed in Chapter 8. If we wanted to calculate the slope of a cantilever tip due to a moment applied there, we can see from Eqs. (8.16) and (8.11) that

$$\theta_L = \frac{L}{EI} M_0 \tag{9.22}$$

Now, when we connect two different beams in a frame, accounting for their different properties (that is, lengths, moduli, and second moments of the area), we assume that the joint remains rigid, so that both beams carry the same moment while rotating through the same angles relative to their far ends.

What happens if the far ends of two rigidly connected beams are not both clamped? A similar analysis starting with beams pinned at both ends and subjected to the same moment would produce a slope change differing only by a number, which would cancel out in the ratio of the two stiffnesses. Thus, the "transmitting" behavior will always be governed by a ratio of the form of Eq. (9.21). This can be verified by examining other moment-loaded beams, as well as by looking at other frame problems (e.g., the frame of Fig. 9.1 with differing leg properties).

Example 9.3. Find the sidesway of the frame shown in Fig. 9.2(a), that is, the lateral deflection of the frame's top beam.

We can use Castigliano's second theorem to calculate the frame sidesway. First we update the complementary energy (9.20) with the known value of the redundant (9.22):

$$U^* = \frac{1}{2}\left(\frac{L_c^3}{12 E_c I_c}\right)(2 + k_r) P^2 \tag{9.23}$$

Then we apply Castigliano's second theorem to find the sidesway to be

$$\delta = \frac{\partial U^*}{\partial P} = (2 + k_r) \frac{P L_c^3}{12 E_c I_c} \tag{9.24}$$

■

The result found in Example 9.3 has no obvious interpretation that jumps off the page. Therefore, in order to get a better sense of what Eq. (9.24) means we will examine two limiting cases. We assume that the properties of the columns are arbitrarily fixed, and then we will see what happens as we vary the stiffness of the top beam making it first more flexible and then stiffer than the columns.

From Eq. (9.21) we see that the case where the beam is, relatively speaking, quite flexible can be modeled by taking the limit $k_r \to \infty$. In this instance we see from

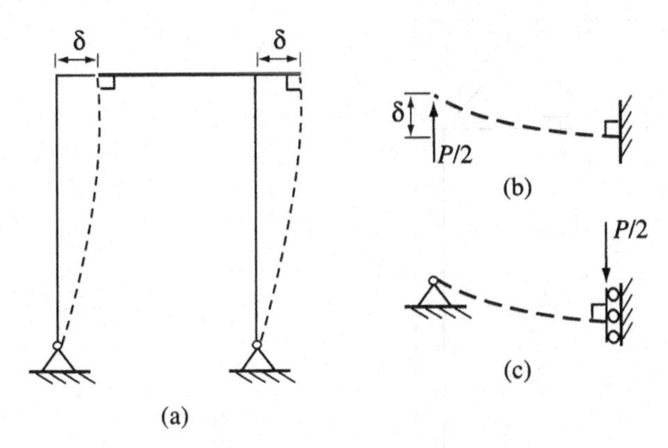

Figure 9.3. Sketches of the deflected shapes of: (a) a frame with a rigid horizontal crossbar, (b) a tip-loaded cantilever, and (c) an equivalent cantilever.

Eq. (9.24) that the lateral sidesway becomes indefinitely large. This makes sense because it amounts to stating that the columns are effectively disconnected from one another when the stiffness of the top beam is negligibly small compared to the stiffness of the columns, and the two individual columns cannot stand in stable equilibrium as tip-loaded beams on single pin supports at their base ends.

The case of the beam being much stiffer than the columns corresponds to the limit $k_r \to 0$. This model produces a deflection limit of

$$\delta_{k_r \to 0} = (2)\frac{PL_c^3}{12E_cI_c} = \frac{PL_c^3}{6E_cI_c} = \frac{(P/2)L_c^3}{3E_cI_c} \tag{9.25}$$

Equation (9.25) appears to suggest that when the beam is very stiff, the sidesway of the frame is equivalent to the deflection of a cantilever beam having the same properties of the columns, but having only one-half the lateral load acting. In fact, this result is both reasonable and right, as can be seen from the deflected shapes shown in Fig. 9.3. The reasoning is as follows. As the horizontal beam becomes quite stiff, it loses its ability to bend and deform as in the sketch in Fig. 9.2(b). The inflection point disappears and the bar ensures that the ends of the columns displace the same amount and that the slopes of the column tops will be increasingly flat and perpendicular to the horizontal beam at points B and C. Note also that while this develops, the columns cannot develop inflection points because of their pinned reaction supports. Thus, the columns must look as sketched in Fig. 9.3(a).

By comparison, we show a sketch of a bent, tip-loaded cantilever beam in Fig. 9.3(b). The only difference between its shape and those of the frame's columns is that the support moves relative to the (now supported) tip (cf. Fig. 9.3(c)). Thus, a rigid top beam ensures that each column transmits one-half the lateral load, and both columns exhibit the same net deflection at their tips due to that one-half load, hence the result of Eq. (9.25).

As we noted earlier, such a circumstance is often found in practice because there are many buildings that require significant open floor spaces, for example, airplane hangars and large manufacturing facilities. The design response is to support the roof on a set of deep trusses that act as very efficient (that is, light-weight), long-span beams. We can see

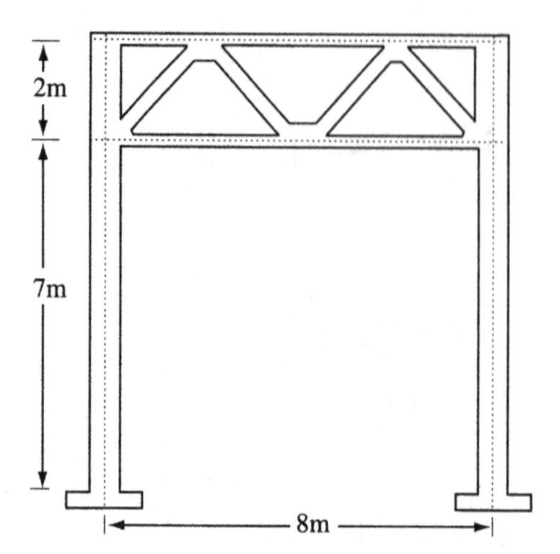

Figure 9.4. A portal frame with a deep truss as its top beam.

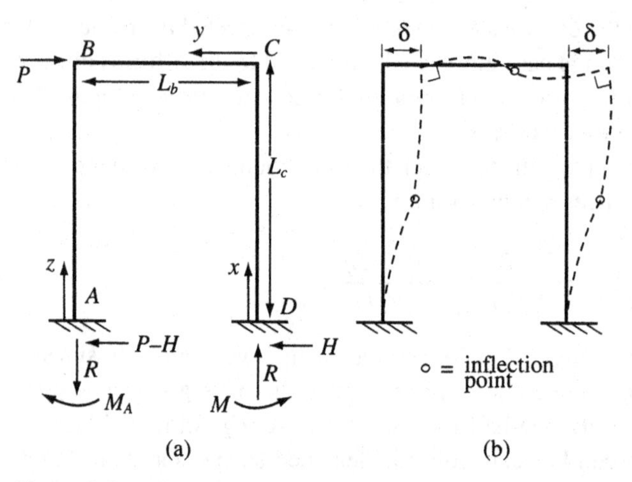

Figure 9.5. A laterally loaded portal frame on fixed supports: (a) overall equilibrium and (b) the deflected shape (greatly exaggerated).

in Fig. 9.4 that such a portal frame will produce exactly the same response that we have described for the limiting case $k_r \to 0$.

Example 9.4. Find the sidesway due to the lateral load P for the portal frame shown in Fig. 9.5.

This frame is indeterminate to the third degree because of the fixed supports at the foot of each column, so analyzing it will be more difficult than were the two frames we have already done. Consequently, we will not provide all of the details here, but we will do enough to establish some points for discussion.

We choose the reactions at the right-hand support as the redundants and, as before, we choose: for member CD an x-coordinate starting at D, for member BC a y-coordinate starting at C, and for member AB a z-coordinate starting at A. Then we can calculate the resultants needed to maintain equilibrium, and here again we report the only the moments:

$$\begin{aligned} M_{DC}(x) &= M - Hx & 0 \le x \le L_c \\ M_{CB}(y) &= M + Ry - HL_c & 0 \le y \le L_b \\ M_{AB}(z) &= M_A + (P - H)z & 0 \le z \le L_c \end{aligned} \tag{9.26}$$

where P is the applied load and the redundants at the support D are as follows: H is the horizontal reaction, R is the vertical reaction, and M is the moment. To simplify the algebra somewhat, we also identify the moment at the support A as

$$M_A = M + RL_b - PL_c \tag{9.27}$$

Note that setting both M and M_A to zero reduces the moments (9.26) to their counterparts (9.15) for the frame on pinned supports, a fact that will also be true of the results that follow. We are also continuing to use the convention of identifying the member for which a moment is valid with a subscript pair, the first of which is the point from which the relevant coordinate is measured. We now substitute the moments (9.26) into the complementary energy (9.9), modified to account for distinct column and beam properties, to find

$$U^* = \int_0^{L_c} \frac{(M - Hx)^2 dx}{2E_c I_c} + \int_0^{L_b} \frac{(M + Ry - HL_c)^2 dy}{2E_b I_b} + \int_0^{L_c} \frac{(M_A + (P - H)z)^2 dz}{2E_c I_c} \tag{9.28}$$

After integration and a fair bit of algebra, we find that the complementary energy becomes (and again we use the relative stiffness parameter k_r)

$$U^* = \frac{1}{2}\left(\frac{L_c^3}{3E_c I_c}\right) \times \left[\begin{array}{l} 3(2 + k_r)\left(\frac{M}{L_c}\right)^2 - 6(1 + k_r)\left(\frac{M}{L_c}\right)H + 3(2 + k_r)\left(\frac{M}{L_c}\right)\left(\frac{RL_b}{L_c}\right) \\ -3\left(\frac{M}{L_c}\right)P + (2 + 3k_r)H^2 - 3(1 + k_r)\left(\frac{RL_b}{L_c}\right)H + HP \\ + (3 + k_r)\left(\frac{RL_b}{L_c}\right)^2 - 3P\left(\frac{RL_b}{L_c}\right) + P^2 \end{array}\right] \tag{9.29}$$

Equation (9.29) can be written more compactly. However, since our primary interest is using the least work principle to find the three redundants, we just set

$$\frac{\partial U^*}{\partial M} = 0, \qquad \frac{\partial U^*}{\partial H} = 0, \qquad \frac{\partial U^*}{\partial R} = 0 \tag{9.30}$$

as a result of which we obtain the following linear equations for the redundants:

$$\begin{bmatrix} 6(2 + k_r) & -6(1 + k_r) & 3(2 + k_r) \\ -6(1 + k_r) & 2(2 + 3k_r) & -3(1 + k_r) \\ 3(2 + k_r) & -3(1 + k_r) & 2(3 + k_r) \end{bmatrix} \begin{Bmatrix} \dfrac{M}{L_c} \\ H \\ \dfrac{RL_b}{L_c} \end{Bmatrix} = \begin{Bmatrix} 3 \\ -1 \\ 3 \end{Bmatrix} P \tag{9.31}$$

These linear equations can be solved in many ways to yield the solution

$$\begin{Bmatrix} \dfrac{M}{L_c} \\ H \\ \dfrac{RL_b}{L_c} \end{Bmatrix} = \begin{Bmatrix} \dfrac{(3+k_r)P}{2(6+k_r)} \\ \dfrac{P}{2} \\ \dfrac{3P}{(6+k_r)} \end{Bmatrix} \tag{9.32}$$

Having found the three redundants, we can substitute them into the complementary energy, Eq. (9.29) and do some algebra to find a simpler and more elegant form:

$$U^* = \frac{1}{2}\left(\frac{L_c^3}{12E_cI_c}\right)\left(\frac{3+2k_r}{6+k_r}\right)P^2 \tag{9.33}$$

Now we can calculate the sidesway of this frame by applying Castigliano's second theorem, that is,

$$\delta = \frac{\partial U^*}{\partial P} = \left(\frac{3+2k_r}{6+k_r}\right)\left(\frac{PL_c^3}{12E_cI_c}\right) \tag{9.34}$$

■

The result found in Example 9.4, like that found in Example 9.3, also lacks an obvious interpretation, so once again we look at limiting cases. We model a highly flexible top beam by taking the limit $k_r \to \infty$, in which case we see from Eq. (9.34) that

$$\delta_{k_r\to\infty} \cong \left(\frac{2k_r}{k_r}\right)\frac{PL_c^3}{12E_cI_c} = \frac{PL_c^3}{6E_cI_c} = \frac{(P/2)L_c^3}{3E_cI_c} \tag{9.35}$$

Thus, the sidesway of the frame is the same as the deflection of a cantilever beam having the same properties as the columns, but acted upon by only one-half the lateral load. This makes sense because when the stiffness of the top crossbar is negligibly small, the columns are (again) disconnected from one another, and so each column acts as a cantilever bent by its half of the lateral load.

We model the case of the top beam being much stiffer than the columns by examining the limit $k_r \to 0$, which here produces the following deflection limit:

$$\delta_{k_r\to 0} = \left(\frac{1}{2}\right)\frac{PL_c^3}{12E_cI_c} = \frac{PL_c^3}{24E_cI_c} = \frac{(P/2)L_c^3}{12E_cI_c} \tag{9.36}$$

As with the similar limit (cf. Eq. (9.25)) for the frame of Fig. 9.2(a), this result is also quite reasonable, as can be seen from the deflected shapes shown in Fig. 9.6. The reasoning is the same, that is, the horizontal beam becomes quite stiff, it loses its ability to bend and deform, and its inflection point disappears. This rigid beam ensures that the columns' ends displace the same amount and that the columns' top slopes are flat and perpendicular to the horizontal beam at points B and C. Thus, the columns will act as sketched in Fig. 9.6.

Now what we have done to analyze the frame shown in Fig. 9.5 is correct and unexceptional. Further, virtually all of the grinding details could have been done with a symbolic algebra computer program, and so we could have been spared most of the algebraic agony that ensues once the moments are identified. However, as we will show in the next section, we could also have given some more thought to the nature of the particular combination of a frame and its loading *before* doing what we have done. That is, we could have done better.

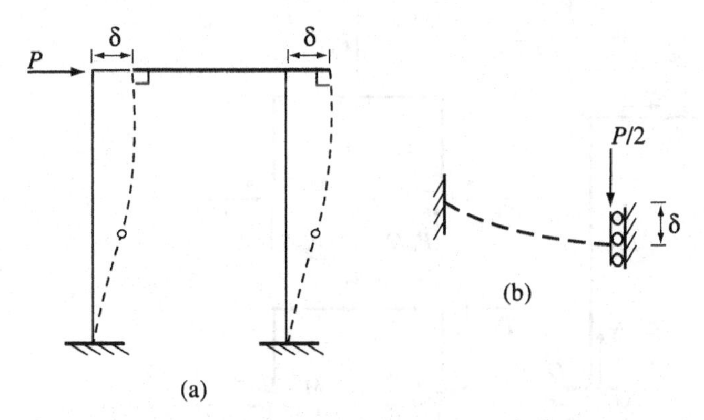

Figure 9.6. Sketches of the deflected shapes of: (a) the laterally loaded frame of Fig. 9.5 with a rigid horizontal crossbar and (b) an equivalent fixed-ended beam.

9.3 Symmetry and Antisymmetry in Frame Analysis

Consider the loaded frame shown in Fig. 9.7(a), which is clearly the frame shown in Fig. 9.5, but now supporting both a vertical load P_v at the center of the beam and a lateral load P_h. We would like to analyze the sidesway resulting from the combined loading on the frame.

We could do straightforwardly what we have just done for this frame under the lateral load alone by modifying the moments of Eqs. (9.26) to reflect the addition of the vertical load, so that they become

$$\begin{aligned} M_{DC}(x) &= M - Hx && 0 \leq x \leq L_c \\ M_{CB}(y) &= M + Ry - HL_c && 0 \leq y \leq L_b/2 \\ &= M + Ry - HL_c - P_v(y - L_b/2) && L_b/2 \leq y \leq L_b \\ M_{AB}(z) &= M_A + (P_h - H)z && 0 \leq z \leq L_c \end{aligned} \tag{9.37}$$

We could develop the appropriate complementary energy integrals with these moments and proceed from there. However, the equations for the redundants would look very much like Eqs. (9.31) because the flexibility matrix for the structure is unchanged, and we only have to superpose the effect of the vertical load P_v on the right-hand side of Eqs. (9.31). With this done, the equations for the redundants under the combined loading appear as

$$\begin{bmatrix} 6(2+k_r) & -6(1+k_r) & 3(2+k_r) \\ -6(1+k_r) & 2(2+3k_r) & -3(1+k_r) \\ 3(2+k_r) & -3(1+k_r) & 2(3+k_r) \end{bmatrix} \begin{Bmatrix} \dfrac{M}{L_c} \\ H \\ \dfrac{RL_b}{L_c} \end{Bmatrix}$$

$$= \begin{Bmatrix} 3 \\ -1 \\ 3 \end{Bmatrix} P_h + \begin{Bmatrix} 6(4+k_r) \\ -6(2+k_r) \\ (24+5k_r) \end{Bmatrix} \left(\frac{P_v L_b}{8L_c} \right) \tag{9.38}$$

Again, these equations can be solved with traditional longhand algebra, by using matrix manipulation, or by exercising a symbolic algebra computer program. However, we recall

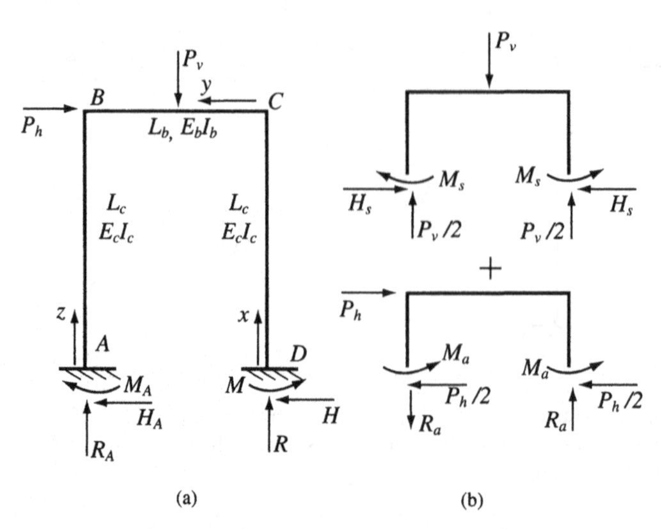

Figure 9.7. The portal frame on fixed supports with two loads: (a) overall equilibrium and (b) a decomposed view.

that we have used superposition to extend Eq. (9.31) into the form (9.38) that reflects both the loads. This suggests looking at the problem posed in Fig. 9.7(a) with a broader view wherein we intend to use superposition and symmetry to formulate the problem.

We show in Fig. 9.7(b) a view of the structure decomposed according to its loads. For the case of the vertical load, which is here placed at the center of the top beam, we see that both the loading and the structure are symmetric. This suggests that the two base moments and the two horizontal reactions (appearing here as shear forces) will be equal in magnitude and oppose each other in direction, and that the two vertical loads (which are axial loads) will be equal in magnitude (to $P_v/2$) and direction. Can we make a similar claim for the laterally loaded frame?

The answer is affirmative. First of all, looking back at the detailed results for the laterally loaded frame, we see that the two vertical reactions (the axial forces) are equal in magnitude and opposite in direction, and by substituting the solution (9.32) into Eq. (9.26) we can confirm that this is also true of the base moments. Further, the two horizontal reactions (the shear forces) in this case have identical magnitudes, here $P_h/2$, and direction.

What's the point? The point has two parts. The first is that we can identify the moment $M(x)$ and the axial force $N(x)$ as symmetric resultants acting on an element or section of a beam because of the sign conventions we use for positive axial forces and positive moments, while the shear force $V(x)$ can be identified as an antisymmetric function (cf. Fig. 7.1). This distinction between moment and shear is also evident in the definition of the shear as the first derivative of the moment, which action causes a symmetric function to become antisymmetric. The second part of the point is the general statement that we expect a symmetric load applied to a symmetric structure to yield a symmetric response, as we also expect that an antisymmetric load applied to a symmetric structure will produce an antisymmetric response.

Thus, with regard to the now decomposed problem at hand, with the reactions as indicated in Fig. 9.7(b), the vertically loaded frame has only two unknown reactions, M_s and H_s, while the laterally loaded frame also has only two reactions, but for this case they are M_a and R_a. So, another way of solving this problem is to analyze two structures, each of which is indeterminate only to the second degree. While we leave the details for you to confirm, we present, first, the following symmetric redundants for the vertically loaded frame,

$$M_s = \frac{k_r}{8(1+2k_r)}(P_v L_b), \qquad H_s = \frac{3k_r}{8(1+2k_r)}\left(\frac{L_b}{L_c}\right)P_v \tag{9.39}$$

and, second, the antisymmetric redundants for the laterally loaded frame,

$$M_a = \frac{(3+k_r)}{2(6+k_r)}(P_h L_c), \qquad R_a = \frac{3}{6+k_r}\left(\frac{L_c}{L_b}\right)P_h \tag{9.40}$$

The antisymmetric set should look quite familiar (see Eq. (9.32)!).

With the results (9.39) and (9.40) in hand, we can find the redundants for the loaded frame of Fig. 9.7(a). With the sign conventions shown there, those redundants, which are also the solutions to Eqs. (9.38), are simply as follows:

$$\begin{aligned} \{M_A, M\} &= \{M_s \mp M_a\}, \qquad \{H_A, H\} = \left\{\frac{P_h}{2} \mp H_s\right\} \\ \{R_A, R\} &= \left\{\frac{P_v}{2} \mp R_a\right\} \end{aligned} \tag{9.41}$$

This example demonstrates that it is important to identify situations where reasoning about symmetry can be helpful in formulating and solving a problem. This is true even when we are using a computer programs for doing structural work because reasoning about symmetry can both reduce the "number of cycles" and support our ability to understand and interpret numerical results.

In the same vein, it is also useful to explore the locations of inflection points in the deflected shapes of the above frames, as we will in the next section. This is very important for some of the approximations we will develop in Section 9.5.

However, there is one more symmetry related issue that we should mention. The problems we have solved so far suggest that we will have lateral motion or sidesway in a frame only when there is an asymmetry in the loading, on the frame or, perhaps, in the frame's construction. The sidesway of a symmetrical frame has been shown to be directly proportional to the magnitude of the lateral load, that is, we find no sidesway if we set the lateral load P to zero in either of the deflection results in Eqs. (9.24) or (9.34). Further, even if it is asymmetric in loading or layout, a frame can be supported against sidesway by providing lateral support near or at the top of the frame. Thus, we need to look carefully at a problem before deciding which considerations can be applied or ignored.

9.4 Inflection Points in Frames

An inflection point in a bent beam or column is a point at which the moment vanishes during a sign change along the member in question. Knowing the locations of inflection points is useful because they serve as additional points about which we can sum moments. This means that every inflection point we identify provides us with an additional point about which we can write a moment equilibrium equation to help determine the redundant reactions of a structure – or, if you prefer, to make the structure less indeterminate. Thus, if we can suggest some ways to do this even approximately, we can provide a useful basis for the approximate analysis for structures we are trying to design.

9.4.1 Inflection Points in Laterally Loaded Portal Frames

Consider the two laterally loaded portal frames we analyzed in Section 9.3. In the first frame, shown in Fig. 9.2(a), both supports are pinned reactions. From the anticipated deflected shape shown in Fig. 9.2(b) and from the equations that defined the moments in the frame legs (Eqs. (9.15)), we expect to find only the single inflection point in the top beam. Note that the moments described in the first and last of Eqs. (9.15) can only vanish at the pinned supports, not for some intermediate value of their respective coordinates. For the top beam, on the other hand, the inflection point occurs when $M_{CB}(y) = 0$, which means when

$$y_{\text{infl}/CB} = y_{\text{infl/beam}} = \frac{HL_b}{P} \tag{9.42}$$

Given our evaluation of the value of the redundant force H in Eq. (9.19), we now know that the inflection point occurs at the center of the beam, and it does so for all values of the stiffness ratio k_r.

For the laterally loaded portal frame with fixed supports (cf. Fig. 9.5), we should anticipate from the deflected shape shown in Fig. 9.5(b) and the moments given in Eqs. (9.26) that all three members will have inflection points whose locations can be expressed in terms of the redundants as follows:

$$\begin{aligned} x_{\text{infl}/DC} &= \frac{M}{H} \\ y_{\text{infl}/CB} &= \frac{HL_c - M}{R} \\ z_{\text{infl}/AB} &= \frac{M_A}{(H - P)} \end{aligned} \tag{9.43}$$

We have given the redundants for this frame in Eqs. (9.32) (and as the antisymmetric components of Eqs. (9.41)). By substituting them into Eqs. (9.43) we can identify the specific locations of the inflection points, that is,

$$\begin{aligned} \frac{x_{\text{infl}/DC}}{L_c} &= \frac{x_{\text{infl/column}}}{L_c} = \frac{3 + k_r}{6 + k_r} \\ \frac{y_{\text{infl}/CB}}{L_b} &= \frac{y_{\text{infl/beam}}}{L_b} = \frac{1}{2} \\ \frac{z_{\text{infl}/AB}}{L_c} &= \frac{x_{\text{infl/column}}}{L_c} = \frac{3 + k_r}{6 + k_r} \end{aligned} \tag{9.44}$$

Thus, for the laterally loaded frame with fixed supports, the inflection point in the beam is at its midpoint and independent of the relative stiffness ratio, just as it is for the frame on pinned supports. The column inflection points, on the other hand, are the same in both columns and dependent on the stiffness ratio. How important is that dependence?

The upper limit of $k_r \to \infty$ corresponds to the evanescence of the connecting beam, in which case both columns act as tip-loaded cantilevers that have no inflection points. The moments are zero only at the cantilever tips, which is reflected in the corresponding limits of the first and third of Eqs. (9.44).

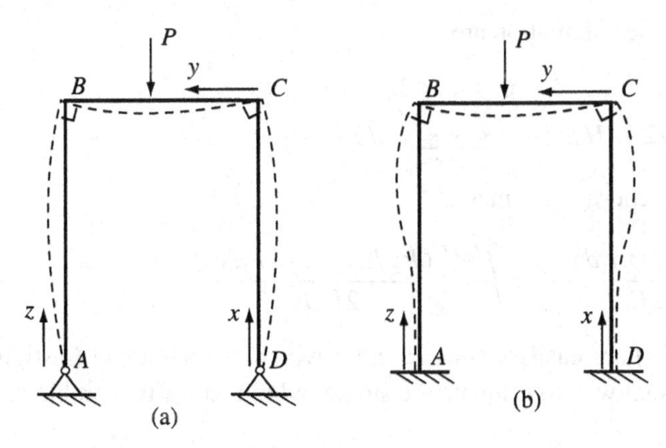

Figure 9.8. Two portal frames under a vertical load: (a) with pin supports and (b) with fixed supports.

For the lower limit of the stiffness ratio, $k_r \to 0$, we find from Eqs. (9.44) that

$$\left.\frac{x_{\text{infl}/DC}}{L_c}\right|_{k_r \ll 1} = \left.\frac{z_{\text{infl}/AB}}{L_c}\right|_{k_r \ll 1} \approx \frac{1}{2}\left(1 + \frac{k_r}{6}\right) \tag{9.45}$$

Thus, it appears that for the class of problems wherein the columns are more flexible than the beam, which turns out to be rather common in structural practice, the inflection points in the columns of laterally loaded portal frames tend to occur at the centers of the columns.

9.4.2 Inflection Points in Vertically Loaded Portal Frames

A situation similar to that described in the previous section also obtains for vertically loaded frames, although the more useful aspect will be the location of multiple inflection points in the beams. In fact, for the first of the two portal frames we have considered, where we have applied a concentrated load at the center and the frame supports are pins, there will be no inflection points in the columns. The columns in this frame will simply bow outward to accommodate the bending of the beam supporting the vertical load, with the column moments vanishing only at the supporting pins (see Fig. 9.8(a)). However, the beam will have two inflection points, symmetrically distributed, under this loading. How do we know this so readily?

There are two answers. First, and qualitatively, in the frame with pin supports the beam will behave as if it were something between a simply supported beam and a fixed-ended beam. These two extremes are inherent because the columns provide rotational stiffness at the ends of the beam – recall the role of the relative stiffness ratio, as well as our analyses of spring-supported beams in Chapter 8 – and so we recognize that our extreme limits provide either no rotational restraint (that is, simple supports) or complete rotational restraint (that is, clamped supports). From the result (8.80) of Example 8.13, we can demonstrate that beams that are rotationally restrained will have inflection points, so we should expect inflection points when such beams form part of a frame.

For our second answer we can actually calculate the location of the inflection points in the beam. We can specialize Eqs. (9.37) to the case of the pin-supported frame subject only to the vertical load by setting $M = 0$ and $R = P/2$. Then, accounting for symmetry, the moments needed for the calculation are

$$\begin{aligned} M_{DC}(x) &= -Hx & 0 \le x \le L_c \\ M_{CB}(y) &= Py/2 - HL_c & 0 \le y \le L_b/2 \end{aligned} \tag{9.46}$$

Then the complementary energy becomes

$$U^* = 2\int_0^{L_c} \frac{(-Hx)^2 dx}{2E_c I_c} + 2\int_0^{L_b/2} \frac{(Py/2 - HL_c)^2 dy}{2E_b I_b} \tag{9.47}$$

The integrals in Eq. (9.47) are easily evaluated, after which we can apply Castigliano's second theorem in the usual way to evaluate the single redundant for this problem:

$$H = \frac{3k_r L_b P}{8(2 + 3k_r)L_c} \tag{9.48}$$

Now we can locate one of the pair of symmetrically located inflection points in the beam by substituting Eq. (9.48) into the second of the moments in Eqs. (9.46), which we in turn set to zero. The result is that

$$\frac{y_{\text{infl}/CB}}{L_b} = \frac{y_{\text{infl/beam}}}{L_b} = \frac{3}{4}\left(\frac{k_r}{2 + 3k_r}\right) \tag{9.49}$$

The two limiting cases of Eq. (9.49) confirm our qualitative estimate of this frame's behavior quite accurately. For the case of vanishingly small rotational stiffness ratio, the columns provide no rotational restraint, the (right-handed) inflection point occurs at the support $y = 0$, and the beam thus can be regarded as simply supported. For an extremely large value of that ratio, the columns provide near-perfect rotational restraint, the (right-handed) inflection point occurs at the beam's quarter point $y = L_b/4$, and so the beam can be thought of as having fixed ends.

For the second case, where the reactions are fixed, we also expect inflection points in the columns, as shown in Fig. 9.8(b). We have already given in Eqs. (9.39) the redundants for this frame, so by substituting them into the moment expressions (9.26) we can find complete expressions for both types of inflection points. Taking the columns first, we find that

$$\frac{x_{\text{infl}/DC}}{L_c} = \frac{x_{\text{infl/column}}}{L_c} = \frac{M_s}{H_s L_c} = \frac{1}{3} \tag{9.50}$$

Thus, the column inflection points are independent of k_r, and they occur closer to the supports than they do under lateral loading. Recall that for laterally-loaded frames the inflection points migrate upward from the midpoints of the two columns (cf. Eqs. (9.44) and (9.45)).

On the other hand, the inflection points in the beam behave very much like that of the beam in the pin-supported frame, that is,

$$\frac{y_{\text{infl}/CB}}{L_b} = \frac{y_{\text{infl/beam}}}{L_b} = \frac{2(M_s - H_s L_c)}{PL_b} = \frac{k_r}{2(1 + 2k_r)} \tag{9.51}$$

where the inflection point migrates from the quarter point of the beam for very large values of the relative stiffness ratio k_r (which we could perhaps also refer to as the case of very stiff supports for the beam) to the end of the beam for very small values of k_r (which we might consider very flexible beam supports).

Now suppose that the vertical load is something other than a concentrated load at the center. Similar behavior would follow, except that the precise location of the inflection points in the beam would shift somewhat. For example, we know from prior work that a fixed-ended beam under a uniform load has inflection points at a distance of approximately $0.211L_c$ from each end, which is not very different from where they are located when the load is concentrated at the center. Shouldn't we expect to get similar behavior in a frame with a uniform load applied along the top beam?

What about the inflection points in the columns? When the supports are pinned, as in Fig. 9.8(a), the change from a (vertical) concentrated load to a uniform load will not affect the column's behavior. They will simply bow out as they did for the concentrated load. If the support reactions are clamps, as in Fig. 9.8(b), will the inflection points under a uniform load also be found one-third of the way up the columns? In fact, that is just how it turns out. We once again leave the details for you, but we can show that if we replace the concentrated load in Fig. 9.8(a) with a uniform load, for pinned column supports the beam's inflection points occur at the solutions of

$$\left(\frac{y_{\mathrm{infl}/CB}}{L_b}\right)^2 - \left(\frac{y_{\mathrm{infl}/CB}}{L_b}\right) + \frac{k_r}{2(2+3k_r)} = 0 \tag{9.52}$$

This quadratic equation has two solutions

$$\frac{y_{\mathrm{infl}/CB}}{L_b} = \frac{y_{\mathrm{infl/beam}}}{L_b} = \frac{1}{2}\left[1 \pm \sqrt{1 - \frac{2k_r}{3+2k_r}}\right] \tag{9.53}$$

For the second case, where the reactions are fixed, we also expect inflection points in the columns, as shown in Fig. 9.8(b). Taking the columns first, we find that

$$\frac{x_{\mathrm{infl}/DC}}{L_c} = \frac{x_{\mathrm{infl/column}}}{L_c} = \frac{M_s}{H_s L_c} = \frac{1}{3} \tag{9.54}$$

while the beam inflection points are found by solving the following quadratic equation:

$$\left(\frac{y_{\mathrm{infl}/CB}}{L_b}\right)^2 - \left(\frac{y_{\mathrm{infl}/CB}}{L_b}\right) + \frac{k_r}{3(1+2k_r)} = 0 \tag{9.55}$$

which has the following two solutions:

$$\frac{y_{\mathrm{infl}/CB}}{L_b} = \frac{y_{\mathrm{infl/beam}}}{L_b} = \frac{1}{2}\left[1 \pm \sqrt{1 - \frac{4k_r}{3(1+2k_r)}}\right] \tag{9.56}$$

Note that we have very similar results for both uniformly loaded frames. We have quadratic equations for the location of the beam inflection points, which are a direct result of the form of the moment equations written for a uniformly loaded beam, and we get the same limiting behavior from Eqs. (9.53) and (9.56). For a beam connected to a very flexible column, that is, for small values of k_r, the inflection points occur at the ends of the beams in both frames, that is, we see the behavior of two simply supported beams under uniform loads. When the beam is connected to a stiff column, that is, for $k_r \to \infty$, the inflection points occur at $y = 0.211L_b$ and at $y = 0.789L_b$, and as we know, these are exactly the inflection points of a fixed-ended beam subjected to a uniform load.

9.4.3 Some Comments on Inflection Points

What have we learned from the foregoing calculations? As a practical matter, we have learned that for laterally loaded frames, columns have inflection points only when they have fixed supports. Further, these inflection points occur somewhere between the midpoints and the tops of each column, depending on the relative bending stiffness of the columns and the top beam. For a beam connected to a relatively strong or stiff column, the inflection points move closer to the tops of the columns, while for relatively weak columns the inflection points migrate to the centers of the columns. The beams typically have only single inflection points, located at their midpoints and independent of the bending strength of the columns to which they are connected.

For the vertically loaded frames, column inflection points again occur only for clamped support reactions. They are found at the columns' lower-third points, regardless of the strength of the column–beam stiffness ratio. The inflection points in the beam occur in pairs, at points along the beam that correspond to the inflection points of equivalently loaded, fixed-ended beams. In the cases we have considered, the loads were symmetrically distributed about the centers of the beams (or frames), and, as a consequence, so were the inflection points.

We should also note that the locations of all of the inflection points are independent of the *magnitude* of the loads applied. Their precise location is coupled to the *precise distribution* of the load, but even here the dependence is not strong. Recall that for symmetric vertical loads, the fixed-ended inflection point limit ranges from a location of $y = 0.211L_b$ for a uniform load to $y = 0.250L_b$ for a concentrated load.

Now, while it is difficult to generalize these results to all loadings and all frame configurations, there are more than enough situations where we can use these results to great practical effect, as we will now demonstrate in Section 9.5.

9.5 Approximate Methods for Frames

We have only dealt with relatively simple structures so far, that is, frames having only one portal (which is often also called a *bent*), typically being symmetric in layout, and being loaded with a either single lateral load, a single symmetric vertical load, or the sum of these two. However, the real world is rather more complicated, and so we must give some thought about how to manage more realistic problems. And it is not enough to say that we will simply do our structural analyses on a computer, thereby finessing any hard work.

It is important to remember that the analysis of indeterminate structures is intimately connected to their configurations, including their layout, the sizes of individual members, and their relevant material properties. Thus, to analyze a particular indeterminate structure, even with the most sophisticated hardware and software available, we must have a pretty good idea of how that structure is put together. Remember that in our description of the structural design cycle (cf. Section 1.3 and Fig. 1.8), we pointed out that we do preliminary layouts of structural types and we estimate the sizes and shapes of the members that will become our columns and beams. Does this sound like a chicken-and-egg problem, wherein we can't do indeterminate analysis without knowing the structure, and we can't know the structure without analyzing how it works?

The answer is that we do preliminary analysis as part of preliminary design, and for complicated frames and similar structures we try to estimate the content of the structure based on an approximate, preliminary analysis. We have not had much need to deal with this particular set of issues so far because we have been concerned only with relatively simple configurations and behaviors–trusses and beams–wherein there isn't nearly as much variability in the types or properties of members as we are likely to see in complex framed structures, e.g., multistory buildings. Further, the approximate analyses we will do are based squarely on what we have learned by examining inflection points. In short, we assume the existence of inflection points at the locations suggested by the analyses performed in Section 9.4, which thus helps us render determinate a given structure. Then we can analyze the distribution of forces and moments in our approximate, determinate structure, independent of its geometric and material properties. Having identified the forces and moments, even roughly, we can then "size" the various members, as a result of which we can then begin to do a proper indeterminate analysis to refine our approximate estimates into a final, detailed design.

9.5.1 The Portal Approximation

As noted, portal frames or bents are often the basic building blocks of framed structures, whether single storied or multi storied. As such, an understanding of their response to lateral loads is quite important. The portal approximation is used to estimate the sidesway of framed structures. It is based on the sidesway analysis we did in Section 9.4.1, but is made even simpler. Recall that we found that the inflection points in the beams, which in real structures are likely to be floor *girders* or roof *joists*, always occur at the center of these girders or joists. In the columns, the inflection points can move up their lengths from the column midpoints, depending on the relative strength of the column and girder stiffnesses. Since we want to *find* or estimate the structural content with our approximate analysis, we will assume that all of the column inflection points are placed at their centers. Thus, for the portal approximation, the a priori assumption is that *all members have inflection points at their midpoints.*

In addition, since we are effectively using superposition to create framed structures by adding portals one to another, we need to decide how to treat the internal (horizontal) shear reactions when "individual portals" are arrayed adjacent to one another, as in Fig. 9.9. If we regard these arrays as literal superpositions, it makes sense to treat each internal column as the sum of two columns, one from the portal to its left, the second from the portal to its right. Thus, we will assume that *an internal column will carry twice the shear reaction as the two end columns*, which are effectively only single columns. Thus, for the n individual portal frames shown in Fig. 9.9(a), lateral equilibrium would dictate that the shear reactions, which are also equal to the shears at the column inflection points, would be found as

$$H + (n-1)(2H) + H = P \tag{9.57}$$

Example 9.5. Find the reactions that support the two-storied frame shown in Fig. 9.10(a).

We start by inserting inflection points at the midpoints of all of the columns and frames, as a result of which we can readily find the base shear reactions to be $H_A = H_H = 10$ kN and $H_G = 20$ kN. Note that there are still six unknown reactions (three moments and three vertical forces), but only two more equations that we can apply to the structure as a whole. Thus, to find all of the reactions and the various member resultants, we draw

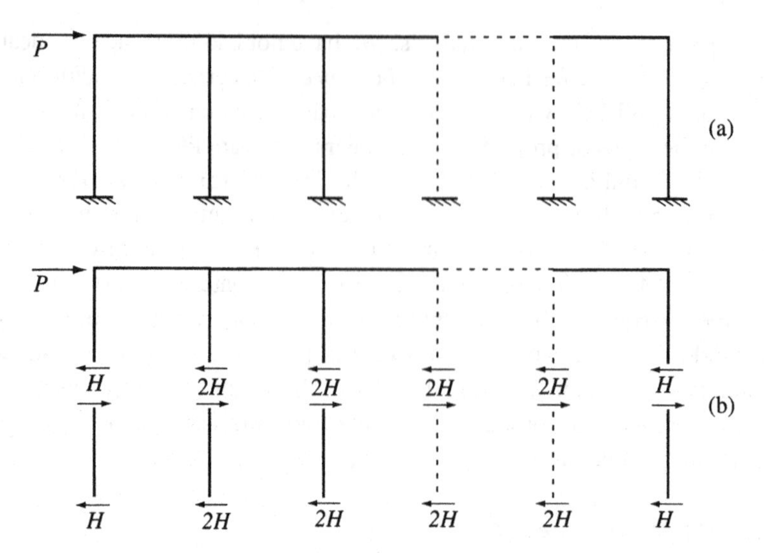

Figure 9.9. A frame made up of n individual portal frames: (a) the complete frame and (b) its shear reactions.

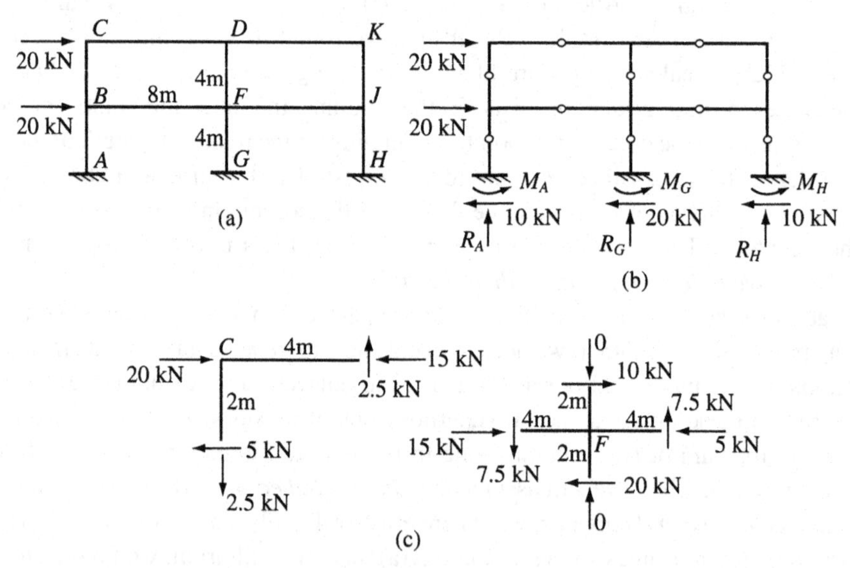

Figure 9.10. A two-story frame structure: (a) the frame, (b) with inflection points, and (c) some sample joint free-body diagrams.

free-body diagrams of each joint as bounded by its nearest member inflection points and write equilibrium equations for that joint. This is, of course, very similar to the idea of the method of joints we use in trusses, although here it is enabled by defining points away from the joints to be moment free. The process proceeds most efficiently if we begin with the uppermost joint at which a lateral load is applied, and we should keep in mind the shear results (cf. Eqs. (9.57)). Some of the joint free-body diagrams required just for determining the reactions are shown in Fig. 9.10(c). After doing the necessary joint

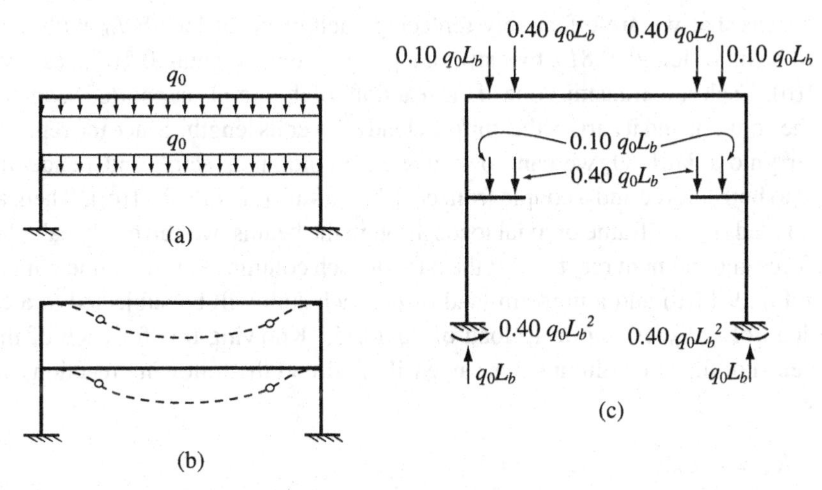

Figure 9.11. A two-story vertically loaded frame: (a) the frame, (b) with beam inflection points, and (c) the independent columns.

analyses, we find the values of the remaining reactions to be

$$\begin{aligned} R_A &= -10 \text{ kN}, & R_G &= 0 \text{ kN}, & R_H &= 10 \text{ kN} \\ M_A &= 20 \text{ kN-m}, & M_G &= 40 \text{ kN-m}, & M_H &= 20 \text{ kN-m} \end{aligned} \tag{9.58}$$

■

Note that in Example 9.5 we were able to calculate all of the reactions without specifying any of the material or geometric properties of the beams and columns that make up this frame. This is because the portal approximation has rendered our frame into a determinate structure. On the other hand, we would have to estimate or know member properties to calculate the sidesway for this frame. Remember that we need member characteristics to formulate the complementary energy as a prelude to applying Castigliano's second theorem.

9.5.2 Approximating Vertically Loaded Beams in Frames

A similar kind of approximation can be useful for preliminary analysis and design of frames subjected to vertical loading. This approach is based on locating the inflection points in the beams that are carrying the vertical loads.

Example 9.6. Find the reactions required to support the uniform load q_0 applied to both beams in the two-storied frame shown in Fig. 9.11(a).

We know that there will be inflection points in the two beams and they will each be within a distance of $0.211L_b$ of the ends of each beam. The precise location, of course, depends on how the stiffnesses of the beams and the columns compare, that is, on the relative stiffness ratio k_r. However, we will assume that the inflection points will be located at a distance of $0.10L_b$ from each of the supports. That is, we will approximate the locations of the inflection points by taking them to be at the average of their range of possible distances. In addition, we will also assume that the replacement beams do not transmit any axial forces between the columns.

This means that we are effectively replacing each beam of length L_b with a simply supported beam of length $0.8L_b$ that sits on a pair of stubs of length $0.10L_b$, as shown in Fig. 9.11(b). Each stub transmits one of the reactions of the simply supported beam (as a tip load) to the column, and it carries the uniform load q_0 over its length. Since the replacement beams carry no axial load, we can now view each column as acting independently, but subjected to both a force and a couple from each beam stub (cf. Fig. 9.11(c)). Then, absent any lateral loads on the frame or axial loads through the beams, we can easily calculate the vertical force and moment reactions at the base of each column. For the frame dimensions shown in Fig. 9.11(a) and a uniform load of q_0, each stub will be subjected to a central vertical load of $q_0L_b/10$ and a tip load of $2q_0L_b/5$. Knowing this for each of the two stubs on each of the two columns, we can easily find find the following reactions at their supports:

$$\begin{aligned} R_A &= q_0L_b, \qquad M_A = 0.09q_0L_b^2 \\ R_G &= q_0L_b, \qquad M_G = 0.09q_0L_b^2 \end{aligned} \tag{9.59}$$

Note that the column lengths do not appear in these results and that we haven't needed member properties because our approximate structure is determinate. ■

9.5.3 Some Comments on Frame Modeling and Approximations

The models and approximations presented in Sections 9.5.1 and 9.5.2 are not the only ways of approximating and estimating frame behavior, nor have we exhausted here the list of reasons for doing these kinds of calculations. One particularly interesting aspect is that lateral motion of frames and other structures occurs in response to ground motion such as that produced by earthquakes. Thus, the response of a portal frame can be modeled as the response of an elastic spring which, augmented by some modeling of its mass distribution, can also be used to estimate the natural frequency of the frame as a guide to its sensitivity to seismic disturbances. For the fixed-ended portal frame of Fig. 9.5(a), for example, we can use the deflection result of Eq. (9.34) to estimate the portal's effective spring stiffness as

$$k_{\text{portal}} = \frac{P}{\delta} = \left(\frac{6+k_r}{3+2k_r}\right)\left(\frac{12E_cI_c}{L_c^3}\right) \tag{9.60}$$

so that the natural frequency (usually expressed in radians per second or hertz) of the portal in lateral motion can be estimated to be

$$\omega = \sqrt{\frac{k_{\text{portal}}}{m_{\text{portal}}}} = \sqrt{\left(\frac{6+k_r}{3+2k_r}\right)\left(\frac{12E_cI_c}{L_c^3 m_{\text{portal}}}\right)} \tag{9.61}$$

where m_{portal} reflects the mass distribution of the frame.

Similarly, there is a cantilever frame model that can be used to estimate the responses – both static and dynamic – of tall, framed buildings to lateral forces. Now, siesmic and other dynamic analyses are clearly beyond our current scope, but the simple results in Eqs. (9.60) and (9.61) are intended to present a bird's-eye view of how one might quickly and usefully estimate important structural behavior. This is something that experienced engineers and designers do all of the time, notwithstanding all of the computer power at their fingertips!

Now, as we've said before, the work in this chapter is based on applying the force or flexibility method, including Castigliano's second theorem. Yet it remains true that most real-world structural calculations are based on numerical implementations of the displacement or stiffness method. So why do all of these frame analyses with Castigliano's second theorem? For two reasons. The first is that the kinds of problems we have posed here and at the end of the chapter can all be done with the force method, and by taking this approach we learn a lot about structural behavior. For example, we can locate inflection points and use them to facilitate our calculations, and we can model and estimate the relative bending stiffnesses of the beams and columns in a frame – and we can do this virtually all by hand. This means we can look at the results and, while sketching the associated deflected shapes, we can interpret and explain what is happening in meaningful physical terms. And, as frequently noted, being able to do this is very useful because the underlying physics must work for all models, whether analytically tractable or computationally intensive.

The second reason is related to our ability to anticipate (and sketch up front) the deflected shape of a structural element or a complex structural device, on the one hand, and a complete equilibrium state of a complex structure, on the other. In short, it is easier to estimate continuous and compatible deflected shapes than it is to lay out an entire equilibrated system of forces and moments in a general way. As a consequence, the computational tools most widely used are those based on the direct displacement approach (see Section 10.3.3), and so we will pay more attention to their formulation in Chapter 10.

Finally, the foregoing models of frames should also convey a style of modeling and thinking about structures and their behavior. The detailed results are less important than knowing what sorts of things to look for, including the right dimensions, the right dimensional ratios, and the presence and absence of terms embodying specific kinds of behavior. Certainly getting the magnitudes right is important, which is where the numerical results come in. All of the results we have given can be done for specific cases on a computer with one program or another. And there are certainly problems (including some below) where you can do it all numerically. However, the kinds of approaches we have emphasized here – and in our discussions of beams and of axially loaded bars – are intended to convey a flavor of what we always need to look for whenever we are doing any numerical work, but most especially when we are using computer-based tools, sometimes called "black boxes." Remember, a computer can't tell you whether the axial force *should* be greater than or smaller than the moment, it can only give you numbers. *You* have to apply some engineering judgement to see if you want to accept those numbers.

Problems

9.1 Find the horizontal and vertical deflections of the tip C of the frame shown in Fig. 9.1, assuming that the column has properties E_c, I_c, and L_c, while the horizontal beam has properties E_b, I_b, and L_b.

9.2 Find the horizontal and vertical deflections of the joint B of the frame shown in Fig. 9.1, assuming that the column has properties E_c, I_c, and L_c, while the horizontal beam has properties E_b, I_b, and L_b. How do these deflections compare with those found for point C (cf. Problem 9.1)?

9.3 For the frame shown In Fig. 9.P3, calculate the horizontal displacement and the slope of rolling support A due to the moment applied as shown. (*Hint*: Apply a force and a moment there and, in applying Castigliano's second theorem, set their values to unity at an appropriate point.)

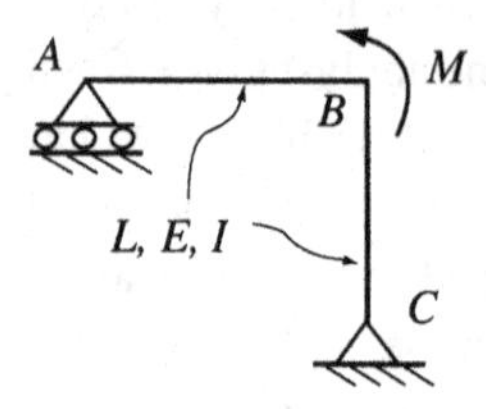

Figure 9.P3. Figure for Problem 9.3.

9.4 You want to use the frame configuration shown in Fig. 9.P4 to support a freeway sign ($W = 4.45$ N, $R = 7.50$ m) and to support a living room lamp ($W = 0.45$ N, $W = 2.00$ m), but you are concerned that there will be too much deflection at the tip. Assume that the semicircular frame will be made of steel in both cases ($E = 200$ GPa $= 20{,}000$ kN/cm^2). Therefore, what moments of inertia should be prescribed to limit their respective displacements to $\delta_{\text{sign}} = 5.00$ cm and $\delta_{\text{lamp}} = 1.00$ cm, respectively?

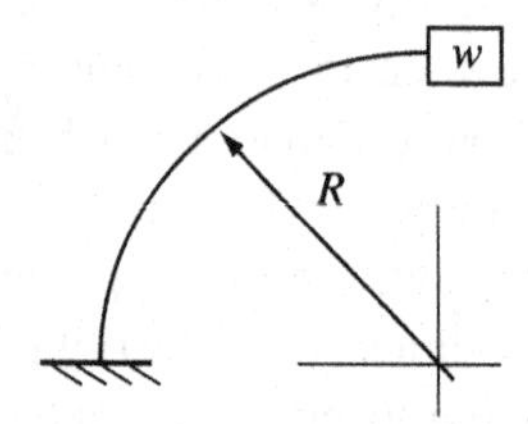

Figure 9.P4. Figure for Problem 9.4.

9.5 Complete the calculations for reactions for the frame shown in Fig. 9.10 and confirm that Eqs. (9.58) are correct. (*Hint*: You may wish to use a program that does symbolic calculations to ease your way.)

9.6 For the frame shown in Fig. 9.P6, calculate the horizontal displacements of joint C and of the rolling support D due to the two loads shown. (*Hint*: Apply appropriate forces and in applying Castigliano's second theorem, set their values to unity at an appropriate point.)

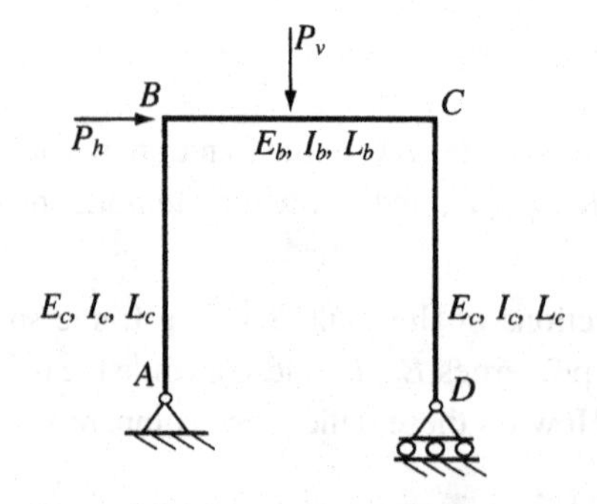

Figure 9.P6. Figure for Problem 9.6.

9.7 For a bay of the type shown in Fig. 9.9 wherein two portal frames are used, calculate the reactions and the sidesway for the entire bay.

9.8 For a bay of the type shown in Fig. 9.9 wherein three portal frames are used, calculate the reactions and the sidesway for the entire bay.

9.9 For a bay of the type shown in Fig. 9.9 wherein n portal frames are combined, estimate the sidesway for the entire bay.

9.10 For the frame shown in Fig. 9.P10, find the reactions, the sidesway, and the deflection of the top beam under the concentrated load. How do these results compare with those of known elementary beams?

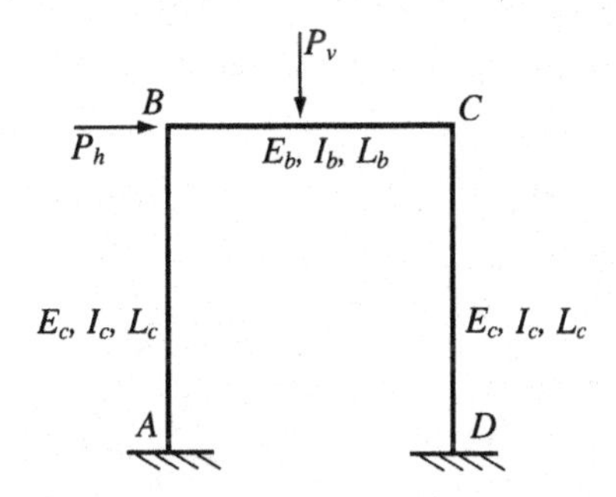

Figure 9.P10. Figure for Problem 9.10.

9.11 Calculate the sidesway for the frame shown in Fig. 9.P11.

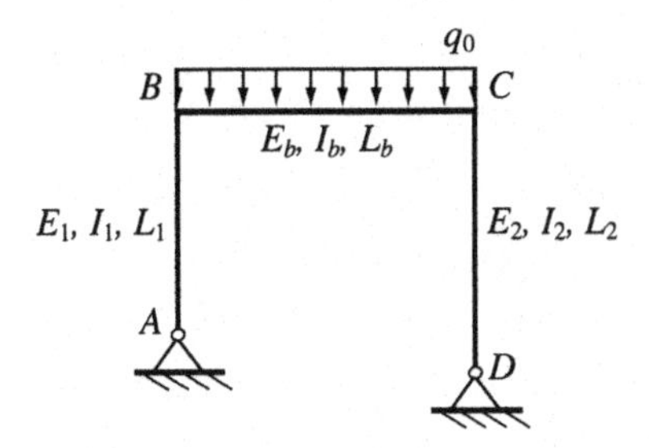

Figure 9.P11. Figure for Problem 9.11.

9.12 Calculate the horizontal and vertical deflections of the point D of the aluminum frame loaded as shown in Fig. 9.P12. Young's modulus for aluminum is ($E \simeq 70$ GPa $=$ 7,000 kN/cm^2).

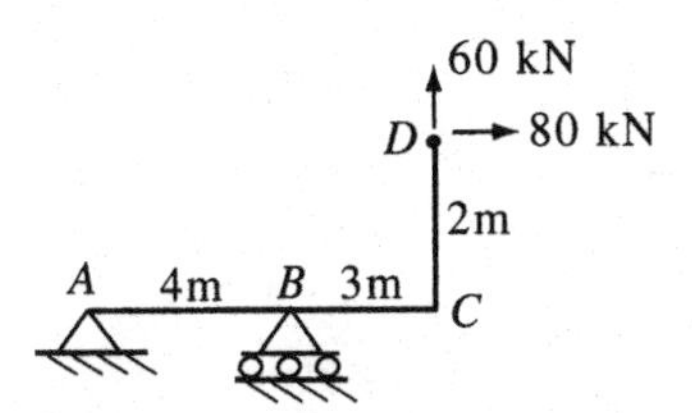

Figure 9.P12. Figure for Problem 9.12.

9.13 Find the redundant force in the frame shown in Fig. 9.2(a) assuming that the two legs have different bending stiffness, that is, assume that for leg AB the bending stiffness is $E_{c1}I_{c1}$ and for leg CD it is $E_{c2}I_{c2}$. Compare your results with the answer found in Example 9.2.

9.14 Find the sidesway of the frame shown in Fig. 9.2(a) assuming that the two legs have different bending stiffness, that is, assume that for leg AB the bending stiffness is $E_{c1}I_{c1}$ and for leg CD it is $E_{c2}I_{c2}$. Compare your results with the answer found in Example 9.3.

The Pontypridd Bridge in Pontypridd, Wales, built in 1756. The holes or "voids" in the haunches of the arch are intended to reduce weight in a region where there is ample cross section to carry the compressive arch forces. Such weight-saving ideas have also been used in modern steel bridges. (British Crown Copyright: Royal Commission on the Ancient and Historical Monuments of Wales.)

10

Force and Displacement Methods

In this last chapter we will pull together the pieces in a way that opens the door to using computer-based tools such as FEM packages. However, we will not do finite element methods in any detail, nor will we give programming instruction, nor will we provide a user's manual (or a disk) for a computer package. Rather, we will express the underlying principles common to these tools in matrix notation to provide both conciseness of language and a representation that is well suited to computer use.

10.1 Recapitulation and Overview

We have presented two fundamental paradigms for formulating and solving structures problems, the force (or flexibility) paradigm and the displacement (or stiffness) paradigm. We have used the stiffness paradigm when we found approximate solutions for axially loaded bars (Section 4.4) and for bent beams (Section 7.3.4) using the direct displacement (Rayleigh–Ritz) method. We have used the flexibility paradigm more extensively, particularly when we applied the principle of consistent deformation and various statements of Castigliano's second theorem to indeterminate axially loaded bars (Sections 4.5 and 4.6), trusses (5.2), determinate (Section 8.1.1) and indeterminate (Section 8.1.2) beams, and frames (Section 9.2–9.5). In fact, we have used the force paradigm rather more often than we have used the displacement paradigm. In this closing chapter, however, we will clearly and consciously do just the reverse. Why?

Well, on several occasions we have noted that most computational analyses of structures are done in the displacement or stiffness paradigm. One reason for that is the relative ease of visualizing and sketching the deflected shape of a structure and of expressing that shape in a set of suitable trial functions. It is much harder to estimate – never mind quantify – the distribution of forces and moments in a beam, a frame, or a more complex structure. In addition, even if we could generate approximate representations for generalized force distributions, we would still be left with a set of results that need to be integrated further to calculate the corresponding generalized displacement quantities. In the same vein, it is also easier to specify and enforce the boundary conditions when they are expressed in terms of displacements (recall how we simplified the solution to the three-bar problem in Section 4.6), and every structure has boundary conditions that restrict its displacements – both deflections and slopes – at some points, if only to keep it from moving! Castigliano's second theorem clearly demonstrates the usefulness of the force paradigm for calculating displacements at discrete, specified points. However, global pictures of displacements and forces at all of the boundary and interior points of an elastic solid or structure are generally found with methods based on the displacement or stiffness paradigm.

Notwithstanding these remarks, the force method is still widely used to analyze trusses and frames. This is the case because, for these structures, we can identify displacement and force boundary conditions and loads at a countable set of discrete points. Further, in designing such structures, we focus first on determining member forces during preliminary design because we need to estimate member sizes for the loads each is expected to carry. So we will now provide a relatively brief discussion of the force method. We follow that with a discussion of the displacement method that builds on the Rayleigh–Ritz method, which in turn results in the total potential energy of a linear elastic structure being approximated as a quadratic form of discrete nodal amplitudes.

10.2 Force (Flexibility) Methods: Matrix Formulation

The force methods are so identified because force is the basic analysis variable (cf. Table 6.1). Energy formulations in the force method depend on the stored complementary energy, which is itself expressed in terms of generalized forces. They are also called flexibility methods because the resulting equations, whose satisfaction ensures compatibility, are written in terms such that displacements are expressed in terms of products of forces and flexibility coefficients. Again, these force or flexibility methods worked well for us when we could express features of a particular problem in terms of discrete loads.

10.2.1 General Formulation

As we noted in Section 6.2.3, the matrix form of the total complementary energy can be written as (with the notation of Δ for structural displacements)

$$\Pi^* = \frac{1}{2}\{P\}^T[f]\{P\} - \{P\}^T\{\Delta\} \tag{10.1}$$

In accord with the principle of minimum total complementary energy, the first variation that establishes the compatibility requirements is

$$\delta^{(1)}\Pi^* = \{\delta P\}^T[[f]\{P\} - \{\Delta\}] \tag{10.2}$$

Thus, the compatible displacements are determined by

$$\{\Delta\} = [f]\{P\} \tag{10.3}$$

where $[f]$ is the symmetric flexibility matrix. This form is by now familiar as we have seen it in our truss (cf. Section 5.2) and frame (e.g., Section 9.2) analyses. However, we want to extend one of our solved beam problems to make two points that become more important for large-scale, more complex structures. In Example 8.2 we formulated the problem of calculating the tip rotation and the deflections at two points of the cantilever beam shown in Fig. 8.1. By applying Castigliano's second theorem we found that the displacements in question are (from Eq. (8.24), in their original notation)

$$\begin{Bmatrix} \delta_a \\ \delta_L \\ \theta_L \end{Bmatrix} = \begin{bmatrix} f_{11} & f_{12} & f_{13} \\ f_{21} & f_{22} & f_{23} \\ f_{31} & f_{32} & f_{33} \end{bmatrix} \begin{Bmatrix} P_1 \\ P_2 \\ M_0 \end{Bmatrix} \tag{10.4}$$

We now use this result to calculate a slope of an indeterminate beam that is clearly an extension of that shown as Fig. 8.2(a).

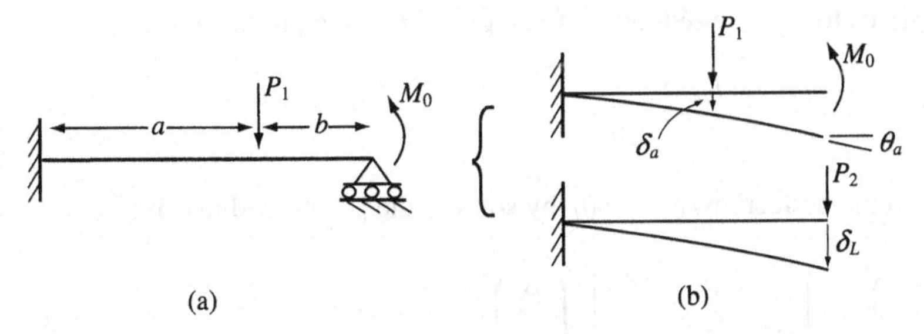

Figure 10.1. A clamped-pinned beam: (a) the full beam and (b) a decomposition.

Example 10.1. Find the slope at the pinned end of a clamped-pinned beam due to the moment applied over the pin (Fig. 10.1).

We begin this new problem by rearranging the flexibility matrix (10.4) so that we can then apply the principle of consistent deformations. In order to get to the new problem from the one for which Eq. (8.24) was derived, we will have to set the deflection at the pin to zero to find the redundant P_2. Thus, we rearrange Eq. (10.4) so that the equation for the deflection of the pin is the last row of the matrix set, rather than the middle row:

$$\begin{Bmatrix} \delta_a \\ \theta_L \\ \delta_L \end{Bmatrix} = \begin{bmatrix} f_{11} & f_{13} & f_{12} \\ f_{31} & f_{33} & f_{32} \\ f_{21} & f_{23} & f_{22} \end{bmatrix} \begin{Bmatrix} P_1 \\ M_0 \\ P_2 \end{Bmatrix} \tag{10.5}$$

Observe that while making this row switch, we also rearranged the column vector of the discrete generalized forces by changing the order of its elements (viz., compare Eq. (10.5) with Eq. (10.4)). We now rewrite Eq. (10.5) using the *partitioned form* of the flexibility matrix:

$$\begin{Bmatrix} \delta_a \\ \theta_L \\ --- \\ \delta_L \end{Bmatrix} = \left[\begin{array}{cc:c} f_{11} & f_{13} & f_{12} \\ f_{31} & f_{33} & f_{32} \\ \hdashline f_{21} & f_{23} & f_{22} \end{array}\right] \begin{Bmatrix} P_1 \\ M_0 \\ --- \\ P_2 \end{Bmatrix} \tag{10.6}$$

Note further that Eqs. (10.5) and (10.6) both show that the symmetry of the flexibility matrix has been preserved during this rearrangement.

As a result of the rearrangement, we are now in a position to calculate the redundant P_2 produced by requiring the tip deflection to vanish, which means that we find the redundant from a reduced version of the matrix (10.6),

$$\begin{Bmatrix} \\ \\ --- \\ \delta_L \end{Bmatrix} = \left[\begin{array}{cc:c} & & \\ & & \\ \hdashline f_{21} & f_{23} & f_{22} \end{array}\right] \begin{Bmatrix} P_1 \\ M_0 \\ --- \\ P_2 \end{Bmatrix} = 0 \tag{10.7}$$

or

$$\{\delta_L\} = [\, f_{21} \quad f_{23}\,] \begin{Bmatrix} P_1 \\ M_0 \end{Bmatrix} + [f_{22}]\{P_2\} = f_{21}P_1 + f_{23}M_0 + f_{22}P_2 = 0 \tag{10.8}$$

which allows us to find the redundant force exerted by the pin to be

$$P_2 = -\frac{f_{21}P_1 + f_{23}M_0}{f_{22}} \tag{10.9}$$

Then we find the deflections δ_a and θ_L by solving the partitioned matrix:

$$\left\{\begin{array}{c} \delta_a \\ \theta_L \\ \hdashline \\ \end{array}\right\} = \left[\begin{array}{cc:c} f_{11} & f_{13} & f_{12} \\ f_{31} & f_{33} & f_{32} \\ \hdashline & & \\ \end{array}\right] \left\{\begin{array}{c} P_1 \\ M_0 \\ \hdashline P_2 \end{array}\right\} \tag{10.10}$$

or

$$\left\{\begin{array}{c} \delta_a \\ \theta_L \end{array}\right\} = \left[\begin{array}{cc} f_{11} & f_{13} \\ f_{31} & f_{33} \end{array}\right] \left\{\begin{array}{c} P_1 \\ M_0 \end{array}\right\} + \left[\begin{array}{c} f_{12} \\ f_{32} \end{array}\right] \{P_2\} \tag{10.11}$$

which we can write out in extenso to find

$$\begin{aligned} \delta_a &= f_{11}P_1 + f_{13}M_0 + f_{12}P_2 \\ \theta_L &= f_{31}P_1 + f_{33}M_0 + f_{32}P_2 \end{aligned} \tag{10.12}$$

Thus, after substituting from Eq. (10.9) for the redundant P_2 we find:

$$\begin{aligned} \delta_a &= \frac{1}{f_{22}}[(f_{11}f_{22} - f_{12}f_{21})P_1 + (f_{13}f_{22} - f_{12}f_{23})M_0] \\ \theta_L &= \frac{1}{f_{22}}[(f_{31}f_{22} - f_{32}f_{21})P_1 + (f_{33}f_{22} - f_{32}f_{23})M_0] \end{aligned} \tag{10.13}$$

■

The matrix formulation of the flexibility method applied to a simple indeterminate structure shows that we can set the order for solving the problem by carefully ordering and labeling those forces that will be considered redundants and by partitioning all of the matrices in the compatibility equation accordingly. We solve first for the redundants of the indeterminate structure and then for the deflections of interest by appropriately partitioning the force, flexibility, and displacement matrices. Stated in more general terms, we begin by removing all of the redundants from the structure to be analyzed to obtain the so-called *primary structure*, which, by definition, is statically determinate. The forces and displacements of the primary structure are identified by a subscript p, and the redundants and their related displacements by the subscript r. Thus, the deflection and force vectors are, respectively,

$$\{\Delta\} = \left\{\begin{array}{c} \Delta_p \\ \hdashline \Delta_r \end{array}\right\}, \qquad \{P\} = \left\{\begin{array}{c} P_p \\ \hdashline P_r \end{array}\right\} \tag{10.14}$$

Now we can write a general compatibility equation after Eq. (10.3), being careful to order the individual compatibility equations to reflect the pattern of primary and redundant quantities shown in Eq. (10.14). That compatibility equation is

$$\begin{Bmatrix} \Delta_p \\ --- \\ \Delta_r \end{Bmatrix} = \left[\begin{array}{c:c} [f]_{p,p} & [f]_{p,r} \\ \hdashline [f]_{r,p} & [f]_{r,r} \end{array}\right] \begin{Bmatrix} P_p \\ --- \\ P_r \end{Bmatrix} \tag{10.15}$$

or, in expanded form, we have an equation for the displacements produced on the primary structure:

$$\{\Delta_p\} = [f]_{p,p}\{P_p\} + [f]_{p,r}\{P_r\} \tag{10.16}$$

and an equation for the displacements that would be produced at points on the structure where the redundants are identified:

$$\{\Delta_r\} = [f]_{r,p}\{P_p\} + [f]_{r,r}\{P_r\} \tag{10.17}$$

Inasmuch as the flexibility matrix is symmetric, the two diagonal submatrices of Eq. (10.15) are symmetric. The off-diagonal submatrices of Eq. (10.15) are not necessarily symmetric, but one is clearly the transpose of the other, that is,

$$[f]_{p,r} = [f]^T_{r,p} \tag{10.18}$$

We then determine the redundants by setting Eq. (10.17) to zero by virtue of the fact that they occur at points where the generalized displacements are zero:

$$\{P_r\} = -[f]^{-1}_{r,r}[f]_{r,p}\{P_p\} \tag{10.19}$$

And now we are positioned to calculate the deflections of interest in the primary structure, which are also clearly the realizable deflections of the actual (indeterminate) structure, by substituting Eq. (10.19) into Eq. (10.16):

$$\{\Delta_p\} = \left[[f]_{p,p} - [f]_{p,r}[f]^{-1}_{r,r}[f]_{r,p}\right]\{P_p\} \tag{10.20}$$

While stated rather abstractly, Eqs. (10.19) and (10.20) embody the force or flexibility paradigm. Equation (10.19) shows us how to calculate the redundants of an indeterminate structure, while Eq. (10.20) shows how we find compatible displacements at a set of specified points in our structure. In both equations we see clearly the role and importance of the the flexibility or influence coefficients. From here on out, "it's all in the details" – although we emphasize that they are serious, difficult details that require study well beyond this conceptual outline.

10.2.2 Some Discretization Aspects of the Flexibility Method for Frames

We have applied the force method to rather simple, monolithic structures, that is, beams and simple frames. More complex, real-world structures such as high-rise and industrial buildings, are comprised of a large number of structural elements – beams, joists, girders, truss-beams, columns, etc. – for which a *global flexibility matrix* must be assembled from building blocks based on *local* or *member* flexibility matrices for each of the structural elements (or even sub-elements). Further, even for a single element, the matrices require thought because in writing matrix equations for a structure we are developing a *discretized model of the structure*. We will illustrate that with a simple frame for which we will use beam members or elements.

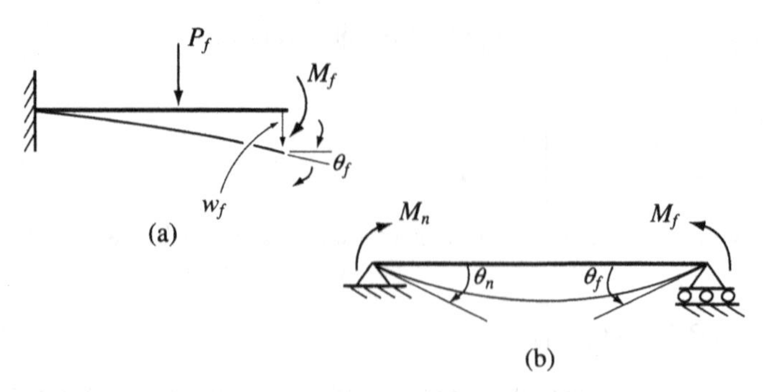

Figure 10.2. Two beam elements used in the analysis of plane frames: (a) moment-loaded cantilever and (b) pinned-pinned.

First of all we need to make a modeling decision about which physical effects we want to incorporate into our model (e.g., shear strain, axial deformation). Recall our force-method analysis of the simple right-angled frame of Fig. 9.1 We found that the flexibility matrix describing the deflections of the tip of that frame was first given in Eq. (9.12):

$$\begin{Bmatrix} \delta_{Cv} \\ \delta_{Ch} \end{Bmatrix} = \begin{bmatrix} \frac{4}{3} + \frac{1}{AL^2}\left(1 + \frac{E}{\kappa G}\right) & -\frac{1}{2} \\ -\frac{1}{2} & \frac{1}{3} + \frac{1}{AL^2}\left(1 + \frac{E}{\kappa G}\right) \end{bmatrix} \begin{Bmatrix} \frac{PL^3}{EI} \\ \frac{QL^3}{EI} \end{Bmatrix} \tag{10.21}$$

If we had chosen to assume that the members were long and thin, we would have neglected both the axial deformation and the shear deformation. In that case the flexibility matrix for that frame would be

$$\begin{Bmatrix} \delta_{Cv} \\ \delta_{Ch} \end{Bmatrix} \cong \begin{bmatrix} \frac{4}{3} & -\frac{1}{2} \\ -\frac{1}{2} & \frac{1}{3} \end{bmatrix} \begin{Bmatrix} \frac{PL^3}{EI} \\ \frac{QL^3}{EI} \end{Bmatrix} \tag{10.22}$$

This is clearly a simpler flexibility matrix, which is obviously of great benefit.

Thus, if we are prepared to ignore axial and shear deformation, we will assemble our frame with beam members or elements that can have only two kinds of generalized forces (transverse force, moment) and two generalized displacements (bending deflection, slope). On the generalized force side, there are a total of four per beam member, but only two of these are independent because there are two equilibrium equations that apply to each beam element. And among the choices, there are basically two bending element models that are used, each of which is similar to the springlike analyses we did in Section 8.2.3. The first is a cantilever beam with a concentrated force and a moment at the tip (Fig. 10.2(a)), and the second, which is more commonly used in the current kind of analyses, a simply supported beam with moments applied over each pin (Fig. 10.2(b)). Thus, using the same techniques we applied in Section 8.2.3, we can write the flexibility matrices that relate generalized displacements at a given end to the generalized forces that produce them – although note that *we are using a different sign convention for the positive angles* shown

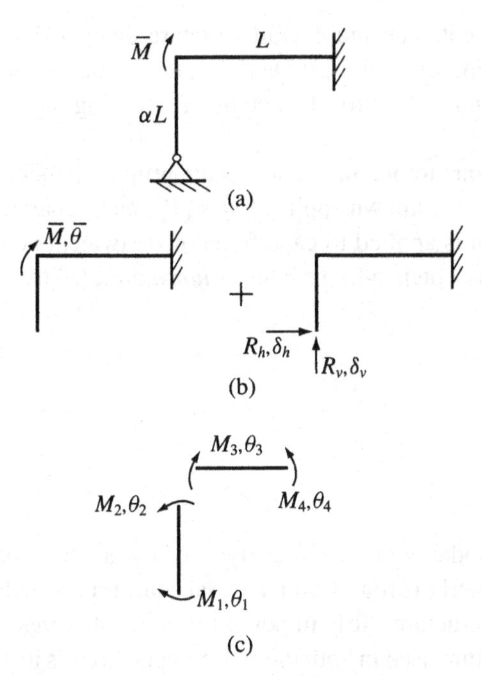

Figure 10.3. An indeterminate frame: (a) the entire frame, (b) its primary structures, and (c) its member (element) decompositions.

in Fig. 10.2. For the cantilever of Fig. 10.2(a)

$$\begin{Bmatrix} w_f \\ \theta_f \end{Bmatrix} = \left(\frac{L}{EI}\right) \begin{bmatrix} L^2/3 & L/2 \\ L/2 & 1 \end{bmatrix} \begin{Bmatrix} P_f \\ M_f \end{Bmatrix} \tag{10.23}$$

while for the simply supported beam of Fig. 10.2(b)

$$\begin{Bmatrix} \theta_n \\ \theta_f \end{Bmatrix} = \left(\frac{L}{6EI}\right) \begin{bmatrix} 2 & 1 \\ 1 & 2 \end{bmatrix} \begin{Bmatrix} M_n \\ M_f \end{Bmatrix} \tag{10.24}$$

We are using subscripts n and f to denote *near* and *far*, respectively. As already noted, the pinned-pinned model is used more often, in large part because a common factor can be pulled out of every term in the flexibility matrix, leaving each matrix element dimensionless. When combining these members, we can then incorporate member properties into scalar multipliers of each matrix, rather than in individual matrix elements. It is also true in many framed structures that each element is tied at its ends to perpendicular elements for whose axial deflection we are ignoring. Hence, that element is effectively restrained to zero transverse motion at its ends, just as in the element of Fig. 10.2(b).

10.2.3 Assembling Discrete Elements in the Flexibility Method for Frames

Now we turn to representing a frame as an assemblage of discrete elements, here the pinned-pinned beams of Fig. 10.2(b), and expressing its behavior in terms of variables at identifiable *nodes* at the ends of each of element. Our goal is to relate both the applied and reaction forces acting on the frame to the set of *generalized member forces* $\{M\}$ that are the (nodal) moments on the elements and the overall structure deformation to the displacements of the nodes of the elements. We locate nodes at the frame supports and at member intersections (and see Example 10.2 for the details of modeling the frame in Fig. 10.3(b)).

We begin by removing the redundants to obtain a determinate primary structure, and then we load the primary structure with the known applied loads $\{P_p\}$ and, separately, the redundants $\{P_r\}$. Moment equilibrium is applied to each frame node twice, once under each load vector, as a result of which we obtain a frame *equilibrium matrix* $\{b\}$ that relates $\{M\}$ to $\{P_p\}$ and $\{P_r\}$:

$$\begin{aligned} \{M\} &= [b]\{P\} \\ &= \left[[b]_p \,\vdots\, [b]_r\right] \begin{Bmatrix} P_p \\ --- \\ P_r \end{Bmatrix} \\ &= [b]_p\{P_p\} + [b]_r\{P_r\} \end{aligned} \tag{10.25}$$

Here the $[b]_p$ matrix results from a nodal equilibrium analysis of the applied loads on the primary structure, while the $[b]_r$ matrix is found from a nodal equilibrium analysis of the redundant forces on the primary structure. It is important to remember that the $[b]$ matrices exist only if the primary structure used in both these decompositions is itself both stable and statically determinate.

Now that we have related the member forces to the applied loads and to the redundants, how do we match up member forces and displacements with the structure's actual forces and displacements when we assemble members to form a frame? There are two steps. First, we need a flexibility matrix that represents all of the members in the frame or structure. One simple way to do this is by constructing a composite matrix that reflects the flexibilities of each member by adding the individual member flexibility terms. We would write this as

$$\{\theta\} = [f]_c\{M\} \tag{10.26}$$

where we use $\{\theta\}$ for the generalized displacements of our pinned-pinned beam elements, and $[f]_c$ is the *unassembled flexibility matrix* (also called the *composite member flexibility matrix*).

Then all we need to complete the formulation is to relate the member force vector $\{M\}$ to the general force vector $\{P\}$ that we had used in our general formulation of the force method (cf. Section 10.2.1). We accomplish this by equating two different forms of the complementary or strain energy stored in a structure. This will, in turn, allow us to recognize and assemble a global flexibility matrix for the structure. Thus, the stored energy resulting from the work done by the externally applied loads must equal the stored energy resulting from the work done by the member nodal forces:

$$\frac{1}{2}\{P\}^T\{\Delta\} = \frac{1}{2}\{M\}^T\{\theta\} \tag{10.27}$$

By virtue of Eq. (10.26) we can write

$$\{P\}^T\{\Delta\} = \{M\}^T[f]_c\{M\} \tag{10.28}$$

We now use Eq. (10.25) to rewrite Eq. (10.28) in terms of the equilibrium matrix and, in so doing, replace the member nodal forces by the actual forces, that is,

$$\begin{aligned}\{P\}^T\{\Delta\} &= ([b]\{P\})^T[f]_c[b]\{P\}\\ &= \{P\}^T[b]^T[f]_c[b]\{P\}\end{aligned} \tag{10.29}$$

from which it follows that

$$\{\Delta\} = [b]^T[f]_c[b]\{P\} \tag{10.30}$$

Finally, we compare the last result with Eq. (10.3) to find the *assembled flexibility matrix* as

$$[f]_{assembl} = [b]^T[f]_c[b] \tag{10.31}$$

It is now useful to partition the assembled flexibility matrix, based on the prior partitioning of the equilibrium matrices, that is, by invoking the different forms of Eqs. (10.25). Thus, substituting the primary and redundant equilibrium matrices into Eq. (10.31) yields

$$\begin{aligned}[f]_{assembl} &= \begin{bmatrix}[b]_p^T\\ [b]_r^T\end{bmatrix}[f]_c\begin{bmatrix}[b]_p & [b]_r\end{bmatrix}\\ &= \left[\begin{array}{c:c}[b]_p^T[f]_c[b]_p & [b]_p^T[f]_c[b]_r\\ \hdashline [b]_r^T[f]_c[b]_p & [b]_r^T[f]_c[b]_r\end{array}\right]\end{aligned} \tag{10.32}$$

This means that the redundants are determined by the equation

$$0 = [b]_r^T[f]_c[b]_p\{P_p\} + [b]_p^T[f]_c[b]_r\{P_r\} \tag{10.33}$$

and the displacements under the external loads are determined by

$$\{\Delta_p\} = [b]_p^T[f]_c[b]_p\{P_p\} + [b]_p^T[f]_c[b]_r\{P_r\} \tag{10.34}$$

Example 10.2. For the the indeterminate frame shown in Fig. 10.3(a), find the rotation $\bar{\theta}$ of the rigid joint over which a moment $\bar{M}$ is applied.

We choose the two support reactions at the pin, R_h and R_v, as our redundants, and $\bar{M}$ is the known, externally applied load. After writing equations of moment equilibrium for the four frame nodes we find

$$\{M\} \equiv \begin{Bmatrix}M_1\\ M_2\\ M_3\\ M_4\end{Bmatrix} = \left[\begin{array}{c:cc}0 & 0 & 0\\ 0 & -\alpha L & 0\\ 1 & -\alpha L & 0\\ 1 & -\alpha L & L\end{array}\right]\begin{Bmatrix}\bar{M}\\ \hdashline R_h\\ R_v\end{Bmatrix} \tag{10.35}$$

The 4×3 matrix on the right-hand side of Eq. (10.35) is our frame equilibrium matrix, appropriately partitioned to distinguish between the applied load and the two redundants. The explicitly partitioned equilibrium matrices and the load vectors in Eq. (10.35), as defined by Eqs. (10.25), take the following forms:

$$[b]_p = \begin{bmatrix} 0 \\ 0 \\ 1 \\ 1 \end{bmatrix}, \quad [b]_r = \begin{bmatrix} 0 & 0 \\ -\alpha L & 0 \\ -\alpha L & 0 \\ -\alpha L & L \end{bmatrix}, \quad \{P\} = \begin{Bmatrix} P_p \\ --- \\ P_r \end{Bmatrix} = \begin{Bmatrix} \bar{M} \\ --- \\ R_h \\ R_v \end{Bmatrix} \tag{10.36}$$

The composite matrix of member flexibility relations for this frame – the counterpart of Eq. (10.26) for the elements of Eq. (10.24) – is collected as

$$\begin{Bmatrix} \theta_1 \\ \theta_2 \\ \theta_3 \\ \theta_4 \end{Bmatrix} = \left(\frac{L}{6EI}\right) \begin{bmatrix} 2\alpha & \alpha & 0 & 0 \\ \alpha & 2\alpha & 0 & 0 \\ 0 & 0 & 2 & 1 \\ 0 & 0 & 1 & 2 \end{bmatrix} \begin{Bmatrix} M_1 \\ M_2 \\ M_3 \\ M_4 \end{Bmatrix} \tag{10.37}$$

where $[f]_c$ is the frame's unassembled flexibility matrix:

$$[f]_c \equiv \left(\frac{L}{6EI}\right) \begin{bmatrix} 2\alpha & \alpha & 0 & 0 \\ \alpha & 2\alpha & 0 & 0 \\ 0 & 0 & 2 & 1 \\ 0 & 0 & 1 & 2 \end{bmatrix} \tag{10.38}$$

Now we can insert the partitioned equilibrium matrices (Eqs. (10.36)) and the unassembled flexibility matrix (Eq. (10.38)) into Eq. (10.30) to find the complete discrete displacement vector for the frame:

$$\{\Delta\} = \begin{Bmatrix} \bar{\theta} \\ \delta_h \\ \delta_v \end{Bmatrix} = \left(\frac{L}{6EI}\right) \left[\begin{array}{c|cc} 6 & -6\alpha L & 3L \\ \hline -6\alpha L & 2(3+\alpha)\alpha^2 L^2 & -3\alpha L^2 \\ 3L & -3\alpha L^2 & 2L^2 \end{array}\right] \begin{Bmatrix} \bar{M} \\ R_h \\ R_v \end{Bmatrix} \tag{10.39}$$

This result shows the partitioned assembled flexibility matrix defined by Eq. (10.32), and we can use it to formulate the counterpart of Eq. (10.33) to find the redundants for our frame problem to be

$$\{R_h \;\; R_v\} = \begin{Bmatrix} R_h \\ R_v \end{Bmatrix}^T = \left[\frac{3}{(3+4\alpha)\alpha} \quad \frac{-6\alpha}{(3+4\alpha)}\right] \left(\frac{\bar{M}}{L}\right) \tag{10.40}$$

Then, by substituting Eq. (10.40) into Eq. (10.34), along with the appropriate equilibrium and flexibility matrices, we finally find the desired angular rotation of the joint at which the moment is applied:

$$\{\delta_p\} = \{\bar{\theta}\} = \left(\frac{\alpha}{(3+4\alpha)}\right) \begin{Bmatrix} \dfrac{\bar{M}L}{EI} \end{Bmatrix} \tag{10.41}$$

■

Now for our last discretization issue. While we are now quite familiar with the removal of external redundants to obtain the determinate primary structure, it is worth mentioning how we might handle *internal indeterminacies* in the force method. In fact, we have already done this in Section 5.2.2 for internally indeterminate trusses. Recall that we inserted a

Table 10.1. Frame member releases for the force paradigm.

Type of release:	The release supports:	It requires compatibility of:
Moment release (hinge)	$N(x)$, $V(x)$	$\theta = w'(x)$
Axial force release	$V(x)$, $M(x)$	$\delta_{\text{axial}} = u(x)$
Shear force release	$N(x)$, $M(x)$	$\delta_{\text{transverse}} = w(x)$

cut of magnitude Δ in place of a redundant bar and then applied there a redundant force T. The value of T was then determined by a compatibility condition (Eq. (5.22)) that ensured that such a cut would not be permissible for the truss to remain intact. In the more general case, particularly for structural frames, we extend this notion by introducing *releases*, by which we mean one or more imaginary cuts or discontinuities in displacement quantities, somewhere along a member, as listed in Table 10.1. These imaginary cuts or discontinuities are then closed by introducing corresponding redundants, which are the appropriate force *duals* or counterparts. Finally, we require that these dual forces satisfy compatibility conditions that force the relevant displacement variables to be continuous by closing the cut(s). (Also, remember that in a continuous member without any cuts, the force resultants $N(x)$, $V(x)$, and $M(x)$ are supported; thus, there are no releases, and so we require no added or special compatibility conditions.)

This is as far as we will go with the matrix formulation or representation of the force or flexibility method. The principles are as outlined, although even matrix notation will not obviate the complexity of analyzing a structure of significant size. The basic style of calculation and the kinds of answers we generally get should be familiar looking extensions of the simpler problems we have tackled previously with the force approach, a point to be borne in mind when we are evaluating someone else's computer analysis of a structure.

10.3 Displacement (Stiffness) Methods: Matrix Formulation

We now present a parallel discussion of displacement or stiffness methods of structural analysis. These methods use displacement as the basic analysis variable (cf. Table 6.1) and their energy formulations depend on the stored strain energy, which is itself expressed in terms of generalized displacements. They are also called stiffness methods because the resulting equations, whose satisfaction ensures equilibrium, are written in terms such that forces are expressed in terms of displacements and stiffness coefficients.

10.3.1 General Formulation

Recall from Section 6.2.3 that the matrix form of the total potential energy can be written as (with the notation of Δ for structural displacements)

$$\Pi = \frac{1}{2}\{\Delta\}^T[k]\{\Delta\} - \{P\}^T\{\Delta\} \tag{10.42}$$

In accord with the principle of minimum total potential energy, the vanishing of the first variation of Eq. (10.42) establishes the following equilibrium equation:

$$\{P\} = [k]\{\Delta\} \tag{10.43}$$

where $[k]$ is the symmetric stiffness matrix. This form is also familiar by now, and we will again extend a beam problem (from Sections 8.1.2 and 10.2.1) to make some points about this formulation that are important for more complex structures. In Section 8.1.2 we calculated the tip rotation and the deflections at two points of the cantilever beam shown in Fig. 8.1. In Section 10.2.1 we used the results of that analysis to find the redundant reaction at the pinned end of the clamped-pinned beam shown in Fig. 10.1 and to calculate the slope due to the moment applied over the pin. Here we will re-examine that problem, but we begin the an inverse formulation of Eq. (8.24), that is,

$$\begin{Bmatrix} P_1 \\ P_2 \\ M_0 \end{Bmatrix} = \begin{bmatrix} k_{11} & k_{12} & k_{13} \\ k_{21} & k_{22} & k_{23} \\ k_{31} & k_{32} & k_{33} \end{bmatrix} \begin{Bmatrix} \delta_a \\ \delta_L \\ \theta_L \end{Bmatrix} \tag{10.44}$$

Here we assume that the elements of the stiffness matrix k_{ij} can be calculated from the elements of the flexibility matrix by matrix algebra (see the Appendix) as

$$k_{ij} = \frac{(-1)^{i+j} M_{ji}^f}{|f|} \tag{10.45}$$

where $|f|$ is the value of the determinant of the flexibility matrix and the M_{ij}^f are the minors of that matrix. We assume here that the inverses of all matrices exist and can be calculated, although we will discuss some restrictions later.

We begin our beam analysis by rearranging Eq. (10.44) so that the equation for the reaction force at the pin is in the last – rather than the middle – row of the matrix (as we did with Eq. (4.73) in Example 4.6 for the three-segment bar):

$$\begin{Bmatrix} P_1 \\ M_0 \\ P_2 \end{Bmatrix} = \begin{bmatrix} k_{11} & k_{13} & k_{12} \\ k_{31} & k_{33} & k_{32} \\ k_{21} & k_{23} & k_{22} \end{bmatrix} \begin{Bmatrix} \delta_a \\ \theta_L \\ \delta_L \end{Bmatrix} \tag{10.46}$$

As we have done before, while making the row switch we have also rearranged the column vector of the discrete generalized forces by changing the order of its elements (viz., compare Eq. (10.46) with Eq. (10.44)). We now rewrite Eq. (10.46) using the *partitioned form* of the stiffness matrix,

$$\begin{Bmatrix} P_1 \\ M_0 \\ \hline P_2 \end{Bmatrix} = \left[\begin{array}{cc|c} k_{11} & k_{13} & k_{12} \\ k_{31} & k_{33} & k_{32} \\ \hline k_{21} & k_{23} & k_{22} \end{array}\right] \begin{Bmatrix} \delta_a \\ \theta_L \\ \hline \delta_L \end{Bmatrix} \tag{10.47}$$

Equations (10.46) and (10.47) show that the symmetry of the stiffness matrix has been preserved during this rearrangement. Further, as in the flexibility approach, we have the partitioned force and displacement vectors to enable us to distinguish between known applied forces (and their consequent nodal displacements) and unknown redundants (and their required nodal displacements, which are typically – but not always – set equal to zero).

How do we determine the redundant? The pin at the right end of the beam requires that the displacement vanishes there, that is, $\delta_L = 0$. Thus, the reaction force needed to keep the pin in equilibrium is determined from

$$\left\{\begin{array}{c} \\ \hline P_2 \end{array}\right\} = \left[\begin{array}{cc:c} & & \\ \hdashline k_{21} & k_{23} & k_{22} \end{array}\right]\left\{\begin{array}{c} \delta_a \\ \theta_L \\ \hdashline \delta_L \end{array}\right\} = 0 \tag{10.48}$$

or

$$\{P_2\} = [k_{21} \quad k_{23}]\left\{\begin{array}{c} \delta_a \\ \theta_L \end{array}\right\} + [k_{22}]\{\delta_L\} = k_{21}\delta_a + k_{23}\theta_L \tag{10.49}$$

Thus, if we know the nodal displacements at those points where loads are applied, we can calculate the redundant(s) directly from Eq. (10.49). In this case the nodal displacements are determined by inverting the partitioned matrix:

$$\left\{\begin{array}{c} P_1 \\ M_0 \\ \hline \\ \end{array}\right\} = \left[\begin{array}{cc:c} k_{11} & k_{13} & k_{12} \\ k_{31} & k_{33} & k_{32} \\ \hdashline & & \end{array}\right]\left\{\begin{array}{c} \delta_a \\ \theta_L \\ \hdashline 0 \end{array}\right\} \tag{10.50}$$

or

$$\left\{\begin{array}{c} \delta_a \\ \theta_L \end{array}\right\} = \begin{bmatrix} k_{11} & k_{13} \\ k_{31} & k_{33} \end{bmatrix}^{-1}\left\{\begin{array}{c} P_1 \\ M_0 \end{array}\right\} \tag{10.51}$$

The redundant P_2 can now be calculated by substituting Eq. (10.51) into the middle form of Eq. (10.49), from which we find

$$\begin{aligned} \{P_2\} &= [k_{21} \quad k_{23}]\left\{\begin{array}{c} \delta_a \\ \theta_L \end{array}\right\} \\ &= [k_{21} \quad k_{23}]\begin{bmatrix} k_{11} & k_{13} \\ k_{31} & k_{33} \end{bmatrix}^{-1}\left\{\begin{array}{c} P_1 \\ M_0 \end{array}\right\} \end{aligned} \tag{10.52}$$

Thus, we see that the redundant can be easily found by a series of matrix manipulations that depend only on the properties of a stiffness matrix for the structure. Further, by properly formulating and partitioning the stiffness matrix, we can calculate the deflections at the nodes or points that do move without having determined the value of the redundant because that movement depends only on the stiffness elements and the given applied loads (cf. Eq. (10.51)).

It is important to recognize that we have not mysteriously or magically turned an indeterminate problem into a determinate problem. It is that by representing and organizing the problem differently we have found a different sequence of calculations needed to do a complete analysis. In the force (flexibility) method we found the redundant first by requiring its displacement to assume the (known) value of zero (see Eqs. (10.8) and (10.9)). Then we were able to calculate structural deflections that depended both on given external loads and on values of the redundants, with the flexibility coefficients providing the key relationship links (viz., Eqs. (10.11) and (10.12)). In the displacement (stiffness) method we found the redundant force to be a function of the nodal displacements (Eq. (10.49)) that we could write in terms of the given external loads (Eq. (10.52)), with the stiffness coefficients providing the key relationship links in both instances. It is left as an exercise to show that the final results are – as they must be – exactly the same: Equation (10.52) produces the same value of the redundant as does Eq. (10.9), and the nodal displacements given by Eqs. (10.51) are identical to those given by Eqs. (10.13).

We should point out here that there is a serious complication in the stiffness method that we have yet to address (because we have not yet seen it arise). We will address it in more detail as part of the more general formulation that comes next, but we note that we have assumed that all matrix inversions that we used in the above problem would proceed without fail. However, it is precisely this assumption that we should *not* take as gospel.

We now proceed to a general matrix formulation of the stiffness method in which we set the order for solving the problem by ordering and labeling those displacements that are known or defined by boundary conditions and by partitioning all of the matrices in the equilibrium equation accordingly. We again use the notion of a primary structure for which we identify (1) by a subscript p the set of given applied loads whose corresponding displacements we want to find and (2) by a subscript r the (unknown) redundants and their corresponding (restricted) displacements. The force and deflection vectors are

$$\{P\} = \begin{Bmatrix} P_p \\ \text{---} \\ P_r \end{Bmatrix}, \qquad \{\Delta\} = \begin{Bmatrix} \Delta_p \\ \text{---} \\ \Delta_r \end{Bmatrix} \tag{10.53}$$

Now we can write a general equilibrium equation that reflects the pattern of primary and redundant quantities shown in Eq. (10.53), that is,

$$\begin{Bmatrix} P_p \\ \text{---} \\ P_r \end{Bmatrix} = \left[\begin{array}{c:c} [k]_{p,p} & [k]_{p,r} \\ \hdashline [k]_{r,p} & [k]_{r,r} \end{array}\right] \begin{Bmatrix} \Delta_p \\ \text{---} \\ \Delta_r \end{Bmatrix} \tag{10.54}$$

If we assume that $\{\Delta_r\} = 0$, that is, the known displacements at boundary points are restricted to be zero, we can relate the given external forces to the nodal displacements (or actual structural deflections) that they produce by using the consequences of the partitioning of Eq. (10.54):

$$\begin{aligned} \{P_p\} &= [k]_{p,p}\{\Delta_p\} + [k]_{p,r}\{\Delta_r\} \\ &= [k]_{p,p}\{\Delta_p\} \end{aligned} \tag{10.55}$$

The matrix $[k]_{p,p}$ is called the *reduced stiffness matrix*, and it can be used to find the displacements produced by the external loads by inverting Eq. (10.55):

$$\{\Delta_p\} = [k]^{-1}_{p,p}\{P_p\} \tag{10.56}$$

Now the reduced stiffness matrix does have a unique inverse because it derives from the satisfaction of equilibrium of the primary structure subjected to the known applied loads. That is, the primary structure is taken to be both stable and statically determinate, and so its stiffness matrix $[k]_{p,p}$, is nonsingular.

We can now calculate the redundants by substituting Eq. (10.56) into the expansion of the lower partition of Eq. (10.54), that is,

$$\begin{aligned} \{P_r\} &= [k]_{r,p}\{\Delta_p\} + [k]_{r,r}\{\Delta_r\} \\ &= [k]_{r,p}[k]^{-1}_{p,p}\{P_p\} \end{aligned} \tag{10.57}$$

where we continue to assume that the prescribed boundary displacements vanish. With Eq. (10.57) we can find the redundants of the indeterminate structure, and so in principle our problem is solved. For the record, we note that inasmuch as the stiffness matrix is symmetric, the two diagonal submatrices of Eq. (10.54) are symmetric. The off-diagonal submatrices of Eq. (10.54) are not necessarily symmetric, but one is clearly the transpose of the other, that is,

$$[k]_{p,r} = [k]_{r,p}^{T} \tag{10.58}$$

Equations (10.56) and (10.57) are the (somewhat) abstract statements of the displacement or stiffness paradigm. Equation (10.56) shows us how to calculate the actual equilibrium displacements at a set of specified points of an indeterminate structure, while Eq. (10.57) shows how we find the redundants in our structure. In both equations we see clearly the role and importance of the the stiffness coefficients. As we move on to some details, however, we want to recall one warning that we have just stated.

It does seem clear that we can calculate a unique set of forces from the matrix equilibrium equation, Eq. (10.46), for given a set of nodal displacements (or a known displacement vector $\{\Delta\}$). However, it is not clear that we can find a displacement vector from this equilibrium equation if we are given a force vector $\{P\}$ because, in general terms, the stiffness matrix $[k]$ may be *singular*, that is, its determinant may vanish. Since by definition (see the Appendix),

$$\text{elements of } [k]^{-1} = \frac{(-1)^{i+j} M_{ji}^{k}}{|k|} \tag{10.59}$$

$[k]$ cannot have an inverse if $|k| = 0$. Thus, in general terms, we cannot invert Eq. (10.46). A stiffness matrix is ordinarily singular when it is derived to allow displacements at all node points of a discretized structure because we can add a *rigid-body displacement* to a given displacement vector $\{\Delta\}$ without changing the strains in the structure – and thus without affecting the (nodal) forces in the structure. We eliminate the possibility of rigid-body-motion when we satisfy the geometric or kinematic boundary conditions on the structure, as a result of which the correspondingly modified stiffness matrix then becomes nonsingular. Thus, formulating a proper, nonsingular stiffness matrix is a very important part of applying the displacement (stiffness) method.

10.3.2 Some Discretization Aspects of the Stiffness Method for Frames

Now we turn to some of the issues involved in assembling a stiffness matrix for the entire structure from building blocks based on the stiffness matrices of the structural elements that we use to develop our discretized structural model.

As before we need to make a modeling decision about which physical effects to incorporate into our model (e.g., shear strain, axial deformation). And as before, we will assume that our frame is made up of long, slender beams for which only bending effects will be evaluated. Further, although the details will differ from the flexibility approach to this problem in some interesting ways, we do want to use the same simple members or elements to make up our discretized structure. That is, while our approach may differ, the physics are such that the members still behave as if their endpoint deflections are zero or negligibly small. Thus, the obvious path is to take the pinned-pinned element of Fig. 10.2(b) and invert its flexibility matrix, that is, invert the matrix equation (10.24). If

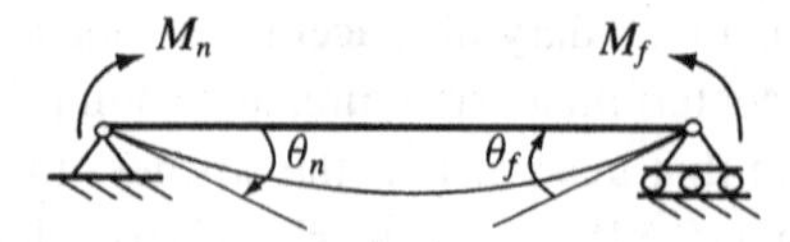

Figure 10.4. A pinned-pinned beam element for stiffness calculations for plane frames.

we did that we would find, in fact, that

$$\begin{Bmatrix} M_n \\ M_f \end{Bmatrix} = \left(\frac{2EI}{L}\right)\begin{bmatrix} 2 & -1 \\ -1 & 2 \end{bmatrix}\begin{Bmatrix} \theta_n \\ \theta_f \end{Bmatrix} \tag{10.60}$$

However, it turns out that practitioners of the art prefer stiffness matrices that have as many positive elements as possible, even for simple elements. Thus, we choose the same simple bending element for our discretization, but we adopt the slightly different sign convention that we show in Fig. 10.4. This convention requires only that we replace M_n by $-M_n$, and θ_n by $-\theta_n$ in Eq. (10.60), which means that our *element* (or *local*) *stiffness matrix* $[k]_e$ is now defined as

$$[k]_e = \left(\frac{2EI}{L}\right)\begin{bmatrix} 2 & 1 \\ 1 & 2 \end{bmatrix} \tag{10.61}$$

Note that Eq. (10.61) and its predecessor (Eq. (10.60)) represent nonsingular stiffness matrices because the kinematic boundary conditions of this problem have been satisfied.

10.3.3 Assembling Discrete Elements in the Stiffness Method for Frames

Now that we have an element stiffness matrix, we can again compose a structure of specified elements in order to analyze the entire structure. For the flexibility method (in Section 10.2.3) we recast the work-energy relation by using: (1) a composite equilibrium equation, $\{M\} = [b]\{P\}$, to cast the forces $\{P\}$ on the structure in terms of the generalized nodal forces $\{M\}$ and (2) a composite flexibility equation, $\{\theta\} = [f]_c\{M\}$, to relate the generalized nodal displacements $\{\theta\}$ to the generalized nodal forces, $\{M\}$ – all to be able to construct a deflection-load relation for the structure, that is, Eq. (10.30). We now follow a similar process for the stiffness method (Example 10.3 gives details for the frame in Fig. 10.3(a)).

We first ensure compatibility by relating the displacement vector for the composite structure, $\{\Delta\}$, to the generalized nodal displacement vector $\{\theta\}$. Thus, we define a *deformation transformation* (or *compatibility*) *matrix* $[T]$:

$$\{\theta\} = [T]\{\Delta\} \tag{10.62}$$

Then the element stiffness matrix (10.61) and the decomposition of a structure into elements, such as that shown in Fig. 10.5 for the aforementioned frame, allow us to relate the moments at the nodes to the set of element rotations and to write a composite relationship for an entire structure as

$$\{M\} = [k]_c\{\theta\} \tag{10.63}$$

where we have introduced a *composite stiffness matrix* $[k]_c$. Equation (10.63) is thus a composite result for the entire structure, and the composite stiffness matrix is clearly analogous to the composite flexibility matrix defined by Eq. (10.26)).

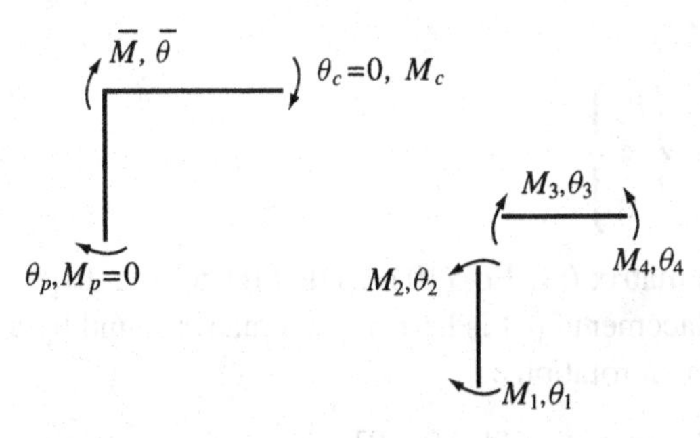

Figure 10.5. Structure (global) and element (local) coordinate systems for the stiffness analysis of the frame of Fig. 10.3.

We now use the energy calculation to relate the structure and element force matrices, much as we constructed the assembled flexibility matrix in Eqs. (10.27) – (10.32). Since the stored energy resulting from the work done by the applied loads must equal the stored energy resulting from the work done by the generalized member forces acting at the member endpoints or nodes,

$$\frac{1}{2}\{P\}^T\{\Delta\} = \frac{1}{2}\{\Delta\}^T\{P\} = \frac{1}{2}\{\theta\}^T\{M\} \tag{10.64}$$

By virtue of Eqs. (10.43) and (10.63) we can write

$$\{\Delta\}^T[k]\{\Delta\} = \{\theta\}^T[k]_c\{\theta\} \tag{10.65}$$

We now use Eq. (10.62) to rewrite Eq. (10.65) in terms of the compatibility or transformation matrix and, in so doing, replace the member nodal displacements by the actual structure displacements, that is,

$$\begin{aligned}\{\Delta\}^T[k]\{\Delta\} &= ([T]\{\Delta\})^T[k]_c[T]\{\Delta\} \\ &= \{\Delta\}^T[T]^T[k]_c[T]\{\Delta\}\end{aligned} \tag{10.66}$$

from which it follows that we can identify the *structure stiffness matrix* as

$$[k]_{structure} = [T]^T[k]_c[T] \tag{10.67}$$

and the *structure equilibrium equation* is written as

$$\{P\} = [k]_{structure}\{\Delta\} = [T]^T[k]_c[T]\{\Delta\} \tag{10.68}$$

Example 10.3. Use the stiffness method to find the rotation of the rigid joint over which the moment $\bar{M}$ is applied for the frame shown in Fig. 10.3(a).

We first identify the rotations at the pinned reaction, θ_p, at the joint under the moment, $\bar{\theta}$, and at the clamped support, θ_c, as generalized structure deflections. The corresponding load and displacement vectors defined by Eqs. (10.53) are, respectively,

$$\{P\} = \begin{Bmatrix} P_1 \\ P_2 \\ P_3 \end{Bmatrix} = \begin{Bmatrix} M_p \\ \bar{M} \\ M_c \end{Bmatrix} = \begin{Bmatrix} 0 \\ \bar{M} \\ M_c \end{Bmatrix} \tag{10.69}$$

and

$$\{\Delta\} = \begin{Bmatrix} \Delta_1 \\ \Delta_2 \\ \Delta_3 \end{Bmatrix} = \begin{Bmatrix} \theta_p \\ \bar{\theta} \\ \theta_c \end{Bmatrix} = \begin{Bmatrix} \theta_p \\ \bar{\theta} \\ 0 \end{Bmatrix} \tag{10.70}$$

The deformation transformation matrix (cf. Eq. (10.62)) that relates the displacement vector $\{\Delta\}$ to the set of nodal displacements $\{\theta\}$ is here a 4×3 matrix found by a visual inspection of the structure and element rotations:

$$\{\theta\} = \begin{Bmatrix} \theta_1 \\ \theta_2 \\ \theta_3 \\ \theta_4 \end{Bmatrix} = \begin{bmatrix} 1 & 0 & 0 \\ 0 & 1 & 0 \\ 0 & 1 & 0 \\ 0 & 0 & 1 \end{bmatrix} \begin{Bmatrix} \theta_p \\ \bar{\theta} \\ \theta_c \end{Bmatrix} = \begin{bmatrix} 1 & 0 & 0 \\ 0 & 1 & 0 \\ 0 & 1 & 0 \\ 0 & 0 & 1 \end{bmatrix} \begin{Bmatrix} \Delta_1 \\ \Delta_2 \\ \Delta_3 \end{Bmatrix} \equiv [T]\{\Delta\} \tag{10.71}$$

The composite stiffness for all of the elements in our frame (recall Eq. (10.63)) is

$$[k]_c \equiv \left(\frac{2EI}{L}\right) \begin{bmatrix} 2/\alpha & 1/\alpha & 0 & 0 \\ 1/\alpha & 2/\alpha & 0 & 0 \\ 0 & 0 & 2 & 1 \\ 0 & 0 & 1 & 2 \end{bmatrix} \tag{10.72}$$

We now substitute Eqs. (10.71) and (10.72) into Eq. (10.68) to obtain the structure equilibrium equation. We write it in partitioned form to maintain the distinctions between the applied loads and their consequent (and unknown) displacements and the redundants and their corresponding displacements:

$$\begin{Bmatrix} 0 \\ \bar{M} \\ \text{---} \\ M_c \end{Bmatrix} = \left(\frac{2EI}{L}\right) \left[\begin{array}{cc:c} 2/\alpha & 1/\alpha & 0 \\ 1/\alpha & 2+2/\alpha & 1 \\ \hdashline 0 & 1 & 2 \end{array}\right] \begin{Bmatrix} \theta_p \\ \bar{\theta} \\ \text{---} \\ 0 \end{Bmatrix} \tag{10.73}$$

which allows us to recognize the counterpart of Eq. (10.55), that is, the reduced stiffness matrix for this problem:

$$\begin{Bmatrix} 0 \\ \bar{M} \end{Bmatrix} = \left(\frac{2EI}{L}\right) \begin{bmatrix} 2/\alpha & 1/\alpha \\ 1/\alpha & 2+2/\alpha \end{bmatrix} \begin{Bmatrix} \theta_p \\ \bar{\theta} \end{Bmatrix} \tag{10.74}$$

We can invert Eq. (10.74) to find both $\bar{\theta}$ and the pinned support rotation θ_p:

$$\{\theta_p \quad \bar{\theta}\} = \begin{Bmatrix} \theta_p \\ \bar{\theta} \end{Bmatrix}^T = \left[\frac{-\alpha}{2(3+4\alpha)} \quad \frac{\alpha}{(3+4\alpha)}\right] \left(\frac{\bar{M}L}{EI}\right) \tag{10.75}$$

The value of $\bar{\theta}$ agrees exactly with the result that we had found using the flexibility method, Eq. (10.41). We can use the lower partitions of Eq. (10.73) to find the redundant reaction at the clamped support M_c, that is,

$$\{M_c\} = \left(\frac{2EI}{L}\right)[0 \quad 2 \,\vdots\, 1] \begin{Bmatrix} \theta_p \\ \bar{\theta} \\ \text{---} \\ 0 \end{Bmatrix} \tag{10.76}$$

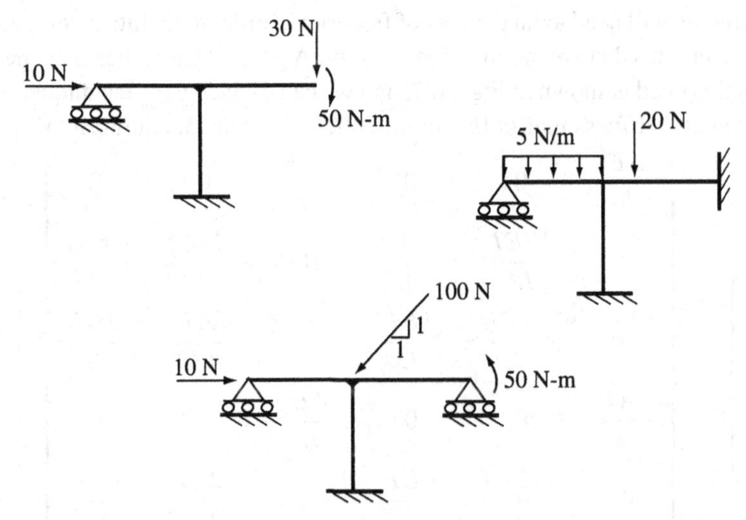

Figure 10.6. Some frame structures for which transverse and axial motion may need to be accounted for in discrete beam element models.

Figure 10.7. A general planar beam element.

which easily leads to the result

$$\{M_c\} = \left(\frac{2\alpha}{3+4\alpha}\right)\left\{\frac{\bar{M}L}{EI}\right\} \tag{10.77}$$

While we did not calculate this redundant in our flexibility analysis of this frame, we can use our prior results to confirm that Eq. (10.77) is correct. By summing moments around the fixed support of the frame of Fig. 10.3(a), we see that

$$M_c = \bar{M} + \alpha L R_h - L R_v \tag{10.78}$$

Then, if we substitute the redundant pin reactions of Eq. (10.40) into Eq. (10.78), we will obtain exact agreement with Eq. (10.77). Thus, we can confirm that the results already obtained with the flexibility method are indeed correct. ■

We have only scratched the surface of stiffness method calculations for real-world structures. One obvious extension has to do with the element or member model that we use to describe any particular structure. Even if we restrict our concerns to frame structures whose elements are, in fact, long and slender, there may well be circumstances in which the simple pinned-beam element (of Fig. 10.4 and Eq. (10.61)) is inadequate.

For example, joints may translate or otherwise move (as in an overhang) to the extent that we cannot ignore transverse displacements, as we can see in the frames in Fig. 10.6. For such cases we should use an element that accommodates both transverse motion and axial motion at each endpoint. Even if the element (or actual frame member) is stiffer axially than in bending, elements will need axial degrees of freedom in order to maintain compatibility with connected elements that are moving transversely. A general planar beam element that accommodates this need is shown in Fig. 10.7, and we can derive by the techniques of Section 8.2.1 an element stiffness matrix $[k]_e$ in the following element equilibrium equation:

$$\begin{Bmatrix} N_n \\ V_n \\ M_n \\ N_f \\ V_f \\ M_f \end{Bmatrix} = \begin{bmatrix} \frac{AE}{L} & 0 & 0 & -\frac{AE}{L} & 0 & 0 \\ 0 & \frac{12EI}{L^3} & -\frac{6EI}{L^2} & 0 & -\frac{12EI}{L^3} & -\frac{6EI}{L^2} \\ 0 & -\frac{6EI}{L^2} & \frac{4EI}{L} & 0 & \frac{6EI}{L^2} & \frac{2EI}{L} \\ -\frac{AE}{L} & 0 & 0 & \frac{AE}{L} & 0 & 0 \\ 0 & -\frac{12EI}{L^3} & \frac{6EI}{L^2} & 0 & \frac{12EI}{L^3} & \frac{6EI}{L^2} \\ 0 & -\frac{6EI}{L^2} & \frac{2EI}{L} & 0 & \frac{6EI}{L^2} & \frac{4EI}{L} \end{bmatrix} \begin{Bmatrix} u_n \\ w_n \\ \theta_n \\ u_f \\ w_f \\ \theta_f \end{Bmatrix} \tag{10.79}$$

And it can be shown that the pinned-beam element stiffness matrix (10.61) is a special case that can be pulled out of Eq. (10.79).

Another consequence of the choice of discretizing element can be seen in the fact that the stiffness approach to our simple frame problem produced some, but not all, of the same results as we obtained with the flexibility approach. Certainly the particular displacements and forces we calculated in common turned out to be the same. However, one important difference was that we found only one redundant in the stiffness approach, while there were two in the flexibility approach. It turned out that way because our choice of pinned-beam elements made it inevitable that the generalized displacements, for both the structure and the elements, would be slopes – as the logical way of enforcing compatibility when constructing the structure stiffness matrix. Consequently, both the structure and the element generalized forces were moments, so it was inevitable that the redundant we would calculate with this model would be the moment at the clamped support.

Still another major issue in the general use of the stiffness method is the construction of the stiffness matrix, and there are two aspects of this issue. One we have already mentioned is that we must be sure that the stiffness matrix is nonsingular, that is, it must be such that its value not vanish, that $|k| \neq 0$. As we have said, stiffness matrices are singular when the kinematic or displacement model used to formulate a problem does not exclude rigid-body displacements because the formulation includes, at least initially, all of the structure's nodes as eligible degrees of freedom. Equilibrium cannot yet be satisfied in this case because the forces are not "grounded," because the structure in question isn't attached to anything. However, once we satisfy the geometric or kinematic boundary conditions, we ground the structure, and we remove the possibility of rigid-body motion – and the stiffness matrix is rendered nonsingular.

The second aspect of assembling a structure stiffness matrix is our wish to make the necessary calculations as straightforwardly algorithmic as possible, that is, we would like them to be done both consistently and automatically on a computer. There is a technique, called the *direct stiffness method*, which sets up an algorithmic assembly of element matrices in a way that guarantees that the assembled stiffness matrix will be nonsingular. However, we leave this and other discretization details for more advanced study, although we note that the process we used to derive the matrix for the simple frame we have just analyzed was equivalent to a simple application of the direct stiffness method.

10.3.4 On the Flexibility Method and the Stiffness Method

Finally, given that we have two powerful methods for analyzing structures, why is one used more often than the other? We have mentioned several times that the displacement or stiffness method is most commonly used. We have seen one reason for this in the simple examples of this chapter, namely, that the stiffness method produced a more direct result for the displacement quantities we wanted to find. The stiffness approach works so well here that it makes the distinction between determinate and indeterminate problems seem much overrated.

The second reason has its roots in the direct displacement or Rayleigh–Ritz method we described in Section 6.5.2 and used in Sections 7.3 and 8.1. In that method we constructed approximate solutions for the displacement(s) or deflection(s) of an elastic structure as a sequence of trial functions. The trial functions, whether finite or infinite in number, were chosen to satisfy at least the geometric or kinematic boundary conditions of the problem at hand. Ideally, of course, the trial functions would satisfy all of the boundary conditions, which would help ensure good approximations not only of the displacements, but also of the strains and stresses (or force resultants). In the direct displacement approach we can more effectively use our intuition to estimate or guess the deflected shape of a structure; in general, it is much harder to predict the spatial distribution of force resultants or stresses. Further, if we do have a good result for the deflected shape, wherein both compatibility and equilibrium are satisfied, we can also easily obtain the stresses or forces by differentiating that known shape, assuming it is a sufficiently good approximation that the differentiations involved do not destroy the "goodness"of our answers.

These remarks also apply to the most often used, computer-based, direct displacement approach, that is, the finite element method. In terms of what we have done in this chapter, it is perhaps useful to think of the FEM as providing the means for solving problems with considerably more complicated geometry than slender beams or frames made up of beams, each of which has a nice symmetric cross section. For example, consider what might be involved in analyzing the complicated roof of Peir Luigi Nervi's Little Sports Palace (Fig. 1.4(c)) or a more modern version of the Pontybridd Bridge in Wales (pictured at the front of this chapter). Supports are considerably more complex – and esthetically likely more interesting – than are simple beam supports, and the analysis details are going to be much complicated than anything we have done here.

Historically the FEM emerged from the application of matrix mathematics to problems similar to those we have seen in this chapter, although certainly for significantly more complicated structures. The computational techniques have advanced in parallel with the advances in computer power of the last three decades, and the range of physical phenomena that can now be numerically modeled is well beyond the static, linear elastic behavior that has been our focus. For example, FEM analysis is now routinely done for problems involving dynamic behavior, instability phenomena that include geometric nonlinearities, several kinds of inelastic behavior, and combinations of these effects.

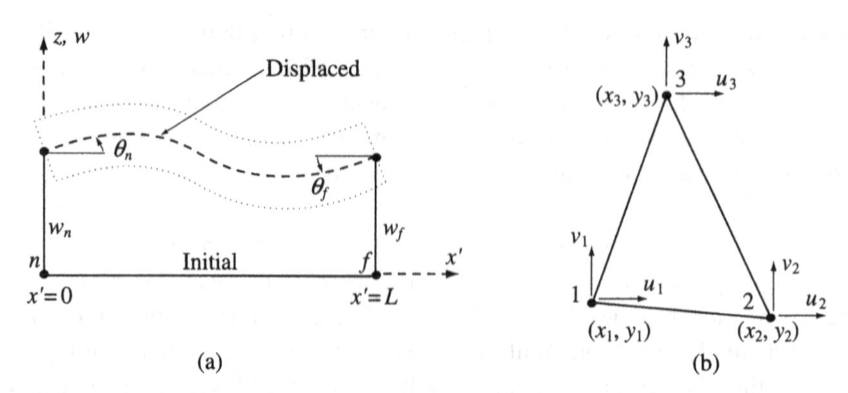

Figure 10.8. Two finite elements: (a) a beam element and (b) a planar, triangular element.

Now, almost all widely used FEM programs are based on the same principles as the direct displacement calculations we have done, with a distinguishing conceptual characteristic that displacement approximations are implemented as elements distributed over the spatial domain of interest, rather than as trial functions, which are typically defined over the entire spatial range. In other words, we are breaking up the structure into tiny finite structures or elements, and writing a trial function for each element that expresses displacements within the element in terms of values of displacements at the element's nodes. In Fig. 10.8 we show two such finite elements, one a *beam element* that is used for elementary thin beams, and the other a *two-dimensional triangular element* that has great utility for modeling two-dimensional problems.

In fact, suppose we are interested in calculating the deflected shape of a beam under a given set of loads. *Instead* of using approximations of the familiar form

$$w(x) = \sum_{n=0}^{N} W_n \phi_n(x) \tag{10.80}$$

where the $\phi_n(x)$ are defined over the entire length of the beam, we *now choose* to represent the beam itself as a collection of beam elements such as the one shown in Fig. 10.8. We would represent bending the displacement *within*, say, the kth element, in the form:

$$w_k = n_1(x')w_{kn} + n_2(x')\theta_{kn} + n_3(x')w_{kf} + n_4(x')\theta_{kf} \tag{10.81}$$

The $n_i(x')$ are *shape functions*, defined over *local coordinates* established for each element, and their coefficients represent the near and far displacement and slope amplitudes at the ends of each element. These shape functions express the variation of the displacement within each element. The generalized nodal displacements are used to ensure compatible connections between elements and to satisfy boundary conditions.

It is clear that this element bears a strong resemblance to the elements shown in Fig. 10.4, whose behavior we defined in Eq. (10.61). But the process of combining elements, while in principle the same as our calculations of Sections 10.3.2 and 10.3.3, is in detail far more complicated than what is done with simple beam elements. When elements are combined, kinematic (geometric) boundary conditions are satisfied to eliminate rigid-body motion, as is compatibility, and they are assembled to form a structure stiffness matrix. The structure stiffness matrix is then used to express equilibrium in terms of the nodal displacements, and then all that remains is to solve the matrix equilibrium equation.

We are not going to get into any more FEM detail here. Suffice it to say that the carefully-crafted, energy-based assembly of these finite elements over a spatial domain is the heart of the FEM, but the underlying principles are the same as those that are the core of this book. Thus, an understanding of the issues developed herein will serve as a more-than-adequate base for the intelligent application of all computational tools for structural analysis, including the FEM.

Problems

10.1 Using the pinned-pinned beam of Fig. 10.2(b) as the basic element, formulate the equilibrium, composite flexibility, and assembled flexibility matrices for the frame shown in Fig. 10.P1.

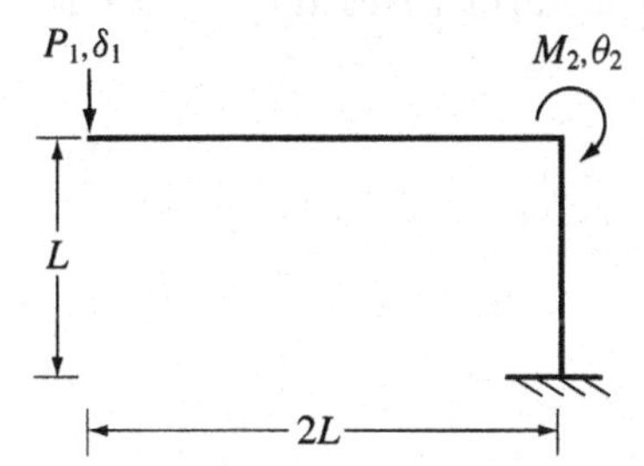

Figure 10.P1. Figure for Problems 10.1 and 10.2.

10.2 Using the pinned-pinned beam of Fig. 10.4 as the basic element, formulate the compatibility, composite stiffness, and structure stiffness matrices for the frame shown in Fig. 10.P1.

10.3 Using the pinned-pinned beam of Fig. 10.2(b) as the basic element, formulate the equilibrium, composite flexibility, and assembled flexibility matrices for the frame shown in Fig. 10.P3.

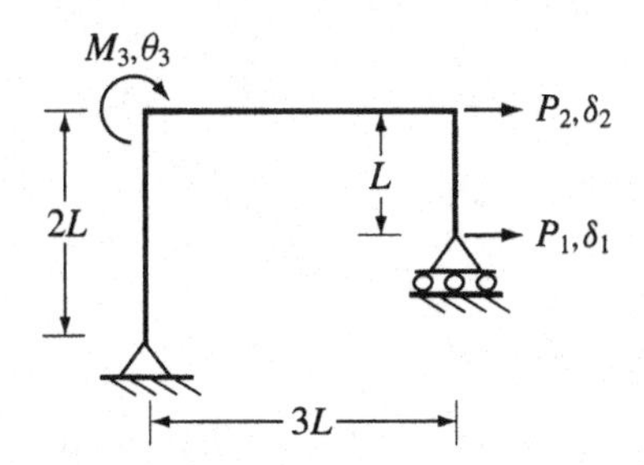

Figure 10.P3. Figure for Problems 10.3 and 10.4.

10.4 Using the pinned-pinned beam of Fig. 10.4 as the basic element, formulate the compatibility, composite stiffness, and structure stiffness matrices for the frame shown in Fig. 10.P3.

10.5 Using the force (flexibility) method, determine all of the reactions of the frame shown in Fig. 10.P5.

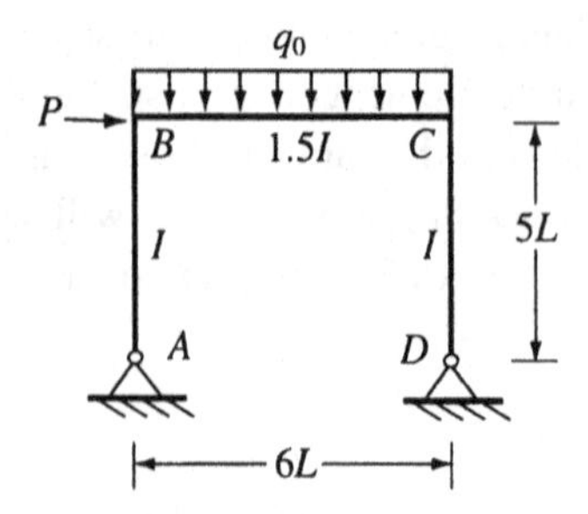

Figure 10.P5. Figure for Problems 10.5 and 10.6.

10.6 Using the displacement (stiffness) method, determine all of the reactions of the frame shown in Fig. 10.P5.

10.7 Using the force (flexibility) method, determine all of the reactions of the frame shown in Fig. 10.P7.

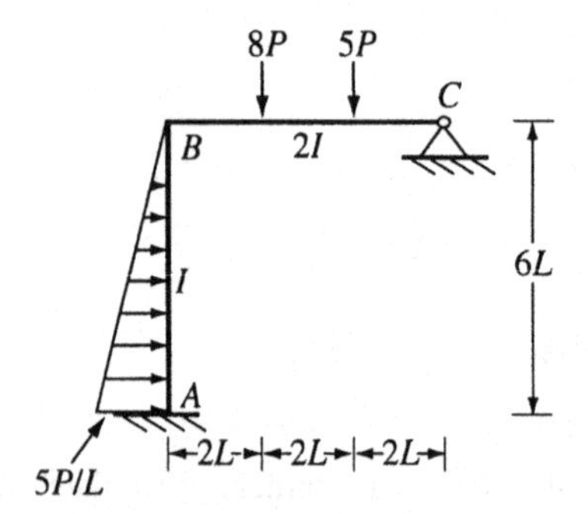

Figure 10.P7. Figure for Problems 10.7 and 10.8.

10.8 Using the displacement (stiffness) method, determine all of the reactions of the frame shown in Fig. 10.P7.

10.9 Using the force (flexibility) method, determine all of the reactions of the frame shown in Fig. 10.P9.

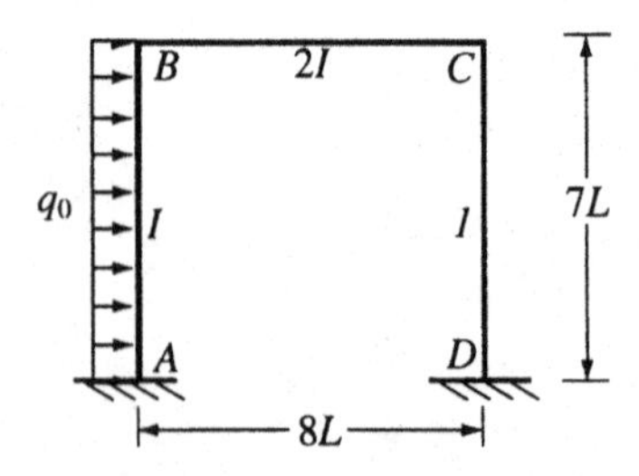

Figure 10.P9. Figure for Problems 10.9 and 10.10.

10.10 Using the displacement (stiffness) method, determine all of the reactions of the frame shown in Fig. 10.P9.

Appendix: A Short Introduction to Matrix Manipulation

In this appendix we outline some basic matrix operations, principally in support of the discussion of the force and displacement methods given in Chapter 8.

A.1 Defining Terms

We begin by defining a *matrix* as a rectangular array of *elements* A_{ij}:

$$[A] = [A_{ij}] = \begin{bmatrix} A_{11} & A_{12} & \cdots & A_{1n} \\ A_{21} & A_{22} & \cdots & A_{2n} \\ \vdots & \vdots & \ddots & \vdots \\ A_{m1} & A_{m2} & \cdots & A_{mn} \end{bmatrix} \tag{A.1}$$

The first subscript of each element refers to the row in which that element occurs, the second to the column. In Eq. (A.1) we show a matrix with m rows and n columns. We characterize it as an $m \times n$ matrix (read as "m by n matrix"). In our work the matrix elements are likely to be numbers or symbols that stand for some properties of a structural model (e.g., force, displacement, stiffness, etc.), although they can also be mathematical operators. There are several special cases of Eq. (A.1) that are of interest, starting with a *column matrix* ($n = 1$):

$$[A] = [A_i] = \begin{bmatrix} A_1 \\ A_2 \\ \vdots \\ A_m \end{bmatrix} \tag{A.2}$$

There is also a counterpart *row matrix* ($m = 1$):

$$[A] = [A_j] = [A_1 \quad A_2 \quad \cdots \quad A_n] \tag{A.3}$$

And, of course, there is the *square matrix*, said to be "a matrix of order m":

$$[A] = [A_{ij}] = \begin{bmatrix} A_{11} & A_{12} & \cdots & A_{1m} \\ A_{21} & A_{22} & \cdots & A_{2m} \\ \vdots & \vdots & \ddots & \vdots \\ A_{m1} & A_{m2} & \cdots & A_{mm} \end{bmatrix} \tag{A.4}$$

Square matrices occur frequently in structural calculations, and often in forms that are of particular interest. We have noted frequently in Chapter 10 (and elsewhere) that stiffness and flexibility matrices are *symmetric matrices*, by which we mean that

$$A_{ij} = A_{ji} \tag{A.5}$$

If the off-diagonal terms are all identically zero, the symmetric square matrix that results is a *diagonal matrix*, that is,

$$[A] = [A_{ij}] = \begin{bmatrix} A_{11} & 0 & \cdots & 0 \\ 0 & A_{22} & \cdots & 0 \\ \vdots & \vdots & \ddots & \vdots \\ 0 & 0 & \cdots & A_{mm} \end{bmatrix} \tag{A.6}$$

And if all of the diagonal terms have a value of unity (or one), we have the *unit matrix*, sometimes called the *identity matrix*:

$$[I] \equiv \begin{bmatrix} 1 & 0 & \cdots & 0 \\ 0 & 1 & \cdots & 0 \\ \vdots & \vdots & \ddots & \vdots \\ 0 & 0 & \cdots & 1 \end{bmatrix} \tag{A.7}$$

Matrices are said to be equal if they are identical arrays, that is, they have the same number of columns and rows, and that each element is exactly the same. In our notation, matrices $[A]$ and $[B]$ are equal if $A_{ij} = B_{ij}$.

Another aspect of square matrices that we will need below requires that we use determinants. A square array or matrix of numbers or symbols or operators has a corresponding *determinant* $|A|$:

$$|A| = |[A_{ij}]| = \begin{vmatrix} A_{11} & A_{12} & \cdots & A_{1m} \\ A_{21} & A_{22} & \cdots & A_{2m} \\ \vdots & \vdots & \ddots & \vdots \\ A_{m1} & A_{m2} & \cdots & A_{mm} \end{vmatrix} \tag{A.8}$$

Unlike its matrix counterpart, the determinant is a *scalar quantity* whose value is calculated by the standard *expansion by minors* (also called Laplace's expansion) that is familiar from basic algebra. We first define the minor M_{ij} corresponding to every element A_{ij} as that determinant obtained by deleting the ith row and the jth column:

$$M_{ij} \equiv \begin{vmatrix} A_{11} & A_{12} & \vdots & A_{1m} \\ A_{21} & A_{22} & \vdots & A_{2m} \\ \cdots & \cdots & \vdots & \cdots \\ A_{m1} & A_{m2} & \vdots & A_{mm} \end{vmatrix} \begin{matrix} \\ \\ i\text{th row} \\ \\ \end{matrix} \tag{A.9}$$

jth column

We then define the *Laplace expansion* for the (scalar) value $|A|$ of a determinant as the sum of the permutation-laden product of the elements of any row or column by its corresponding minor:

$$
\begin{aligned}
|A| &= \sum_{k=1}^{m} (-1)^{i+j} A_{ik} M_{ik} \qquad \text{(by row; } i = 1 \ldots, m\text{)} \\
&= \sum_{k=1}^{m} (-1)^{i+j} A_{kj} M_{kj} \qquad \text{(by column; } j = 1 \ldots, m\text{)}
\end{aligned}
\tag{A.10}
$$

A.2 Matrix Operations: Things We Can Do *to* a Matrix

We now describe two matrix operations wherein we change their appearance or construction by operating *on* them or doing something *to* them. One such operation is *transposition*, in which we interchange the rows and columns of a matrix. We denote the transpose of a matrix by a superscript as

$$
[A]^T = \begin{bmatrix} A_{11} & A_{12} & A_{13} \\ A_{21} & A_{22} & A_{23} \\ A_{31} & A_{32} & A_{33} \end{bmatrix}^T \equiv \begin{bmatrix} A_{11} & A_{21} & A_{31} \\ A_{12} & A_{22} & A_{32} \\ A_{13} & A_{23} & A_{33} \end{bmatrix}
\tag{A.11}
$$

or, in indicial shorthand,

$$
[A]^T = [A_{ij}]^T \equiv [A_{ji}]
\tag{A.12}
$$

The second operation in which we change the appearance of a matrix reflects our need to treat different members of a class of physical elements in different ways. For example, when we wrote the forces on a structure as a column matrix, we often distinguished known forces from redundants, and as a result we drew distinctions about elements in stiffness and flexibility matrices. We expressed such distinctions by writing a *partitioned form* of a matrix:

$$
\left[\begin{array}{cc:c} A_{11} & A_{12} & A_{13} \\ A_{21} & A_{22} & A_{23} \\ \hdashline A_{31} & A_{32} & A_{33} \end{array}\right] \quad \text{or} \quad \left\{\begin{array}{c} A_1 \\ A_2 \\ \hdashline A_3 \end{array}\right\}
\tag{A.13}
$$

A.3 Matrix Operations: Things We Can Do *with* a Matrix

How about matrix operations? We can add two matrices together or subtract one from the other if they are both $m \times n$ arrays, in which case we simply add or subtract the corresponding elements:

$$
[A] \pm [B] = [A_{ij} \pm B_{ij}]
\tag{A.14}
$$

If we multiply (or divide) a matrix by a scalar, each element of that matrix is correspondingly multiplied (or divided):

$$
c[A] = [cA_{ij}]
\tag{A.15}
$$

Can we multiply and divide matrices, and if so, how? We can multiply two matrices, say, $[A]$ and $[B]$, if they are *conformable*, that is, if the number of columns in matrix $[A]$

is the same as the number of rows in $[B]$. Then we can form the product $[A][B]$ in accord with the following definiton:

$$\underbrace{[A]}_{(m\times n)}\underbrace{[B]}_{(n\times p)} \equiv \underbrace{[C]}_{(m\times p)} \tag{A.16}$$

The matrix $[C]$ resulting from this definition is an $m \times p$ array with elements

$$C_{ij} = \sum_{k=1}^{n} A_{ik}B_{kj} \tag{A.17}$$

We also note that the requirement that the two matrices multiplied in Eq. (A.16) are conformable means that a specific order of multiplication is imposed on matrix multiplication. Consider the following simple multiplication example:

$$\underbrace{[A]}_{(2\times 3)}\underbrace{[B]}_{(3\times 1)} = \begin{bmatrix} 2 & 4 & 6 \\ 1 & 3 & 5 \end{bmatrix}\begin{bmatrix} 5 \\ 7 \\ 9 \end{bmatrix} = \underbrace{[C]}_{(2\times 1)} = \begin{bmatrix} 82 \\ 71 \end{bmatrix} \tag{A.18}$$

However, we cannot reverse the order of multiplication, that is, we cannot write

$$\underbrace{[B]}_{(3\times 1)}\underbrace{[A]}_{(2\times 3)} = \begin{bmatrix} 5 \\ 7 \\ 9 \end{bmatrix}\begin{bmatrix} 2 & 4 & 6 \\ 1 & 3 & 5 \end{bmatrix} \quad \Rightarrow \quad \text{undefined} \tag{A.19}$$

Thus, the product $[B][A]$ does not exist just because $[A][B]$ does.

In fact, quite apart from the conformability requirement, we need to make a significantly more general (and more important) statement about matrix multiplication, that is, we cannot interchange or reverse the order of a matrix multiplication, that is, *matrix multiplication is not commutative*:

$$[A][B] \neq [B][A] \tag{A.20}$$

On the other hand, *matrix multiplication is distributive*:

$$[A]([B]+[C]) = [A][B]+[A][C] \tag{A.21}$$

Further, *matrix multiplication is associative*:

$$[A]([B][C]) = ([A][B])[C] \tag{A.22}$$

We have asked before whether we could divide matrices. The answer is that we can't divide one matrix by another because there is no mathematical means of implementing divisors other than scalars. Recall, for example, that the use of tensors in engineering reflects in part an attempt to rationalize the expressions of quantities expressed as ratios of vectors. However, because we are often interested in solving linear systems of equations, that is, of finding the solution $[y]$ to a linear system of equations expressed in matrix form as

$$[A][y] = [x] \tag{A.23}$$

it turns out to be useful to define the *inverse* $[A]^{-1}$ of a matrix $[A]$:

$$[A][A]^{-1} = [A]^{-1}[A] = [I] \tag{A.24}$$

so that the solution of the linear system (A.23) can be written as

$$[y] = [A]^{-1}[x] \tag{A.25}$$

As a formalism, Eq. (A.25) is all well and good, but how do we actually calculate the inverse of a matrix $[A]^{-1}$ so that we can calculate the unknown elements of the unknown matrix $[y]$? The answer is a matrix adaptation of the well-known Cramer's rule of determinants. In particular, the inverse of the matrix $[A]$ is written in terms of the *adjoint matrix* $[A]^*$, which is in turn defined in terms of the minors M_{ij} of the (original) matrix $[A]$:

$$[A]^* \equiv [A_{ij}]^* \equiv [(-1)^{i+j} M_{ji}] \tag{A.26}$$

Note that the elements of the adjoint matrix are the permutation-laden elements of the transpose of the matrix of the minors of the matrix $[A]$. Then the inverse matrix is calculated according to the following, Cramer-like formula:

$$[A_{ij}]^{-1} \equiv \frac{[A]^*}{|A|} = \frac{[(-1)^{i+j} M_{ji}]}{|A|} \tag{A.27}$$

where $|A|$ is the value of the determinant of the matrix $[A]$.

We note in passing one important point. The formula we have given for the inverse of a matrix is quite correct, but in the matrix analysis of structures, whether in their original formulation or in their current applications in FEM programs, we rarely ever invert matrices numerically. It turns out to be significantly faster and safer (in terms of avoiding serious numerical errors) to solve linear systems of equations using other techniques, the most common being some variant of the Gauss elimination scheme. Thus, since our interest in matrix structural analysis is limited here to expressing the kinds of formal relations we have given in Chapter 10 and to recognizing the roots and roles of these formalities in other aspects of structural mechanics, we will not go any further with this very gentle introduction to matrix manipulation.

Bibliography

Classical civil engineering structures. The books listed here are in the tradition of classical structural analysis, explaining the various classical methods of structural analysis and built around the traditional determinate–indeterminate dichotomy. The analysis methods are often presented as separate, unrelated techniques. Also included are a few books that have a clear design emphasis.

A. E. Armenàkas, *Classic Structural Analysis: A Modern Approach*, McGraw-Hill, New York, 1988.

T. Au and P. Christiano, *Structural Analysis*, Prentice-Hall, Englewood Cliffs, NJ, 1987.

J. R. Benjamin, *Statically Indeterminate Structures*, McGraw-Hill, New York, 1959.

S. T. Carpenter, *Structural Mechanics*, John Wiley, New York, 1960.

A. Chajes, *Structural Analysis*, Prentice-Hall, Englewood Cliffs, NJ, 1990.

H. Cross and N. D. Morgan, *Continuous Frames of Reinforced Concrete*, John Wiley, New York, 1932.

R. Englekirk, *Steel Structures: Controlling Behavior Through Design*, John Wiley, New York, 1994.

J. M. Gere and W. Weaver Jr., *Analysis of Framed Structures*, 2nd ed., D. Van Nostrand, New York, 1982.

A. Ghali and A. M. Neville, *Structural Analysis*, Intext, New York, 1972.

A. S. Hall and R. W. Woodhead, *Frame Analysis*, 2nd ed., John Wiley, New York, 1967.

R. C. Hibbeler, *Structural Analysis*, 3rd ed., Prentice-Hall, Englewood Cliffs, NJ, 1995.

M. Hoit, *Computer-Assisted Structural Analysis and Modeling*, Prentice-Hall, Englewood Cliffs, NJ, 1995.

Y.-Y. Hsieh and S. T. Mau, *Elementary Theory of Structures*. 4th ed., Prentice-Hall, Englewood Cliffs, NJ, 1995.

R. L. Ketter, G. C. Lee, and S. P. Prawel Jr., *Structural Analysis and Design*, McGraw-Hill, New York, 1979.

J. S. Kinney, *Indeterminate Structural Analysis*, Addison-Wesley, Reading, MA, 1957.

W. H. Mosley and W. J. Spencer, *Microcomputer Applications in Structural Engineering*, Elsevier, New York, 1984.

B. G. Neal, *Structural Theorems and Their Application*, Pergamon Press, Oxford, 1964.

C. H. Norris, J. B. Wilbur, and S. Utku, *Elementary Structural Analysis*, 3rd ed., McGraw-Hill, New York, 1976.

A. C. Palmer, *Structural Mechanics*, Clarendon Press, Oxford, 1976.

J. I. Parcel and R. B. B. Moorman, *Analysis of Statically Indeterminate Structures*, New York, John Wiley, 1955.

W. R. Spillers, *Introduction to Structures*, Halsted Press, Chichester, UK, 1985.

B. S. Taranath, *Structural Analysis and Design of Tall Buildings*, McGraw-Hill, New York, 1988.

S. P. Timoshenko and D. H. Young, *Theory of Structures*, 2nd ed., McGraw-Hill, New York, 1965.

C. K. Wang, *Indeterminate Structural Analysis*, McGraw-Hill, New York, 1983.

H. H. West, *Analysis of Structures*, 2nd ed., John Wiley, New York, 1989.

R. N. White, P. Gergely, and R. G. Sexsmith, *Structural Engineering*, combined ed. (Vols. 1 and 2), John Wiley, New York, 1976.

Energy methods. The works in this group employ variational techniques to derive structural models and approximate solutions. They are mostly written from the viewpoint of people who do engineering mechanics rather than classical structural analysis.

T. M. Charlton, *Energy Principles in Applied Statics*, Blackie & Son Ltd., London, 1959.
C. L. Dym and I. H. Shames, *Solid Mechanics: A Variational Approach*, McGraw-Hill, New York, 1973.
Y. C. Fung, *Foundations of Solid Mechanics*, Prentice-Hall, Englewood Cliffs, NJ, 1965.
N. J. Hoff, *The Analysis of Structures*, John Wiley, New York, 1956.
H. L. Langhaar, *Energy Methods in Applied Mechanics*, John Wiley, New York, 1962.
J. T. Oden and E. A. Rippenberger, *Mechanics of Elastic Structures*, 2nd ed., McGraw-Hill, New York, 1981.
I. H. Shames and C. L. Dym, *Energy and Finite Element Methods in Structural Mechanics*, Hemisphere Publishing, New York, 1985.
T. R. Tauchert, *Energy Principles in Structural Mechanics*, McGraw-Hill, New York, 1974.

Finite element and matrix methods. This category includes those works that are outgrowth of early work in matrix analysis of structures, the precursor of finite element methods. Their emphasis is much more on computational techniques than on structural behavior.

R. J. Astley, *Finite Elements in Solids and Structures*, Chapman & Hall, London, 1992.
R. D. Cook, *Concepts and Applications of Finite Element Analysis*, 2nd ed., John Wiley, New York, 1981.
C. Desai and J. F. Abel, *Introduction to the Finite Element Method*, D. Van Nostrand, New York, 1972.
R. H. Gallagher, *Finite Element Analysis, Fundamentals*, Prentice-Hall, Englewood Cliffs, NJ, 1975.
T. J. R. Hughes, *The Finite Element Method, Linear Static and Dynamic Finite Element Analysis*, Prentice-Hall, Englewood Cliffs, NJ, 1987.
H. I. Laursen, *Matrix Analysis of Structures*, McGraw-Hill, New York, 1966.
R. Levy and W. R. Spillers, *Analysis of Geometrically Nonlinear Structures*, Chapman & Hall, New York, 1995.
R. K. Livesley, *Matrix Methods of Structural Analysis*, Pergamon Press, Oxford, 1962.
H. C. Martin, *Introduction to Matrix Methods of Structural Analysis*, McGraw-Hill, New York, 1966.
H. C. Martin and G. F. Carey, *Introduction to Finite Element Analysis*, McGraw-Hill, New York, 1973.
W. McGuire and R. H. Gallagher, *Matrix Structural Analysis*, John Wiley, New York, 1979.
J. L. Meek, *Matrix Structural Analysis*, McGraw-Hill, New York, 1971.
R. J. Melosh, *Structural Engineering Analysis by Finite Elements*, Prentice-Hall, Englewood Cliffs, NJ, 1990.
V. J. Meyers, *Matrix Analysis of Structures*, Harper & Row, New York, 1983.
J. S. Przemieniecki, *Theory of Matrix Structural Analysis*, McGraw-Hill, New York, 1968.
M. F. Rubinstein, *Matrix Computer Analysis of Structures*, Prentice-Hall, Englewood Cliffs, NJ, 1966.
I. H. Shames and C. L. Dym, *Energy and Finite Element Methods in Structural Mechanics*, Hemisphere Publishing, New York, 1985.
O. Ural, *Finite Element Method*, Intext, New York, 1973.
O. C. Zienkiewicz, *The Finite Element Method*, 3rd ed., McGraw-Hill, New York, 1977.

Special topics. This category is represented here only sparsely, as the inclusion of all of the specialized work on all aspects of structural mechanics and analysis would be in itself an overwhelming task. Thus, this list includes just a sampling of specialized works, each of which includes a reasonable bibliography of other works in that specialty.

R. W. Clough and J. Penzien, *Dynamics of Structures*, McGraw-Hill, New York, 1975.

M. Z. Cohn (Editor), *An Introduction to Structural Optimization*, University of Waterloo, Waterloo, Canada, 1970.

C. L. Dym, *Introduction to the Theory of Shells*, updated ed., Hemisphere Publishing, New York, 1990. (Original edition published by Pergamon Press, Oxford, 1974.)

C. L. Dym, *Stability Theory and Its Applications to Structural Mechanics*, Noordhoff International Publishing, Leyden, The Netherlands, 1974.

R. Englekirk, *Steel Structures: Controlling Behavior Through Design*, John Wiley, New York, 1994.

F. Fahy, *Sound and Structural Vibration*, Academic Press, London, 1985.

M. R. Horne, *Plastic Theory of Structures*, MIT Press, Cambridge, MA, 1971.

H. M. Irvine, *Cable Structures*, MIT Press, Cambridge, MA, 1981.

A. Kalnins and C. L. Dym (Editors), *Vibration: Beams, Plates and Shells*, Dowden, Hutchinson and Ross, Stroudsburg, PA, 1976.

B. G. Neal, *Structural Theorems and Their Application*, Pergamon Press, Oxford, 1964.

T. Sarpkaya and M. Isaacson, *Mechanics of Wave Forces on Structures*, Van Nostrand Reinhold, New York, 1981.

E. Simiu and R. H. Scanlan, *Wind Effects on Structures: An Introduction to Wind Engineering*, John Wiley, New York, 1978.

D. B. Steinman, *A Practical Treatise on Suspension Bridges*, John Wiley, New York, 1929.

B. S. Taranath, *Structural Analysis and Design of Tall Buildings*, McGraw-Hill, New York, 1988.

General structural behavior and the history of structures. Most of the books in this category are intended to provide a flavor for how structural elements and structures behave. Generally written either for architectural students or for general audiences with little background in mathematics, their emphasis is on trying to give a "feel" for how and why structures behave as they do. Also included here are several works on structural history and philosophy.

W. Addis, *Structural Engineering: The Nature of Theory and Design*, Ellis Horwood, Chichester, UK, 1990.

D. P. Billington, *Robert Maillart's Bridges*, Princeton University Press, Princeton, NJ, 1979.

D. P. Billington, *The Tower and The Bridge*, Basic Books, New York, 1983.

H. J. Cowan, *Architectural Structures: An Introduction to Structural Mechanics*, 2nd ed., Elsevier, New York, 1976.

E. DeLony, *Landmark American Bridges*, American Society of Civil Engineers, New York, and Little, Brown and Company, Boston, 1992.

A. J. Francis, *Introducing Structures*, Pergamon Press, Oxford, 1980.

J. E. Gordon, *Structures: Or, Why Things Don't Fall Down*, Plenum Press, New York, 1978.

M. Hayden, *The Book of Bridges*, Galahad Books, New York, 1976.

J. Heyman, *Coulomb's Memoir on Statics: An Essay in the History of Civil Engineering*, Cambridge University Press, Cambridge, UK, 1972.

J. Heyman, *The Stone Skeleton: Structural Engineering of Masonry Architecture*, Cambridge University Press, Cambridge, UK, 1995.

M. Levy and M. Salvadori, *Why Buildings Fall Down*, W. W. Norton, New York, 1992.

T. Y. Lin and S. D. Stotesbury, *Structural Concepts and Systems for Architects and Engineers*, 2nd ed., Van Nostrand Reinhold, New York, 1988.

National Geographic Society, *The Builders: Marvels of Engineering*, National Geographic Society, Washington, DC, 1992.

P. L. Nervi, *Structures* (Translated by G. and M. Salvadori), F. W. Dodge, New York, 1956.

C. O'Connor, *Roman Bridges*, Cambridge University Press, Cambridge, UK, 1993.

M. Overman, *Roads, Bridges, and Tunnels*, Doubleday, Garden City, NY, 1968.

N. Rosenberg and W. G. Vincenti, *The Britannia Bridge: The Generation and Diffusion of Technological Knowledge*, MIT Press, Cambridge, MA, 1978.

T. Ruddock, *Arch Bridges and Their Builders 1735–1835*, Cambridge University Press, Cambridge, UK, 1979.

M. Salvadori, *Why Buildings Stand Up*, McGraw-Hill, New York, 1980.

M. Salvadori and R. Heller, *Structure in Architecture*, Prentice-Hall, Englewood Cliffs, NJ, 1963.

M. Salvadori and M. Levy, *Structural Design in Architecture*, Prentice-Hall, Englewood Cliffs, NJ, 1981.

D. L. Schodek, *Structures*, Prentice-Hall, Englewood Cliffs, NJ, 1980.

D. B. Steinman and S. R. Watson, *Bridges and Their Builders*, Dover, New York, 1957.

E. Torroja, *Philosophy of Structures* (Translated by J. J. and M. Polivka), University of California Press, Berkeley and Los Angeles, 1958.

N. Upton, *An Illustrated History of Civil Engineering*, Heinemann, London, 1975.

Design specifications and design codes. A sampler of some of the principal design codes and design specifications that are used in modern structural engineering practice.

Aluminum Assoociation (AA), *Aluminum Construction Manual: Specifications for Aluminum Structures*, 5th ed., AA, New York, 1986.

American Association of State Highway and Transportation Officials (AASHTO), *Standard Specifications for Highway Bridges*, AASHTO, Washington, DC, 1983.

American Concrete Institute (ACI), *Building Code Requirements for Reinforced Concrete*, ACI, Detroit, MI, 1989 (revised 1992).

American Institute of Steel Construction (AISC), *Manual of Steel Construction, Load & Resistance Factor Design*, AISC, Chicago, 1986.

——, *Manual of Steel Construction, Allowable Stress Design Design*, 9th ed., AISC, Chicago, 1989.

American Institute of Timber Construction (AITC), *Timber Construction Manual*, John Wiley, New York, 1994.

American National Standards Institute (ANSI), *Building Code Requirements for Minimum Design Loads in Buildings and Other Structures*, ANSI, New York, 1982.

American Railway Engineering Association (AREA), *Specifications for Steel Railway Bridges*, AREA, Chicago, 1965.

American Society for Testing and Materials (ASTM), *Specification for Heat-Treated Steel Structural Bolts, 150 ksi Minimum Tensile Strength*, ASTM, Philadelphia, 1985.

Building Officials and Code Administrators International, Inc., (BOAC), *Basic Building Code*, 7th ed., BOAC, Country Club Hills, IL, 1994.

International Conference of Building Officials (ICBO), *Uniform Building Code*, ICBO, Whittier, CA, 1991.

National Fire Protection Association (NFPA), *Life Safety Code*, NFPA, Quincy, MA, 1985.

Prestressed Concrete Institute (PCI), *PCI Design Handbook*, PCI, Chicago, IL, 1992.

Structural Engineers Association of California (SEAOC), *Recommended Lateral Force Requirements and Commentary*, SEAOC, San Francisco, 1974.

Index

Page number entries followed by a "p" indicate photographs.